Second Edition

SOIL EROSION

RESEARCH METHODS

Second Edition

SOIL EROSION RESEARCH METHODS

R. Lal, Editor
Associate Professor
Department of Agronomy
Ohio State University
Columbus

CRC Press
Taylor & Francis Group
Boca Raton London New York

CRC Press is an imprint of the
Taylor & Francis Group, an **informa** business

CRC Press
Taylor & Francis Group
6000 Broken Sound Parkway NW, Suite 300
Boca Raton, FL 33487-2742

First issued in hardback 2017

ISBN-13: 978-1-8840-1509-0 (pbk)
ISBN-13: 978-1-1384-6860-3 (hbk)

Visit the Taylor & Francis Web site at
http://www.taylorandfrancis.com

and the CRC Press Web site
http://www.crcpress.com

Library of Congress Cataloging-in-Publication Data

Soil Erosion Research Methods / edited by Rattan Lal.—2nd ed.
p. cm.
Includes bibliographical references

1. Soil erosion—Research—Methodology. 2. Soil conservation—Research—Methodology. I. Lal, R.
S626.S65 1994
631.4'5'072—dc20 93-45352
CIP

CONTRIBUTORS

D. V. Armbrust
Soil Scientist, Wind Erosion Research Unit, Agricultural Research Service, U.S. Department of Agriculture, Manhattan, Kansas

A. A. Durar
Research Assistant, Department of Agronomy, Kansas State University, Manhattan, Kansas

W. Elliot
Project Leader, Intermountain Research Station, USDA, Forest Service, Idaho

G. R. Foster
Director, National Sedimentation Laboratory, Agricultural Research Service, U.S. Department of Agriculture, Oxford, Mississippi

D. W. Fryrear
Research Leader, Conservation Systems Research Laboratory, Agricultural Research Service, U.S. Department of Agriculture, Big Spring, Texas

L. J. Hagen
Agricultural Engineer, Wind Erosion Research Unit, Agricultural Research Service, U.S. Department of Agriculture, Manhattan, Kansas

J. M. Laflen
Research Leader, National Soil Erosion Research Laboratory, Agricultural Research Service, U.S. Department of Agriculture, West Lafayette, Indiana

R. Lal
Associate Professor, Department of Agronomy, Ohio State University, Columbus

L. J. Lane
Hydrologist, Aridland Watershed Management Research Unit, Agricultural Research Service, U.S. Department of Agriculture, Tucson, Arizona

V. L. Lopes
Assistant Professor, School of Renewable Natural Resources, University of Arizona, Tucson, Arizona

D. K. McCool
Agricultural Engineer, Washington State University, Agricultural Research Service, U.S. Department of Agriculture, Pullman, Washington

K. C. McGregor
Agricultural Engineer, National Sedimentation Laboratory, Agricultural Research Service, U.S. Department of Agriculture, Oxford, Mississippi

L. D. Meyer
Supervising Agricultural Engineer, National Sedimentation Laboratory, Agricultural Research Service, U.S. Department of Agriculture, Oxford, Mississippi

C. E. Murphree
Agricultural Engineer, National Sedimentation Laboratory, Agricultural Research Service, U.S. Department of Agriculture, Oxford, Mississippi

C. K. Mutchler
Supervisory Research Hydraulic Engineer, National Sedimentation Laboratory, Agricultural Research Service, U.S. Department of Agriculture, Oxford, Mississippi

M. A. Nearing
Agricultural Engineer, National Erosion Laboratory, Purdue University, West Lafayette, Indiana

F. J. Pierce
Associate Professor, Crop and Soil Sciences Department, Michigan State University, East Lansing, Michigan

K. N. Potter
Soil Scientist, Soil and Water Research Laboratory, Agricultural Research Service, U.S. Department of Agriculture, Temple, Texas

K. G. Renard
Research Hydraulic Engineer, Agricultural Research Service, U.S. Department of Agriculture, Tucson, Arizona

C. W. Rose
Foundation Professor, School of Australian Environmental Studies, Griffith University, Brisbane, Queensland

E. L. Skidmore
Soil Scientist, Northern Plains Area, Agricultural Research Service, U.S. Department of Agriculture, Manhattan, Kansas

M. A. Stocking
Consultant, Food and Agriculture Organization, United Nations, and Soil Scientist, Overseas Development Group, University of East Anglia, Norwich, England

L. E. Wagner
Agricultural Engineer, Agricultural Research Service, U.S. Department of Agriculture, Manhattan, Kansas

D. R. Walling
Hydrologist, Department of Geography, University of Exeter, Exeter, England

T. M. Zobeck
Soil Scientist, Wind Erosion Research Unit, Agricultural Research Service, U.S. Department of Agriculture, Manhattan, Kansas

CONTENTS

FOREWORD

The second edition of "Soil Erosion Research Methods" is a result of the enthusiastic reception of the first edition by soil erosion researchers around the world. This new version is a response to the continued and growing need to promote the use of appropriate methodology to evaluate soil erosion.

Both the first edition of 1986 and this second edition are sponsored by Subcommission C on Soil Conservation and Environment of the International Society of Soil Science (ISSS), and by the World Association of Soil and Water Conservation (WASWC). This sponsorship is a recognition by ISSS that the problems of soil erosion are increasing throughout the world. This is partially due, especially in developing countries, to lack of appropriate evaluations of the processes and causes of soil erosion in specific situations, and to the generalized use of empirical approaches to select soil conservation practices.

The main objective of Subcommission C of ISSS is to promote studies and research leading to practices and systems of soil conservation and environmental protection. One of the main steps toward reaching this objective is the identification, selection, and use of methodologies adequate to the required evaluation of soil erosion. Selection and use of the appropriate methodologies is a prerequisite for the development of soil conservation practices that are adapted to specific agroecological and socioeconomic conditions, for sustainable agriculture in countries around the world.

Again, we are grateful to the authors and editors, and to the Soil and Water Conservation Society, for making the publication of the second edition possible.

Ildefonso Pla Sentis
Chairman of Subcommission C of the ISSS

PREFACE

The second edition of "Soil Erosion Research Methods" retains the themes and layout of the first edition. However, most chapters have been revised and some additional chapters have been added. Much of the necessary updating was needed in Chapter 1 regarding current statistics available on the global magnitude of soil erosion by water and wind and on denudation rates. New chapters have also been added on modeling wind and water erosion. Extensive revisions and updating have been done in chapters dealing with assessment of erosivity and erodibility, erosion-crop productivity, measuring sediment yield from river basins, and on field plot techniques. Several new authors have made significant improvements in revising and updating the available information.

All authors responded enthusiastically in promptly revising and updating appropriate chapters, and manuscripts were revised on schedule. However, delays encountered in prompt publication of the second edition were unforeseen. In this regard, I express my thanks to all authors for their cooperation and understanding. As usual, the cooperation and support from the editorial staff of the Soil and Water Conservation Society is greatly appreciated.

Rattan Lal, Editor
October 1993

Chapter 1

SOIL EROSION by WIND and WATER: PROBLEMS and PROSPECTS

R. Lal

That accelerated soil erosion is a serious global problem is widely recognized. What is difficult to assess reliably and precisely, however, are the dimensions—the extent, magnitude, and rate—of soil erosion and its economic and environmental consequences. Information readily available in the literature often is based on reconnaissance surveys and extrapolations based on sketchy data.

Judson (*12*) estimated that river-borne sediments carried into the oceans increased from 10 billion Mg a year before the introduction of intensive agriculture, grazing, and other activities to between 25 billion and 50 billion Mg thereafter (*1*). Dudal (*4*) reported that the current rate of agricultural land degradation worldwide by soil erosion and other factors is leading to an irreversible loss in productivity on about six million ha of fertile land a year. Oldeman (*21*) estimated that human-induced soil degradation has affected 24 percent of the inhabited land area. The values for individual continents range from 12 percent in North America, 18 percent in South America, 19 percent in Oceania, 26 percent in Europe, 27 percent in Africa and 31 percent of Asia. Large areas of the earth are affected by accelerated erosion by water and wind (Table 1.1). According to the United Nations Environmental Program (*27*), crop productivity on about 20 million ha each year is reduced to zero or becomes uneconomic because of soil erosion and erosion-induced degradation. On the basis of information available worldwide on soil erosion, Higgins and associates (*9*) reported that crop yields in rainfed areas might decrease 29 percent over the next 25 years. Since the beginning of settled agriculture, soil erosion has destroyed about 430 million ha of productive land (*14*). Buringh (2) esti-

Table 1.1 Global extent of land affected by wind and water erosion (*21*).

Region	Land Area Affected by Erosion (10^6 ha)	
	Water Erosion	Wind Erosion
Africa	227	186
Asia	441	222
South America	123	42
Central America	46	5
North America	60	35
Europe	114	42
Oceania	83	16
World	1094	548

mated that the annual global loss of agricultural land is three million ha due to soil erosion and two million ha due to desertification. Of the total annual sediment load of one billion Mg carried by rivers from the continental United States, about 60 percent is estimated to be from agricultural land (*19, 20*). The off-site damages caused by sediment in the United States are estimated at $6 billion annually (*3*), including $570 million for dredging several million cubic meters of sediment from U.S. rivers, harbors, and reservoirs (*23*).

Equally ruinous are the adverse effects of wind erosion. Although wind erosion is less than water erosion on a world scale, the problem is severe in many semiarid and arid regions. However, even less research data are available for wind erosion than for water erosion. Wind erosion is the principal factor responsible for desertification. Large proportions of arid and semi-arid regions of the world are affected by desertification (Table 1.2). The basic principles governing wind erosion processes and erosion control are similar to those for water erosion. Nonetheless, the specific cause-and-effect relationships, the magnitude of wind erosion in different ecoregions, and the effectiveness of erosion control practices on management systems for wind erosion have not been fully established.

The need for improved data

These frightening statistics leave many questions unanswered on data sources, methods of data collection and extrapolation, and data accuracy and reliability. At present, the quality of available data is extremely uneven. Land use planning based on unreliable data can lead to costly and gross errors. Soil erosion research is a capital-intensive, time-consuming exercise. Global extrapolation on the basis of few data collected by diverse and unstandardized methods can lead to gross errors. Erroneous and

Table 1.2 Global extent of land affected by desertification (10^6 ha, *27*).

	Irrigated Cropland		Rainfed Cropland		Rangeland		Total Agriculturally Used Dryland	
Continent	Total	Degraded	Total	Degraded	Total	Degraded	Total	Degraded
Africa	10.4	1.9	79.8	48.9	1342.4	995.1	1432.6	1045.8
Asia	92.0	31.8	122.3	122.3	1571.2	1187.6	1881.4	1311.7
Australia	1.9	0.3	14.3	14.3	657.2	361.4	701.2	375.9
Europe	11.9	1.9	11.9	11.9	111.6	80.5	145.6	94.3
N. America	20.9	5.9	11.6	11.6	483.1	411.2	578.2	428.6
S. America	8.4	1.4	6.6	6.6	390.9	297.8	420.7	305.8
Total	145.5	43.2	215.5	215.6	4556.4	333.5	5159.7	3592.2

unreliable information is worse than having no information because it can lead to costly mistakes and misjudgments on critical policy issues.

There is an urgent need for standardizing methodologies to increase the reliability and accuracy of data on soil erosion. The information should be factual and devoid of emotional rhetoric. Some basic problems with data collection are as follows:

Data extrapolation

Extrapolations based on limited data and to regions outside the ecological limits in which experiments were conducted are often misleading. There is also a major problem with scaling. Soil erosion processes are scale-dependent. It is difficult to extrapolate data measured at plot scale to that of the watershed scale. Extrapolations on the basis of limited data can lead to gross errors. Two examples are the attempts to compile suspended sediment yield maps for Africa and South America (Figure 1.1). Fournier (*5, 6*) determined sediment yield in Africa south of the Sahara Desert on the basis of extrapolation of relationships between specific sediment yield and catchment relief in other regions, for example, continental Europe. In another case, Strakhov (*25*) used meager data and extrapolated it to the entire African continent. These two maps differ both in trend and in sediment yields by several orders of magnitude.

Walling (*29*), in collaboration with UNESCO, compiled the most up-to-date map based on sediment yield data for Africa and other regions of the world. This map, based on existing data, differs from both the Fournier and Strakhov maps. Walling indicates the dangers of insufficient data and points out that his map is still tentative, subject to further refinements. The magnitude of error caused by using Strakhov's map or Fournier's map, if

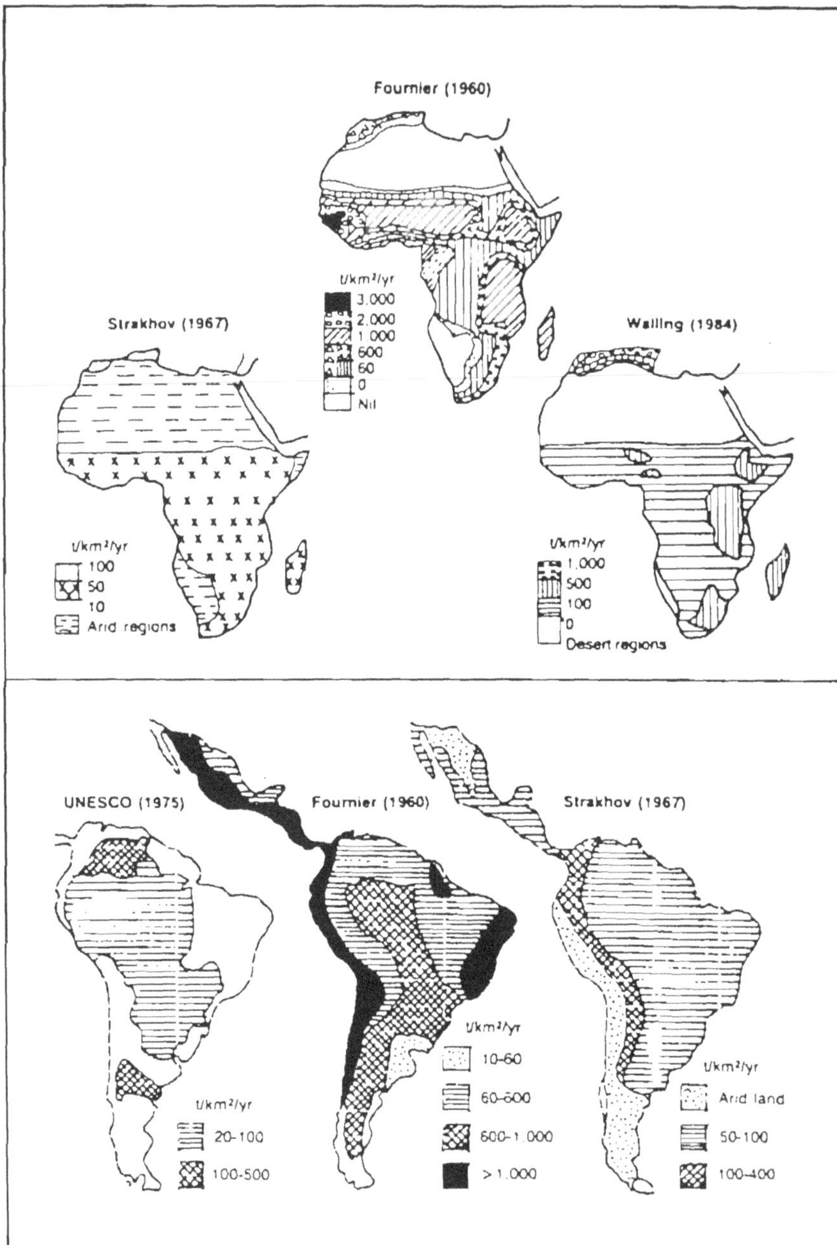

Figure 1.1 Comparison of suspended sediment yield maps of Africa (top) and South America (bottom) prepared by Fournier (*6*), and Strakhov (*13*), UNESCO (*14*) and Walling (*16*).

Walling's is, in fact, more reliable, is striking indeed. Discrepancy among estimates of global denudation rates is also shown by the data in Table 1.3. Estimates of global denudation rates range from 0.088 mm/yr to 0.30 mm/yr.

Rather than quantitatively measuring soil loss, field assessment often involves a qualitative evaluation of soil erosion. Soil movement is visually judged to be none, slight, moderate, or severe. Assessment of soil erosion by this technique is rather selective and depends upon personal judgment. It is also difficult to relate quantitatively erosion-caused effects on soil properties and crop yields to soil erosion. Information of this nature, though useful, is only of relative importance. Unless supported by data on soil profile characteristics, such qualitative information is hard to generalize. Slight erosion may have severe effects on crop yields in a marginal soil, whereas severe erosion may have little effect on crop yields in a deep, fertile soil.

Standardization of methodology and data reliability

There continues to be a lack of standard methods to measure and quantify soil erosion and its impact on productivity and environment. A wide range of plot dimensions and runoff and erosion collection systems are used to monitor the severity of erosion in different ecoregions. The equipment used is often insufficient for expected runoff; overflow thus is a common problem. A range of multidivisor systems, storage and siltation tanks and reservoirs, and sediment collection and filtration methods makes soil erosion research more of an art than a science.

In collecting samples for determining suspended sediment and bedload measurements in rivers, Walling (29) listed some important considerations, for example, equipment and procedures used for collecting samples, sampling frequency, sampling location in relation to river bank, depth of sampling, and methods of data calculation and analysis. In fact, data reliability is one of the serious problems in soil erosion research. Erosion rates assessed by an unstandardized methodology are unreliable. Regrettably, the literature is polluted with such data.

It is common to estimate soil erodibility from empirical equations developed elsewhere. The estimated values of erodibility often differ from the measured values by a factor of two to five. Similar problems are encountered in estimating rainfall erosivity without prior knowledge of drop size distribution, intensity, and kinetic energy and the relationships among these variables. Isoerodent maps prepared on the basis of "guestimated" R factors can do more harm than good.

Table 1.3 Estimates of global denudation rates.

Source	Mean Annual Load (from Walling, 1990) (10^9 Mg)	Denudation Rate* (mm)
Fournier (1960)	58.1	0.30
Gilluly (1955)	31.7	0.16
Goldberg (1976)	18.0	0.09
Holeman (1968)	18.3	0.09
Jensen & Painter (1974)	26.7	0.14
Keunen (1950)	32.5	0.17
L'Vovich et al. (1990)	15.0	0.059
Lopatin (1952)	12.7	0.06
Mackenzie & Garrels (1966)	8.3	0.04
Milliman & Meade (1983)	13.5	0.07
Pechinov (1959)	24.2	0.12
Schumm (1963)	20.5	0.10
Walling & Webb (1983)	15.0	0.075
Walling (1987)	17.4	0.088

* Assuming bulk density of 1.5 Mg/m^3

Land area = 13.1×10^9 ha

1 mm loss = 196.5×10^9 Mg

Record duration and data continuity

Data should be obtained with standard equipment and for a reasonable length of time to provide long term trends. Soil erodibility and detachability are time-dependent functions. Rain and wind erosivity assessments should be made based on long term trends in rainfall or wind velocity. The sediment load in rivers is subject to seasonal and annual fluctuations. The interannual variability of these parameters can only be ascertained from long term data obtained using standard methods. Spot measurements of erodibility, erosivity, or sediment load can be misleading.

Toward standardization of research methods

Soil erosion is among the most critical environmental hazards of modern times. Vast areas of land now being cultivated may be rendered unproductive or at least economically unproductive if erosion continues unabated. Although accelerated erosion and erosion-caused soil degradation are unmitigated evils, quantitative data on the rate of soil erosion for different environments and land uses are not available for many regions. Guesses, emotional statements, and qualitative assessments should be replaced by facts and figures supported by verifiable data. There is a need for more basic research to generate data that is accurate, reliable, and obtained by standardized methods. There is much talk about the severity of erosion, yet little is known about quantitative rates of soil erosion. What are the criteria to evaluate the degree of soil degradation caused by erosion? How can one assess the impact of soil erosion in economic terms?

For results to be comparable and for their extrapolation from one region to another, it is important that methodologies be standardized. This volume is an attempt to define standard methods and techniques used for soil erosion and sedimentation research. The emphasis is on basic data collection. Existing models also are explained, and those that can be used as tools to identify the knowledge gaps and to extrapolate the data to other regions are discussed.

REFERENCES

1. Brown, L.R., and E.C. Wolf. 1984. *Soil erosion: Quiet crisis in the world economy*. Paper No. 60. Worldwatch Institute, Washington, D.C. 50 pp.
2. Buringh, P. 1981. *An assessment of losses and degradation of productive agricultural land in the world*. Working Group on Soils Policy, Food and Agriculture Organization, United Nations, Rome, Italy.
3. Clark, E.H. 1985. *The off-site costs of soil erosion*. Journal of Soil and Water Conservation 40: 19–22.
4. Dudal, R. 1981. *An evaluation of conservation needs*. In R.P.C. Morgan (editor) Soil Conservation, Problems and Prospects. J. Wiley & Sons, Chichester, England. pp. 3–12.
5. Fournier, F. 1960. *Climat et erosion*. Presses Universitaires de France, Paris.
6. Fournier, F. 1962. *Map of erosion danger in Africa south of the Sahara*. Commission for Technical Cooperation in Africa, European Economic Community, Paris, France.
7. Gilluly, J. 1955. *Geologic contrasts between continents and ocean basins*. Geol. Soc. Am. Special Pap. No. 62. pp. 7–18.
8. Goldberg, E.D. 1976. *The Health of the Oceans*. UNESCO, Paris.
9. Higgins, G.M., A.H. Kassam, and L. Naiken 1982. *Potential population supporting capacities of land in developing world*. Project INT/75/P13. Food and Agriculture Organization, United Nations. Rome, Italy.
10. Holeman, J.N. 1968. *The sediment yield of major rivers of the world. Water Resources Research*, 4, 737–747.
11. Jensen, J.H.L., and Painter, R.B. 1974. *Predicting sediment yield from climate and topography*. J. Hydrol., Vol. 21, pp. 371–380.
12. Judson, S. 1981. *What's happening to our continents*. In: B.J. Skinner (editor) *Use and Misuse of Earth's Surface*. William Kaufman, Inc., Los Altos, California. pp. 12–139.

13. Kuenen, P.H. 1950. *Marine Geology*. John Wiley, New York.
14. Kovda, V.A. 1983. *Loss of productive land due to salinization*. Ambio 12: 91–93.
15. Larson, W.E., F.J. Pierce, and R.H. Dowdy. 1983. *The threat of soil erosion to long term crop production*. Science 219: 458–465.
16. Lopatin, G.C. 1952. *Detritus in the rivers of the USSR*. Zap. Vses. Geogr. Obsch., Vol. 14, Geografgiz, Moscow.
17. L'Vovich, M.I., N.L. Bratseva, G. Ya Karasik, G.P. Medvedeva, and A.V. Meleshko. 1990. *A map of contemporary erosion of the earth's surface*. Mapping Sciences and Remote Sensing 27: 51–67.
18. MacKenzie, F.T., and R.M. Garrels. 1966. *Chemical mass balance between rivers and oceans* Am. J. Sci., Vol. 264, pp. 507–525.
19. Milliman, J.D. and Meade, R.H. 1983. *Worldwide delivery of river sediment to the oceans*. J. Geol., Vol. 91, pp. 1–21.
20. National Academy of Science. 1974. *Productive agriculture and a quality environment*. Washington, D.C. 189 pp.
21. Oldeman, L.R. 1991–92. *Global extent of soil degradation*. Bi-annual report, International Soil Reference and Information Center, Wageningen, The Netherlands: 19–36.
22. Pechinov, D. 1959. *Vodna eroziya i to'rd ottok. Priroda*. Vol. 8, pp. 49–52.
23. Pimental, D., J. Allen, A. Beers, L. Guirand, R. Linder, P. McLaughlin, B. Meer, D. Mussonds, D. Perdue, S. Poisson, S. Siebert, K. Stoner, R. Salazar, and A. Hawkins. 1987. *World agriculture and soil erosion: Erosion threatens world food production*. BioScience 37(4): 277–283.
24. Schumm, S.A. 1963. *The disparity between present rates of denudation and Orogeny*. US Geological Survey Prof. Paper 454H: H1–H13.
25. Strakhov, N.M. 1967. *Principles of lithogenesis (volume I)*. Oliver & Boyd, Edinburgh, England.
26. United Nations Educational, Scientific and Cultural Organization and International Association of Hydrological Sciences. 1975. *Gross sediment transport into the oceans*. Paris, France.
27. United Nations Environmental Program. 1991. *Status of desertification and implementation of the UN plan of action to combat desertification. UNEP*, Nairobi, Kenya.
28. UNEP. 1990. *Desertification Revisited*. Proc. Ad. Hoc. Consultative: *Meeting on the Assessment of Desertification* UNEP-DC/PAC, Nairobi, Kenya, 344 pp.
29. Walling, D.E. 1984. *The sediment yields of African rivers*. Publication 144. International Association of Hydrological Sciences, Wallingford, England. pp. 265–283.
30. Walling, D.E. and Webb, B.W. 1983. *Patterns of sediment yield*. In: K.J. Gregory (editor), *Background to Palaeohydrology*, pp. 69–100, John Wiley and Sons, Chichester, England.

31. Walling, D.E. 1987. *Rainfall, runoff, and erosion of the land: a global review.* In: K.J. Gregory (editor) *Energetics of Physical Environment*, J. Wiley and Sons, Ltd., England. pp. 89–117.
32. Walling, D.E. 1990. *Sediment yield investigations: A perspective on recent developments and future needs.* Proc. Fourth International Symposium on River Sedimentation Nov. 1–5 1989, Beijing, China.

Chapter 2

LABORATORY and FIELD PLOTS for EROSION RESEARCH

C. K. Mutchler, C. E. Murphree, and K. C. McGregor

Soil erosion research must be based on experimental results of some form. Often, laboratory and field plots are used to obtain experimental data for predicting and evaluating soil erosion and sediment yield.

Laboratory and field plots are important only as tools for acquiring data to accomplish a research objective. Plots must be designed to furnish data that can be analyzed to test hypotheses, to quantify equation and model components, and to provide some knowledge or technique useful for soil conservation.

THE PROJECT OUTLINE

The first step in any research project is to develop a working outline that gives specific instructions for performing and completing the experiment. First is a set of objectives and justification for doing the research. These elements are necessary for two reasons: (a) to identify money and materials needed for the research and (b) to guide the research through the life of the project.

The next part of the outline is a literature review. This ensures that the research does not "reinvent the wheel." The length of time needed for a literature review depends upon the experience of the researcher and the technical guidance provided.

The procedures part of the outline describes in detail the instrumentation required and data collection methods. No single part of the outline can stand alone, except perhaps the objectives and justification. The procedures section in particular must support all other parts of the research outline.

The last part of the outline is a plan for data analysis. Experiments using erosion plots generally are expensive. Therefore, researchers must take care that all data needed to complete the analysis are collected. Also, the plan for analysis helps to ensure that questions in the objectives will be answered.

Careful preparation of a research outline is critical because of the peculiar nature of erosion research. Most procedures tend to destroy the erosion plot. This is especially true in laboratory experiments that require expensive, time-consuming efforts to prepare a plot for application of a simulated rainstorm. In short, all data must be collected correctly the first time to avoid the need to re-prepare the laboratory plot or a long wait to allow plot conditions to be reestablished.

Physical modeling of the soil surface with or without plants must always be done in full scale. Sand, silt, and clay particles cannot be physically modeled for reduced-scale experiments. Agricultural soils have many chemical and physical properties, particularly those of cohesion, that differ for each soil type. Also, land invariably has some runoff channelization tendencies that are modified by cropping patterns. These factors require plots to be full size. Likewise, plants are difficult to use except in a full-size model. Most important, rainfall cannot be scaled. Every raindrop has its own shape and velocity, depending on its size. Consequently, plot size is important to represent properly the soil and plant situation being sampled. The research outline should ensure correct use of the plot.

THE SOIL EROSION PROCESS

The term "erosion" often is used as an all-inclusive word to describe the wearing down of a landscape. However, we prefer to define erosion as the detachment or entrainment of soil particles, thus distinguishing it from deposition or sedimentation and sediment transport. Herein, we use the term sediment yield to refer to the amount of eroded material that passes a designated point at the outflow end of a plot, field, channel, or watershed (catchment). The term soil loss, expressed as quantity per unit area and time, often is used for small plots. Even in the most erodible situations, soil loss or sediment yield is limited by the transport capacity of the runoff. As runoff flows through a watershed, changes in topography, vegetation, and soil characteristics often reduce this transport capacity. There are various opportunities for deposition as a result. Gross erosion in a watershed, as predicted from plot results, must be reduced by a sediment delivery ratio (unique for each watershed) to obtain an expected sediment yield for a watershed.

Because of the surface tension properties of water, runoff concentrates as it flows downhill; this causes rills and possibly gullies, even on a plane slope. The result of this phenomenon can be seen after landscapes have eroded under the flow concentrations. Researchers have devised various systems of stream orders to study the results of runoff concentrations on the land. Although farm operators are more concerned with the total farmable areas between stream channels, the same general effect of flow concentrations is observable as rills in these areas between channels.

Raindrop impact erodes any unprotected land surface between rills and also initiates transport of detached soil particles to the rills. The complete transport system involves the initial movement of soil from the point of raindrop impact to small rills, to larger rills, to ephemeral channels, and to continuously flowing rivers. Erosion and sedimentation can occur at any point in the system. This process is called geologic erosion when it occurs without human influence. Alternatively, the process is called accelerated erosion when human activity causes increased erosion, such as land disturbed for crop production.

During agricultural operations, cultivation, row direction of planting, and water management practices designed to convey runoff without allowing excessive erosion all change the natural rill system. In row-crop agriculture, rills generally form in tillage paths, especially where the crop is cultivated. For up-and-down-hill rows, the rills enlarge as runoff amounts and rill erosion increase. For contour rows, rill enlargement is slower because of slower runoff velocities. However, the capacity of rills parallel to the rows often is exceeded under large storms; runoff overtops the rows, and rills are formed parallel with the slope. Terraces built across a slope interrupt downslope rill patterns and keep rill formation at the rudimentary stage.

Knowledge of the erosion process is necessary because erosion studies use plots that are samples of this "big picture." To be valid, results of such research must be from a representative sample of the landscape. Researchers, of course, use various plot types to study soil erosion. The major concern is that any plot is a sample of the landscape, and it must furnish data that can be analyzed and formulated for use on the land.

In summary, a land area is comprised of rills and channels and the land between them. More specifically, farmland can be described as rill and interrill areas and the erosion processes as interrill erosion, rill transport, rill erosion, and deposition. Plots can represent full-size areas of interrill erosion or rills separately, or they may represent areas that include all of the processes. Thus, erosion plots fall into three categories: small plots, USLE (universal soil loss equation) plots, and large plots.

SMALL PLOTS

Researchers use small plots to study interrill erosion separately. Such plots are typified by the square-meter plots used by Meyer and Harmon (*12*) to study interrill erosion on cropland in row crops. Laboratory plots are also commonly used to isolate a small part of the erosion process for intensive study under controlled conditions (Figure 2.1).

Small plots usually are studied using simulated rainstorms (see Chapter 4).

There is extensive literature in soil physics on the use of small soil plots. Here, we discuss only those principles that are important to erosion research.

Objectives

Generally, researchers use small plots to study basic erosion phases that are difficult to study in detail on larger plots; for example, surface sealing, aggregate stability, raindrop detachment, and splash transport. Note that these topics deal with processes on the interrill area. Experiments with rill processes require plots long enough to develop sufficient runoff. Meyer and Harmon's (*13*) study of rill and interrill erosion illustrates the use of small plots in the field under simulated rainfall.

Small plot experiments should have objectives complementary to field plot experimentation or application. Moldenhauer, for example, used small plots to study the effects of soil characteristics on erodibility (*14*). He used laboratory plots to isolate certain phases of the erosion process, then examined and verified the results with rainfall, climate, and soils in the field.

A currently popular objective is to develop a mathematical model of the erosion process. Small plots then are used to develop or verify basic operating equations that govern the physical processes of soil erosion.

Justification

The greatest justification for small plots is their utility in studying the basic aspects of soil erosion in detail. A researcher can control the parameters of the erosion process more closely on small plots than on larger plots. Experiments using small plots are more justifiable if the anticipated results have a planned application. In most cases, these experiments provide basic concepts and knowledge required for efficient developmental research.

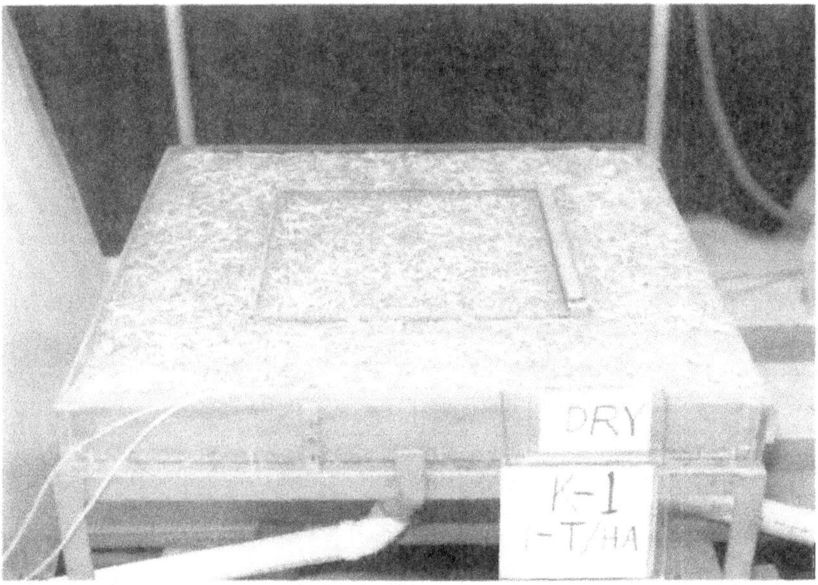

Figure 2.1 A laboratory erosion plot used to study the effects of crop residue placement on soil erosion. The 46-cm x 46-cm plot is surrounded by a border 23 cm wide. Runoff enters a slot at the bottom of the small plot and is collected in a container for measurement and sediment analysis. Rainfall is applied with a nozzle type rainfall simulator mounted 3 m over the plot.

Procedures

Problems that influence study procedures on small plots include those related to size and duplication of the soil as it is found in a field.

Edge effects, which must be considered in all plots, are greatest with small plots. Edge effects include the movement of splashed soil onto or off the plot, the addition of eroded soil to a plot treatment from an erodible soil ridge used as a boundary, a barrier effect on soil water and plant root movement, and other boundary effects caused by the restriction of splash or runoff on a plot. Laboratory plots also have a boundary under the soil plot.

Edge effects can be minimized by collecting all or almost all of the splash, generally suitable for single-drop experiments using a soil target as exemplified by Al-Durrah and Bradford (*1*), or by using a border of thin metal, for example, to delineate the plot and maintain a border area around the plot in the same condition as the plot area. For very small plots in the field or laboratory, a border area is necessary to ensure that as much splash will enter the plot as leaves the plot. Splash from raindrop impact is greatest from a surface covered with a very thin layer of water. Thus, results from any small plot will be biased if the surface water is not maintained in a condition representative of the treatment in the field. This thin layer is best described as the amount of water on an interrill surface when runoff is initiated. The phenomena is demonstrated by examining records of runoff and sediment concentration from a plot under simulated rainfall; invariably, the concentration values at the beginning of runoff are higher than at any other time in the test.

Small laboratory plots must have a bottom to hold the soil. The design of a bottom depends, of course, upon the objective of the experiment. For erosion experiments where runoff and soil loss are the primary indicators of differences in the treatments, an open bottom made of screens overlain by cloth often is used to allow free passage of soil water. This is representative of an extreme condition of layered soil, such as loam underlain with a much coarser material. Moldenhauer (*14*) found that using suction under a plot caused considerable variation, especially with thin layers of soil. Consequently, he used a 15-cm-deep soil layer with free drainage. Use of such a system is supported by considering the events that occur during a simulated rainfall experiment. Most simulator storms are 30 to 60 minutes long and surface saturation must occur before runoff begins. Hydrographs from field plots often show a constant runoff rate (and constant infiltration rate) after about 45 minutes for initial dry runs and after five to 15 minutes for wet runs made four to 24 hours after the dry run. Usually, the time for infiltration and saturation of a 15-cm soil profile is longer than the time used to evaluate erosivity with a rainfall simulator at 6.4 cm/h. Of course, lower rainfall rates change these times, and there are some unique volcanic soils with extremely high infiltration rates.

Another problem in simulating a field soil occurs because soil in the laboratory plot has been disturbed. One solution is to use soil only from the plow layer in the field, air dry it for storage, sieve it through a large screen before placing it in the plot container, and pack it into the plot container in layers (*14*). All soil must be prepared uniformly to minimize differences between replicates.

Many problems with plot preparation can be avoided by covering field locations to allow only simulated rainfall to fall on the plots. Thus, field cultivation, rainfall compaction, and other events need not be simulated.

Figure 2.2 The standard USLE plot is 22.1 m long and, in this case, 4 m wide (four crop row widths). The plot is bordered with steel strips driven into the ground and an endplate at the bottom end of the plot that also holds a runoff collecting trough. Runoff is measured with an H-flume and water-stage recorder. An aliquot sample of storm runoff is taken with a Coshocton wheel sampler.

Analysis

Small plots provide information about infiltration, detachment, and other factors influencing interrill erosion, but small plots do not give complete information about the erosion process. It is difficult to discuss any particular analysis of data from small plots because they are used for a variety of objectives. The same may be said of the objectives and justification statements discussed earlier. Most important, researchers must consider analysis when planning the experiment as well as use of the results either in field research or in application by the conservationist or land-owners.

USLE PLOTS

The second type of erosion research plot is large enough to represent the combined processes of rill and interrill erosion (Figure 2.2). Good examples are the standard erosion plots used to develop the USLE (*26, 27*). Measured erosion from these plots represents field erosion at the top of a slope where runoff begins. Plots are sufficiently wide to minimize edge or border effects, large enough for development of downslope rills, and limited to rows parallel to the slope. The topographical factors of slope length and steepness, as used in the USLE, are determined from plots of different lengths and steepnesses. Slope irregularity effects are studied with concave and convex plots. Surface roughness, residue cover, plant canopy, antecedent cropping effects, and other cropping and management effects are evaluated using specific crops and tillage sequences on standard plots. Predicting soil loss from other locations is made possible by an erodibility evaluation of each soil type and a rainfall erosivity estimate. To expand the use of data collected to field-sized areas, a factor is used to reduce the erosion estimate for contouring, stripcropping, or terracing.

The USLE plot exemplifies a representative slope length. The USLE (*27*) is A = RKLSCP, where A is soil loss per unit area, R is the rainfall factor, K is the soil erodibility factor, L and S are the slope length and steepness factors, C is the cover and management factor, and P is the support practice factor. Only A, R, and K have dimensions. In customary English units, A is tons per acre per year, R is in hundreds of foot-ton inches per acre-hour-year, and K is in ton-acre-hours per hundreds of foot-ton-inch-acres. In SI units[1], A is in t/ha · y, R is in MJ · mm/(ha · h · y), and K is in t · ha · h/(ha · MJ · mm), where t is metric tons. Note that R customarily represents the average annual accumulated storm erosivity index (EI). For a cropstage, the EI terminology is retained, and the soil loss equation is A (cropstage) = (EI)KLS(SLR)P, with the same unit used for A and EI. Also, SLR (soil loss ratio) is used instead of C for time periods not equal to one year. Referring to data from a plot, R is the accumulated rainfall EI that produced the soil loss, A, on a nine percent slope, 22.1 m (72.6 ft) long, in continuous cultivated fallow. K, then, is the ratio of A/R. In practice, slopes of nine percent are not found for all soil types, so the S factor is used to adjust the soil loss value to that expected from a nine percent slope. Likewise, the L factor is used to adjust the soil loss to that from a 22.1-m length if some other slope length is used.

---

[1] The USLE was developed using U.S. customary units. Conversions of units to the International System of Units are used according to Foster and associates (*5*). U.S. customary units are given in parentheses where it is thought helpful to avoid confusion.

Objectives

The general objective for field plots in USLE-related research was to develop a soil loss prediction equation for conservation planning, which requires an average annual soil loss estimate. This is what the unmodified USLE furnishes. The equation, like the plots used for gathering its data base, is suitable only for estimating erosion along an up-and-down-hill profile; that profile must stretch from the top of a hill to the bottom, until the slope steepness decreases to the point that deposition occurs or a drainageway is encountered. Conservationists use the equation to design conservation plans on a worst-case basis. For example, the most erodible portion of a field in question is either cut out of the tillable area, used to control the farming practices in the entire field, or used as the basis for a compromise plan that may include such practices as terraces.

Justification

The main reason for early research leading to development of the USLE was to give working soil conservationists a means of quantifying soil losses on land treated with conservation practices in comparison with untreated land. Landowners are more easily persuaded to use soil conservation methods if, for example, a scientific estimate shows that a soil loss of 40 t/ha · y is occurring which exceeds an allowable 11 t/ha · y, rather than stating that terraces reduce excessive losses. Also, the USLE assures landowners that their respective fields are being evaluated on the same basis. Soil conservation is difficult to promote because significant soil loss rates of 20 to 40 t/ha · y are not noticed easily, even by trained observers. Gross mismanagement of land that causes a soil loss of 80 t/ha · y or more almost invariably results in large (>10 cm deep) cropland rills and badly eroded drainageways. Erosion from channels is not estimated with the USLE, but channels can readily be observed in the field.

Recently, researchers have tried to modify the USLE and thereby extend its use beyond the original intent of providing an average annual soil loss estimate. Williams (23) developed MUSLE (modified USLE), which uses procedures to obtain an average K, L, S, and C for small agricultural watersheds. He replaced the R factor with a runoff factor to give a storm estimate of sediment yield. However, this requires use of a storm estimate of runoff volume and peak runoff rate from the watershed.

Researchers also have employed the USLE to estimate sediment yield; the procedure uses a sediment delivery ratio (SDR) to account for deposition of eroded material as estimated by the USLE. The dissimilarity of watersheds has made determination of sediment delivery ratios difficult.

The CREAMS model, designed to estimate storm sediment yield from small agricultural watersheds (*22*), also uses USLE factors. This model greatly extends the USLE by considering runoff, rill and interrill erosion, sediment sizes and densities, erosion of channels, and deposition of detached soil in parts of the watershed, including areas of ponded water. The justification for this model, in addition to that for the USLE, is to provide a storm model of sediment yield from the field or small agricultural watershed.

Researchers have made many other uses of the USLE parameters to devise estimates of soil loss for purposes other than agriculture. A notable example is the adaptation of the USLE for use on forest land (*3*).

Procedures

As explained earlier, the USLE-type plot samples the combined processes of rill and interrill erosion. Experience has shown that 5 m is about the minimum slope length that will adequately represent a rill system in an up-and-down-hill plot. A better length is at least 10 m. However, for studies to adapt the USLE to an area or country with little erosion data, the standard length of 22.1 m (72.6 ft) is best because the equation is normalized on that length. Mutchler (*16*) provided guidelines for researchers inexperienced in the use of erosion plots.

Anyone beginning experiments to adapt the USLE to a particular situation should start with a set of treatments that includes continuous fallow (for use as a standard for comparisons) and prevailing cropping treatments. Such a set of plots, operated with common procedures, should be established in each major climatic area of a country. These plots will furnish the essential data to adapt and establish confidence in the EI concept (or some other concept). Standard plots and treatments are necessary to correlate the data from different locations and to establish a relationship between rainfall erosivity and soil erodibility. There is no shortcut to such basic data collection, which should be the first endeavor in adapting the USLE to a new geographical area or in developing any erosion equation.

Experiments to acquire data for the L, S, and C factors do not require as much time as the R and K correlations. Such experiments can, therefore, be delayed. Of prime importance is the fact that one set of plots is only a sample of the climate and land conditions at that particular site. Data from a single location must be combined with data from other locations to permit researchers to make erosion predictions over a large area. If such data are not available from other locations, field data must be used to verify a soundly-based theoretical prediction method. Of course, a compromise usually is made. There are never enough funds to sample every specific climate-soil combination, and every theoretical equation should be verified across the range of its use to enable interpolation, which is much safer than extrapolation.

Following are more detailed guidelines for evaluating each of the USLE factors:

Rainfall erosivity and soil erodibility. These variables must be evaluated together because they are essentially the cause and effect of soil erosion.

We leave to others the argument of whether the best parameter to represent rainfall erosivity is EI used in the USLE, kinetic energy, momentum, or another. Some parameter must be used to represent and simplify the highly variable nature of rainfall distribution over even relatively small areas of 25,000 to 50,000 ha. For instance, the R factor (average annual accumulated EI) increases from 5100 MJ · mm/(ha · h) to more than 9300 MJ · mm/(ha · h) over a north-south distance of 480 km in the state of Mississippi (U.S.). In general, R changes even more where major elevation differences occur. Furthermore, rain fall initiates erosion, so erosion can be predicted only as well as one can predict rainfall and its parameters. Thus, a fairly intensive, well-distributed system of rain gauges throughout a country is essential for developing a reliable system of assessing the erosion hazard.

For the USLE, the standard erosion surface for measuring the effect of rainfall is a continuously cultivated fallow plot, 22.1 m (72.6 ft) long and 4.1 m (13.3 ft) wide, that allows development of rill and interrill erosion patterns. This surface was specified in "Instructions for Establishment and Maintenance of Cultivated Fallow Plots," a letter to field workers from D. D. Smith, dated January 1, 1961. The instruction said, "Plow plot to normal depth and *smooth immediately* by disking or cultivating two or more times except for fall plowed plots in areas where wind erosion during the winter may be serious. In such cases delay disking or cultivation until spring. Plowing shall be done each year at the time continuous row crop plots are plowed. Cultivate at row crop planting and cultivating times, in the fall at small grain seeding time, and at other times when necessary to eliminate a serious crust formation. Chemical weed control may also be necessary." The letter also emphasized the requirement that the fallow plot used for computing K be operated according to a "standard plan" by each research location having erosion plots. These plots served to allow national analysis of erosion data from all locations as well as for determining K for the soil on the plot.

Farmers in the United States now plow much less than they did in 1962, and the plowing in Smith's letter is not used at all in some parts of the world. However, the requirements of a standard treatment and plot to establish K factors and allow convenient comparison of erosion data from different locations remains fully justified.

Measurements of runoff and sediment concentrations from aliquot samples are used to compute soil loss. Samples to determine sediment concentration in the runoff are collected after each storm or each group of closely spaced storms if there is not time to collect samples before the next storm.

Concurrent rainfall measurements are made for each segment of erosion data using a recording rain gauge. The rain gauge chart gives rainfall amount per time interval, which is the basis for calculating storm EI.

Slope length (L) and steepness (S). These two factors represent plot topography. For experimental purposes, the plot used for determining the L and S factors, as well as the K factor, should be a plane surface. Such sites may be difficult to find. Also, it is desirable to collect data on slopes of nine percent for L and slope lengths of 22.1 m (72.6 ft) for S to avoid adjustments in slope data from slope lengths different than 22.1 m long, etc. However, such ideal plot sites are seldom available, especially considering the need for uniform soil conditions.

Slope length experiments must be conducted on plots of several different steepnesses because L is a function of slope length, λ, and slope steepness, θ. Because slope lengths (as defined for use in the USLE) in nature are relatively short on steep slopes and longer on less steep slopes, some assessment of the land's topography in an area or country should be made. Terraces and other structures must be used to prevent destruction of soils on slopes exceeding about 10 percent. Therefore, slope length on steep slopes is further restricted. Slope lengths on less steep slopes are much shorter than first appearance would indicate because typical slope irregularities may pond water and invalidate the USLE. Conversely, flatlands often are smoothed or land-planed to a uniform slope. In such cases, slope lengths can exceed several hundred meters.

Farmland slopes generally are restricted to those that can be farmed with machinery or that will not produce excessive soil loss. There are, of course, notable exceptions to this rule, such as mountainside slopes farmed with hand tools and the steep Palouse region in the northwestern United States.

To repeat, slope lengths and steepness of concern must be determined from a study of the location topography and research needs. Plot widths should be the same for all plot lengths or slopes. The difficulty of finding the same soil conditions on different slope steepnesses requires either slope studies within a relatively small range of steepnesses or an adjustment of soil loss measurements using K for the soil on the plot.

Slope factors can be studied with plots using any cover. However, a fallow surface results in larger amounts of soil loss, which are easier to measure accurately. Use of rainfall simulators is convenient for experiments to determine the L and S factors. As is always the case with rainfall simulation, care must be taken that the simulated storms adequately represent rainstorms received throughout the year at the locations where the research results are to be applied.

Cover and management. The C factor is "the ratio of soil loss from an area with specified cover and management to that from an identical area in tilled continuous fallow" (*27*). C factors are measured most accurately using the experimental treatment paired with a standard fallow plot. If previous research has established confidently the values of K, L, and S, the soil loss ratio and C factor can be computed by solving the USLE (P is always unity for a standard plot) and the fallow plot is not needed. However, if the research uses simulated rainfall, the fallow plot should be used to reduce the importance of verifying the accuracy of simulated storms.

Cover, canopy, and tillage differ throughout the year. Dividing the year into cropstages allows the use of an average soil loss ratio to replace a relatively small range of values. As a result, the term "C factor" is reserved for the average annual soil loss ratio, and cropstage values are called soil loss ratios. This procedure requires year-round measurements of cover, canopy, and tillage operations. Also, to allow better use of the soil loss ratio at other locations, researchers should measure the cover and canopy of weeds in the treatment. Effects of weeds then can be introduced at the location of the USLE estimate.

Cropstage soil loss ratios and average annual EI for each cropstage are used to compute expected average annual soil loss. This requires research values of the same parameters, except EI values are measured.

Rainfall simulators can be used to evaluate new cropping systems. However, simulators generally cannot be used to evaluate cold weather cropstages. This may be a handicap, depending upon climate and crop. For instance, in Mississippi, cover for row crops is lowest during the winter and early spring, but EI is nearly as high as in the warm season. Also, runoff is highest from the cool, wet soils during the cool season.

Support practices. These variables cannot be evaluated on the standard USLE plot. All rows on standard plots must be parallel to the slope because plot widths are too narrow to allow the natural runoff concentration and row break-over found in fields.

We have used contour rows on 1000-m^2 plots, 22.1-m slope length, and 46-m row length across the slope. In this case, the topography of the area was such that a 46-m contour length allowed a sufficient distance to approximate the spacing between rills resulting from row break-over. Concrete lining was necessary to prevent deposition in the 46-m-long channel leading to the runoff-measuring apparatus. Four of these plots on a 5 percent slope, two on a 2.5 percent slope, and two on a 10 percent slope were instrumented in 1957—all but the 10 percent plots are still in use. These plots were difficult to locate and required some land shaping to achieve the desired plane surface.

In general, experiments using terraces, stripcropping, or other cropping practices are expensive and difficult to establish. Because of the space required and the variability in topography and soils, research sites defy efforts to make statistical replication. The support practices, in most cases, are evaluated better on small watersheds.

Analysis

Outlining the proposed analysis of plot experiments can raise questions that cause a revision of procedures in a research outline.

R factor. Several equations are available for describing the erosivity of rainfall. McGregor and Mutchler discussed rainfall energy characteristics (*8, 10*) and geographical differences in rainfall (*20*). Wischmeier and Smith (*27*) truncated the energy-intensity equation formulated by Laws and Parsons (*6*) to read

$$E = 916 + 331 \log_{10} I; \text{ for } I \leq 3 \qquad [1]$$

and

$$KE = 1074; \text{ for } I > 3 \qquad [2]$$

where KE is kinetic energy in foot-tons per acre-inch and I is intensity in inches/hour. Thus, the KE values dependent upon rainfall intensity are close to those based on the equation derived by McGregor and Mutchler (*8*):

$$KE = 1035 + 822 \exp(-1.22\ I) - 1564 \exp(-1.83\ I) \qquad [3]$$

Equations 1, 2, and 3 can be converted to the International System of Units (SI) as described by Foster and associates (*5*).

The major point of the Mutchler and McGregor study (*20*) was that no known erosion study had distinguished between the storm types that occur throughout the year at any location. The magnitude of storm variations at a location and among locations is sufficient to justify the use of a lumped annual data set, such as that used by Laws and Parsons (*6*) or McGregor and Mutchler (*8*).

Rainfall relationships provided by Wischmeier and Smith (*27*) in *Agriculture Handbook 537* allow comparison of new research data with data previously collected and used as a basis for the USLE. Our comparison of Equations 1 and 3 using 19 years of data from 15 to 34 rain gauges resulted in almost identical annual average EI values. The restriction of Equation 2 had little effect on average EI for our rainfall conditions.

Because of large variations in storm EI, magnitude, and time of year, EI should be measured for every soil loss measurement. For the same reason, at least three years of field research data should be used to evaluate soil loss from an experimental treatment. If the aggregate EI is not representative in magnitude and distribution of the values in *Agriculture Handbook 537*, the project should continue for another year. At locations without established EI distributions, an analysis of measured rainfall distributions can be made with somewhat less confidence to determine the representativeness of the rainfall during the experiment. This same problem is magnified when using rainfall simulation to evaluate soil erodibility because soil loss is affected by time of year (temperature, antecedent moisture, etc.) in addition to storm EI.

K factor. Soil erodibility is "the soil loss rate per erosion index unit for a specified soil as measured on a unit plot..." (*27*). Measurements on a unit plot mean measurements made over a sufficiently long period to sample adequately the rainfall at the location. Wischmeier and Smith (*27*) suggested using 22-year records to compute R because of the cyclical patterns in rainfall. This restriction, however, is only a guideline. As mentioned earlier, three years of rainfall data usually represent the annual rainfall distribution adequately in northern Mississippi, where annual rainfall averages roughly 1400 mm, ranges from 1000 to 1800 mm, and is distributed fairly evenly throughout the year except during the somewhat drier late summer and fall months. In addition, the rain falls on soil that is seldom frozen during rainfall. The analysis of rainfall (and EI) representation for areas with highly variable amounts of rainfall at different times of the year requires more years of replication. The same is true for areas that have lower rainfall amounts and large portions of the year with frozen soil. Low annual rainfall would be treated more easily and more properly on a storm basis.

The K factor is a measure of the effect of rainfall and runoff on soil erosion, as measured throughout the year. A particular soil may rill more easily than another. Rain may arrive at different times after the prescribed tillage on the standard USLE plot. It may arrive at different antecedent water contents of the soil. It does arrive over a year of varying temperature. All of these factors (and others) affect the amount of erosion resulting from a unit of EI. Thus, K is not entirely a measure of the inherent erodibility of soil. K is properly computed as follows:

$$SL = a + K\,(EI) \qquad [4]$$

where SL is storm soil loss, EI is the storm parameter of energy computed with Equations 1 and 2 and maximum 30-minute intensity, and a is a constant that primarily represents the effect of soil water content and temperature at the beginning of rainfall. The restrictions on E and I given in *Agriculture Handbook 537* (*27*) and outlined below attempt to reduce a to zero to attain the proportionality of A, R, and K.

A storm is defined arbitrarily as one rain separated from another rain by more than six hours with less than 1 mm (0.04 in) of rain. Storms with less than 13 mm (0.5 in) of rain generally do not cause appreciable soil loss, defined as greater than or equal to 0.02 t/ha (0.01 ton/acre). However, short, intense bursts of rainfall less than 13 mm often exceed the infiltration rate of a plot soil and cause measurable soil loss. Therefore, rains of less than 13 mm but with a 15-minute intensity of 25 mm/h (one in/hr) or greater are included in the EI/SL relationship. Also, I30 in EI is limited to 64 mm/h (2.5 in/hr). However, we always use data from storms with low amounts of rainfall and EI if there is any measurable soil loss from the storm.

These criteria are suitable for general application of the USLE. However, they should be verified at any location or situation outside the data base of the USLE.

In our experimentation, we consider the above criteria as factual and use the following equation:

$$K = b\Sigma SL/\Sigma EI \qquad [5]$$

where b represents the other factors of the USLE that are not unity.

Because K-factor evaluation is a time-consuming process, few soils have been evaluated using rainfall. Instead, rainfall-evaluated soils have been used as a basis for evaluations using rainfall simulators. A fallow plot is relatively easy to prepare, but a careful correlation of simulated rainfall data with rainfall must be made to obtain confidence in such measured K factors. Wischmeier and associates (*25*) used a weighting procedure for accumulating the results of rainfall simulator storms of 0.5, 1.0, 1.5, and 2.0 hours to account for storm patterns of natural rainfall and antecedent soil water conditions in the U.S. Corn Belt. Other researchers have used different methods, but any K-factor evaluation with simulated rainfall must consider rainfall distribution at a location.

Using simulated rainfall is complicated by the fact that the K factor is not a constant but varies throughout the year (*17*). The K factor of a Mississippi soil varies from a low of 31 percent of annual average on August 5 to a high of 169 percent on February 4. For this reason, it is desirable to have K-factor plot data obtained under natural rainfall to correlate simulated rainfall results.

L factor. The slope length factor "is the ratio of soil loss from the field slope length to that from a 22.1-m (72.6-ft) length under identical conditions" (*27*). To the researcher, this definition of L also means the same plot width should be used for all plots in a slope-length study.

Past research has shown the L-factor relationship to be

$$L = (\lambda/22.1)^m \quad [6]$$

where λ is slope length in meters and m varies with slope steepness, θ, in degrees. The value of the exponent, m, can be computed using data from a slope-steepness experiment. Formulation of

$$m = f(\theta) \quad [7]$$

requires slope-length plots on several different slopes. Data from a set of slope-length plots can be used to fit

$$A(\lambda) = a\lambda^m \quad [8]$$

where A(λ) is the soil loss from each plot and a is a proportionality constant depending upon all of the USLE factors except slope length.

By definition,

$$L = A(\lambda/A(22.1) \quad [9]$$

However, results from a 22.1-m-long plot are not required because m can be evaluated from Equation 8 with other plots of various lengths.

The exponent m, as given in *Agriculture Handbook 537* for use by field technicians, is a set of discrete values for a range of slopes. Mutchler and Murphree (*18*) used data from Mutchler and Greer (*19*) and from earlier research to derive the following:

$$m = 1.2(\sin\theta)^{1/3} \quad [10]$$

which produced a more accurate exponent for slopes under 2 percent and made little change in the exponents given in *Agriculture Handbook 537* for steeper slopes.

Although crops can be grown on slope-length plots, a fallow surface tilled similarly to the continuous fallow treatment is best. The higher rates of erosion under fallow conditions are advantageous because they are easier to measure accurately. This same observation applies to slope steepness plots.

Foster and Wischmeier developed a procedure (*4, 27*) to evaluate LS for slopes that are slightly irregular.

S factor. The slope-steepness factor "is the ratio of soil loss from the field slope gradient to that from a 9 percent slope under otherwise identical conditions" (*27*). Evaluation or verification of the S factor is done most easily on standard plots with a fallow surface. Erodibility will be near the K value of the soil, C will be near unity, and L will be unity. Thus, data from the slope plots can be used to fit the parameters of

$$A(\theta) = f(\theta) \qquad [11]$$

and

$$S = A(\theta)/A(9\%) \qquad [12]$$

Early slope relationships were given in terms of percent slope because it was and is a convenient field measurement. The relationship given in *Agriculture Handbook 537* is based upon slope in degrees:

$$S = 65.41 \sin^2\theta + 4.56 \sin\theta + 0.065 \qquad [13]$$

This equation evolved from Smith and Wischmeier (*21*):

$$A = 0.43 + 0.30\ S + 0.043\ S^2 \qquad [14]$$

where this S is percent slope. Equation 14 was normalized by dividing by the value of soil loss, A, for S = 9 percent. Later, researchers decided to substitute sinθ for S because percent slope is 100 tanθ and tanθ gave unrealistic values at high angles. This change was primarily theoretical because the sine and tangent are not much different for the slope range where the USLE has validity.

The above slope equation was developed using data generally from three to 18 percent slopes. Murphree and Mutchler (*15*) found that Equation 13 overestimated soil loss from slopes less than three percent. They proposed a slope factor coefficient of

$$S_c = \left(\frac{\beta}{2}\right) + 0.5 \qquad [15]$$

where

$$\beta = 1 - 0.67 \exp(-160 \sin\theta) \qquad [16]$$

to modify the S factor at low slopes. The experimental work was done by shaping 0.2 percent land to steepnesses of 0.1, 0.2, 0.5, 1.0, 2.0, and 3.0 percent while preserving the A horizon, about 20 cm (8 in) deep, on all plots. The rainulator (*11*) was used to produce runoff and erosion from the plots. It is, of course, preferable to use plots without any grading; smoothing is usually a necessity, especially on less steep slopes.

McCool and associates (*7*) developed the following equation for slopes of 9 to 60 percent in the dryland grain region of the northwestern United States:

$$LS = (\lambda/22.13)^{0.3} (s/9)^{1.3} \qquad [17]$$

where s is percent slope. They used data from field observations and erosion plot data to formulate the equation.

Experiments to collect data for a generalized S factor are difficult, primarily because the same soil cannot be found on a wide range of steepnesses.

C factor. The cover and management factor "is the ratio of soil loss from an area with specified cover and management to that from an identical area in tilled continuous fallow" (*27*). Evaluations of cropping systems can be made using erosion plots under natural or simulated rainfall. In both cases, the best method is to compare a cropping treatment and a fallow treatment on identical plots. Thus,

$$SLR = A(\text{treatment})/A(\text{fallow}) \qquad [18]$$

The C factor for a particular cropping system is computed from soil loss ratios (SLR) that are computed using Equation 18 for a cropstage that is a period of time when "cover and management effects may be considered approximately uniform" (*27*). Cropstages are defined in *Agriculture Handbook 537*. The researcher measures SLR for the numbers (n) of cropstage periods, and the USLE application uses the following:

$$C = \sum_{i=1}^{n}(EI_i \cdot SLR_i) \Big/ \sum_{i=1}^{n} EI_i \qquad [19]$$

Equation 19 implies that SLRs are constant for all parts of a country or area of USLE usage and accounts for different EI distributions at different locations within the country.

Standard-size erosion plots under rainfall are used in blocks with one fallow plot and several cropping treatments. In Mississippi, three years or more of replicated data are required, depending upon an evaluation of the rainfall received, to evaluate a cropping system properly. Areas with less rainfall may require more annual replications to sample the expected EI or rainfall distribution properly.

Recently, there has been increased interest in further understanding the C factor using subfactors based on suggestions by Wischmeier (*24*). Consequently, we routinely estimate cover and canopy throughout the year, record observations of surface configuration and tillage application, and record cropping history. Also, we measure crop yield and details of residue management to describe SLRs for the cropping system more properly. Because weed varieties and populations differ by location, we measure cover and canopy separately (*9*) to allow reporting of SLRs without weeds. The ratios can then be adjusted for expected weeds by the USLE user.

Use of a rainfall simulator to determine SLRs requires careful analysis to represent properly the EI and crop growth of a particular cropstage.

Outdoor use of rain simulation is limited to the warm seasons of the year; simulators are difficult to use in cool, wet conditions and impossible in freezing temperatures. We have used a rainulator primarily for slope studies, C-subfactor experiments, and related experiments on the mechanics of erosion.

Use of a fallow plot to evaluate SLRs with Equation 18 requires that the fallow plot be prepared at least two years prior to the comparison with treatment plots to eliminate the effects of previous cropping. Also, fallow plots must not be used too long because the soil will deteriorate from excessive erosion and, hence, will not properly represent the soil used for adjacent cropping treatments. There are two ways to avoid this problem: either use the fallow plot only long enough to determine K for the soil or use the SLRs for a carefully evaluated crop that will not significantly change the soil erodibility.

The SLR can be evaluated using a known K with the following equation:

$$\mathrm{SLR} = \frac{\Sigma \mathrm{SL}}{\mathrm{K\,L\,S\,P}\ \Sigma \mathrm{EI}} \qquad [20]$$

If a standard plot is used, L and P are unity (S is not unity). The SLR is computed by dividing the accumulated soil loss by the accumulated EI for the particular cropstage (replicated by years) and by a constant representing K and S.

If a check plot (with a known C) is used, then the equation is as follows:

$$\mathrm{SLR} = \frac{\Sigma \mathrm{SL(treatment)}}{\Sigma \mathrm{SL(check)}} \cdot \frac{\mathrm{SLR(check)}}{1} \qquad [21]$$

In our research, we use a check plot to increase confidence in our results even if we know the soil erodibility.

P factor. The support-practice factor "is the ratio of soil loss with a support practice like contouring, stripcropping, or terracing to that with straight-row farming up-and-down the slope" (*27*). As discussed earlier, support practices cannot be evaluated using standard USLE plots.

The effectiveness of support practices are difficult to evaluate. Consequently, few data are available for this purpose. P factors must be measured on slope widths that will evaluate row break-over effects. The proposed analysis of our slope-segment plots (Figure 2.3) depends upon data from standard plots with the same crop on both types of plot. Thus,

$$\mathrm{P(contouring)} = \frac{\mathrm{A(contouring)}}{\mathrm{A(up\text{-}and\text{-}down)}} \qquad [22]$$

which follows the definition given earlier because both plots have the same slope length. However, adjustments must be made for any differences in slope and soil.

Figure 2.3 A contour segment plot (46-m contour rows and 22.1-m slope length) used to determine the reduction in soil loss due to tilling and planting on the contour. The plot is bordered on the sides and top by earthen ridges. A concrete-lined channel along the bottom of the plot carries runoff to the runoff collector and H-flume equipped with a water-stage recorder. Sediment concentrations in the aliquot sample taken with the Coshocton sampler are used with measured runoff to compute storm soil loss.

Tillage treatments, ranging from relatively smooth surfaces to row ridges, impose various surface configurations on a field. The combination of these configurations with slope length, steepness, and topography makes the design of a research program to evaluate contouring comprehensively almost impossible. The same is true of stripcropping. Consequently, present-day P-factor values for these practices have been systematized and evaluated primarily by group agreement among researchers and USLE users and with very little data.

A contour plot, like our slope segment, should carry runoff out of the end of the rows (side of the plot) for smaller rains and through the interior of the plot at some low point in the rows for a storm sufficiently large to cause row break-over. After row break-over, most runoff would exit at that point until subsequent tillage rebuilt the row ridges. Solving the row slope design problem would, of course, provide the answer or bias the answer to the contour problem. The best solution to this dilemma is to use well-described watersheds and adequate annual replications to compare contour and up-and-down hill tillage.

LARGE PLOTS—UNIT-SOURCE WATERSHEDS

The third type of erosion research plot is a small catchment or watershed large enough to include at least one natural drainageway. These often are called "unit-source watersheds," which also implies that only a single crop is grown on the entire catchment. This plot would contain interrill areas, rills, and small ephemeral channels that may or may not be cultivated or planted to the same crop as the remainder of the catchment area. Also, terraces or other practices may be included in this large plot. The attractive feature of this plot is that it may combine the results of all the erosion processes and conservation measures in a single measurement. However, use of large plots affords little opportunity to learn about the different parts of the erosion process. Also, natural catchments are not replicated easily to increase confidence in the results.

Some support practices affecting erosion require so much space that they are difficult to evaluate on any plot smaller than a small unit-source watershed of several hectares. Contouring and stripcropping are among such practices. Terraces can be evaluated on watersheds, but because they concentrate and control runoff, they are evaluated better as a simple terrace interval unit. Small watersheds have a major advantage—farmers are more willing to believe results from larger scale experimentation. They also have a major disadvantage—research results on one watershed are difficult to apply to other watersheds.

Small watersheds are best for research that seeks to evaluate conservation systems or to verify a modeling concept. Such research is justified by the need for basic data to give the researcher confidence that formulations are correct and to persuade conservation workers that conservation systems effectively conserve soil and are usable by farmers.

Procedures

Site selection is critical for research with small watersheds because they must be representative of a significant area of land. However, the watershed characteristics must still be measured carefully. In contrast to plots used to study isolated parts of the erosion process, small watersheds allow study of the entire process as affected by topography and climate. Transfer of the results to unmeasured watersheds and verification of models then can be made more easily from a comprehensive description of the research watershed.

As with field plots, a recording rain gauge record of rainfall is needed to compute rainfall intensities and energy and for analysis of rainfall distribution. Limits of a watershed should be established with some type of border. We usually use a low earthen ridge turned up with a farm implement.

Runoff and sediment yield from small watersheds are too great for continuous sampling. On flatland watersheds, we have used a Parshall flume and a water-stage recorder to measure runoff. A pumping sampler can extract sediment concentration samples at preset time intervals, usually five minutes initially, until storm analysis indicates that a longer interval is more appropriate. On small upland watersheds, we have used an H-flume and a water-stage recorder with an intermittent proportional sampler to obtain sediment concentrations. Details of instrumentation can be found in *Agriculture Handbook 224* (2).

The size of the flume should be determined on the basis of the largest runoff rate expected because larger storms produce the largest amounts of sediment. For USLE-type plots, the design runoff rate used is 100 percent of the five-minute-duration rainfall intensity. For larger plots or unit-source watersheds, rainfall duration can be estimated from the estimated time of concentration for the watershed. Some local experience in estimating runoff rates is highly desirable.

All conditions of cropping and management practices should be recorded, including estimates of residue cover, weed cover, and crop canopy, at a sufficient number of times during the year to allow a description of changes. The USLE factors can be used as a guideline for describing the characteristics of the farmable part of a watershed. The efficiency of sediment transport from interrill areas and the erosion of rills and small channels becomes more important for watersheds. Thus, researchers should pay particular attention to drainage and rill pattern characteristics.

Sediment analysis of the concentration samples is done by routine laboratory methods. Both total sediment and particle-size distribution analyses should be made. Particle size is important in sediment transport and deposition.

Most data collection problems arise because the equipment is not ready to operate when runoff occurs. Therefore, gauging stations should be inspected at least once a week. Operability of the samplers should be verified after a storm. However, seemingly unexplainable problems can arise between storms. All of our watershed equipment is automatic, but there is no substitute for the attention of a good technician to keep the equipment operable.

Analysis

Data from small watersheds are analyzed on a storm basis. We use the same storm definition for watershed plots as for USLE plots. To determine storm sediment yield, runoff increments between concentration samples are multiplied by the average sediment concentration to obtain sediment yield increments, which are then accumulated to determine storm sediment yield. Missing data are a particularly difficult problem because watersheds are so difficult to replicate. If a runoff measurement is lost, the storm sediment yield is estimated from examination of the long term record. If all the sediment samples are lost, the runoff can be used to help estimate the storm sediment yield. More commonly, part of the sediment concentration samples may be lost due to sampler malfunction. If most of the sediment samples are available, they can be used to construct a smooth sediment concentration graph to estimate the missing values by interpolation. Of course, if only representative storm results are desired, a storm with too many missing measurements is discarded.

Small watersheds usually have a high SDR (sediment delivery ratio), which means that little eroded material is deposited in depositional areas. Therefore, small watersheds may be used to evaluate factors of the USLE with little error if significant concentrated flow erosion is not present. The specific SDR can be estimated by inspecting the watershed for evidence of deposition. SDR also can be estimated by analyzing the primary particle size of the watershed soil and of the sediment yield; most of the clay fraction of the eroded soil is transported from the watershed.

Researchers can use several procedures on small watersheds to evaluate conservation practices. A single watershed can be operated with a calibration treatment for several years, then subsequently with some desired treatment, such as contouring. The change in sediment yield, adjusted for rainfall differences, then is the treatment effect. Also, two watersheds with different treatments can be paired and operated several more years. Both of these procedures are time-consuming, and some changes in watershed performance due to previous erosion must be accounted for. A better system is to use a single watershed to verify a model. Use of a model is quicker and often enhances use of the experimental results on unmeasured areas. Inherent in the modeling approach is the requirement of some experience to use as the basis for forming a concept of the watershed sediment yield process.

CONCLUSIONS

Erosion research includes studies using plots representative of the landscape or part of the landscape that is of interest. A carefully written research outline ensures that the study focuses on a well-defined part of the erosion process and that an analysis of the data can provide either basic knowledge useful for further research or for application to erosion problems in the field.

Three types of erosion plots are useful—small plots used primarily for studying interrill erosion; USLE standard plots that include both rill and interrill processes; and the small, unit-source watershed that includes rill, interrill, deposition, and small channel processes.

Small plots probably are most useful for establishing principles and accumulating knowledge about rainfall-induced erosion for further use in applied research. Experiments using small plots can be controlled more closely than experiments with large plots. This allows studies of isolated variables. Some problems with research using small plots are created by the large spatial variability of soil and rainfall.

The USLE standard plot is perhaps the most important plot because of the extensive U.S. data base collected generally in a uniform manner and the proven usefulness of the USLE to soil conservationists. Also, USLE plots are convenient for use as a reference in discussing the principles of field-plot experimentation.

Small, unit-source watersheds are the only type of plot large enough to evaluate conservation structures and practices, such as contour rows. Being difficult to replicate, they are most useful for verification of runoff and sediment yield models.

These three types of plots are by no means all those used for the many diverse objectives of research in soil erosion.

REFERENCES

1. Al-Durrah, M., and J.M. Bradford. 1981. *New methods of studying soil detachment due to waterdrop impact.* Soil Science Society of America Journal 45(5): 949-953.
2. Brakensiek, D.L., H.B. Osborn, and W.J. Rawls. 1979. *Field manual for research in agricultural hydrology.* Agriculture Handbook 224. U.S. Department of Agriculture, Washington, D.C. 550 pp.
3. Dissmeyer, G.E., and G.R. Foster. 1980. *A guide for predicting sheet and rill erosion on forest land.* Technical Publication SA-TP 11. Forest Service, U.S. Department of Agriculture, Washington, D.C. 40 pp.

4. Foster, G.R., and W.H. Wischmeier. 1974. *Evaluating irregular slopes for soil loss prediction.* Transactions, American Society of Agricultural Engineers 17(2): 305-309.
5. Foster, G.R., D.K. McCool, K.G. Renard, and W.C. Moldenhauer. 1981. *Conversion of the universal soil loss equation to SI units.* Journal of Soil and Water Conservation 36(6): 355-359.
6. Laws, J.O., and D.A. Parsons. 1943. *The relation of raindrop size to intensity.* Transactions, American Geophysical Union. 24: 452-460.
7. McCool, D.K., W.H. Wischmeier, and L.C. Johnson. 1982. *Adapting the universal soil loss equation to the Pacific Northwest.* Transactions, American Society of Agricultural Engineers 25(4): 928-934.
8. McGregor, K.C., and C.K. Mutchler. 1976. *Status of R-factor in North Mississippi.* In: *Soil Erosion: Prediction and Control.* Soil and Water Conservation Society, Ankeny, Iowa. pp. 135-142.
9. McGregor, K.C., and C.K. Mutchler. 1983. *C factors for no-till and reduced-till corn.* Transactions, American Society of Agricultural Engineers 26(3): 785-788, 794.
10. McGregor, K.C., C.K. Mutchler, and A.J. Bowie. 1980. *Annual R values in North Mississippi.* Journal of Soil and Water Conservation 35(2): 81-84.
11. Meyer, L.D., and D.L. McCune. 1958. *Rainfall simulator for runoff plots.* Agricultural Engineering 39(10): 644-648.
12. Meyer, L.D., and W.C. Harmon. 1979. *Multiple-intensity rainfall simulator for erosion research on row sideslopes.* Transactions, American Society of Agricultural Engineers 22(1): 100-103.
13. Meyer, L.D., and W.C. Harmon. 1985. *Sediment losses from cropland furrows of different gradients.* Transactions, American Society of Agricultural Engineers 28(2): 448-453, 461.
14. Moldenhauer, W.C. 1965. *Procedure for studying soil characteristics using disturbed samples and simulated rainfall.* Transactions, American Society of Agricultural Engineers 8(1): 74-75.
15. Murphree, C.E., and C.K. Mutchler. 1981. *Verification of the slope factor in the universal soil loss equation for low slopes.* Journal of Soil and Water Conservation 36(5): 300-302.
16. Mutchler, C.K. 1963. *Runoff plot design and installation for soil erosion studies.* ARS-41-79. Agricultural Research Service, U.S. Department of Agriculture, Washington, D.C. 26 pp.
17. Mutchler, C.K., and C.E. Carter. 1983. *Soil erodibility variation during the year.* Transactions, American Society of Agricultural Engineers 26(4): 1102-1104, 1108.
18. Mutchler, C.K., and C.E. Murphree. 1981. *Prediction of erosion on flatlands.* In: R.P.C. Morgan (ed.), *Soil Conservation, Problems and Prospects.* John Wiley and Sons, Chichester, England. pp. 321-325.
19. Mutchler, C.K., and J.D. Greer. 1980. *Effect of slope length on erosion from low slopes.* Transactions, American Society of Agricultural Engineers 23(4): 866-869, 876.

20. Mutchler, C.K., and K.C. McGregor. 1979. *Geographical differences in rainfall.* In: *Proceedings, Rainfall Simulator Workshop, Tucson, Arizona, March 7-9, 1979.* ARM-W-10. U.S. Department of Agriculture, Washington, D.C. pp. 8-16.
21. Smith, D.D., and W.H. Wischmeier. 1957. *Factors affecting sheet and rill erosion.* Transactions, American Geophysical Union 38: 889-896.
22. U.S. Dept. of Agriculture. 1980. *CREAMS--A field scale model for chemicals, runoff, and erosion from agricultural management systems. Volume I: Model documentation and Volume II: User manual.* Conservation Research Report No. 26. Washington, D.C.
23. Williams, J.R. 1975. *Sediment-yield prediction with universal equation using runoff energy factor.* In: *Present and Prospective Technology for Predicting Sediment Yield and Sources: Proceedings, Sediment Yield Workshop, USDA Sedimentation Laboratory, Oxford, Mississippi, November 28-30, 1972.* ARS-S-40. U.S. Department of Agriculture, Washington, D.C., pp. 244-252.
24. Wischmeier, W.H. 1975. *Estimating the soil loss equations's cover and management factor for undisturbed area.* In: *Present and Prospective Technology for Predicting Sediment Yield and Sources: Proceedings, Sediment Yield Workshop, USDA Sedimentation Laboratory, Oxford, Mississippi, November 28-30, 1972.* ARS-S-40. U.S. Department of Agriculture, Washington, D.C., pp. 118-124.
25. Wischmeier, W.H., C.B. Johnson, and B.V. Cross. 1971. *A soil erodibility nomograph for farmland and construction sites.* Journal of Soil and Water Conservation 26(5): 189-193.
26. Wischmeier, W.H., and D.D. Smith. 1965. *Predicting rainfall erosion losses from cropland east of the Rocky Mountains.* Agriculture Handbook 282. U.S. Department of Agriculture, Washington, D.C. 47 pp.
27. Wischmeier, W.H., and D.D. Smith. 1978. *Predicting rainfall erosion losses.* Agriculture Handbook 537. U.S. Department of Agriculture, Washington, D.C. 58 pp.

Chapter 3

MEASURING SEDIMENT YIELD from RIVER BASINS

D. E. Walling

INTRODUCTION

The context

Information on the sediment yield at the outlet of a river basin can provide a useful perspective on the rate of erosion and soil loss in the watershed upstream. Many countries undertake such measurements of sediment transport as part of national hydrometric programs or more specific research investigations. Therefore, some background data are often available. In nearly all cases, these measurements refer to the suspended sediment load of the river. The bedload component is not included because of the practical difficulties in obtaining such measurements (*27*). However, in most rivers the suspended sediment component will account for the majority of the total load. Furthermore, this is of most relevance in soil erosion investigations, since most of the bedload will be eroded from the channels. Where data are unavailable or detailed investigations are to be initiated, measurement programs can be established using standard equipment and procedures (*46*). It is, however, essential to realize that there are a number of constraints that must be recognized in attempting to use sediment yield measurements in soil erosion studies. These relate both to measurement problems and associated data reliability and to difficulties of interpreting the resultant data to provide information regarding erosion rates in the upstream watershed.

A global perspective

On a global basis, documented suspended sediment yields include minimum values well below 2.0 t/km^2/yr. For example, Douglas (*25*) cites yields of 1.3 t/km^2/yr for the Brindabella catchment (26.1 km^2) and 1.7 t/km^2/yr for the Queanbeyan River (172 km^2) in the Southern Tablelands and Highlands of New South Wales, Australia. Values of less than 1.0 t/km^2/yr have been reported for several rivers in Poland (*10*). Improved availability and global coverage of sediment yield data in recent years have changed the perception of the upper bounds of sediment yield significantly. Values in excess of 10,000 t/km^2/yr have been reported for several rivers (Table 3.1). The highest value listed in Table 3.1 relates to a drainage basin in the middle reaches of the Yellow River, which drains the gullied loess region, an area that is well known for its severe erosion problems. Converted to local rates of soil loss, the sediment yields of the Huangfuchuan River represent soil erosion rates on the order of 500 t/ha/yr.

Sediment yield data are available for many areas of the world, but the global coverage currently is inadequate to produce a reliable and detailed world map of sediment yields. Figure 3.1 nevertheless represents a tentative map, prepared by the author, of the global pattern of suspended sediment yield from river basins on the order of 1000 to 10,000 km^2 in area. The pattern discussed in detail elsewhere (*116*) reflects the interaction of climatic, topographic, geologic, and other physiographic controls and provides some indication of the global distribution of fluvial erosion.

Problems of interpretation

Sediment yield measurements possess the advantage of providing a spatially integrated assessment of erosion rates in the upstream catchment area and thereby avoid many of the sampling problems associated with direct measurements. Thus, in principle, measurements of sediment yield at a single point at the basin outlet can provide information on *average* rates of erosion within the basin, whereas a large number of plot or similar measurements might be required in order to derive an equivalent average. However, there are several major problems that must be recognized in any attempt to provide meaningful information about on-site rates of erosion and soil loss within a drainage basin. The first relates to the processes of sediment delivery or conveyance interposed between on-site erosion and downstream sediment yield (*63, 85, 107*). Only a fraction, and often a rather small fraction, of the sediment eroded within a drainage basin reaches the basin outlet and is represented in the sediment yield. Deposition and temporary or permanent storage may occur on the slopes, particularly where gradients decline downslope; at the base of a slope; in swales; on the flood-

Table 3.1 Recently reported values of sediment yield in excess of 10,000 t/km²/yr.

Location	River	Drainage Area (km²)	Mean Annual Suspended Sediment Yield (t/km²/yr)
China (mainland)	Huangfuchuan	3199	53500
	Dali	96	25600
	Dali	187	21700
China (Taiwan)	Tsengwen	1000	28000
Kenya	Perkerra	1310	19520
Java	Cilutung	600	12000
	Cikeruh	250	11200
New Guinea	Aure	4360	11126
North Island, New Zealand	Waiapu	1378	19970
	Waingaromia	175	17340
	Hikuwai	307	13890
South Island, New Zealand	Hokitika	352	17070
	Cleddau	155	13300
	Haast	1020	12736

Based on Walling and Webb (*118*)

plain; and in the channel itself. The relative magnitude of this loss tends to increase with increasing basin size. Hadley and Shown (*45*), for example, indicated that only 30 percent of the sediment eroded in many of the small tributary basins (0.5–5.2 km²) of the Ryan Gulch basin in northwestern Colorado reaches the main valley. In turn, only 30 percent of this sediment is transported to the mouth of the 124.8 km² basin. Similarly, a countrywide study of 105 agricultural production areas in the United States by Wade and Heady (*104*) documented a range of sediment output between 0.1 percent and 37.8 percent of gross erosion.

The concept of a sediment delivery ratio (SDR) has been introduced as a means of quantifying these effects (*38, 57, 86*) and of converting measurements of sediment yield (SY) to estimates of gross erosion rates (E) in the upstream basin. The SDR is the ratio of sediment delivered at a basin outlet (t/km²/yr) to the gross erosion within the basin (t/km²/yr) and thus SY = E • SDR. SDR values as low as 0.05, or 5 percent, are frequently cited. However, considerable uncertainty surrounds the methods available for calculating or estimating SDR for a drainage basin.

A second problem in attempting to link downstream sediment yield to upstream erosion rates is the temporal discontinuity that may be involved in sediment delivery. Sediment eroded at one location may be stored temporarily and subsequently remobilized several times before reaching the drainage basin outlet. Sediment transported out of a basin may therefore reflect the recent history of erosion and sediment delivery rather than con-

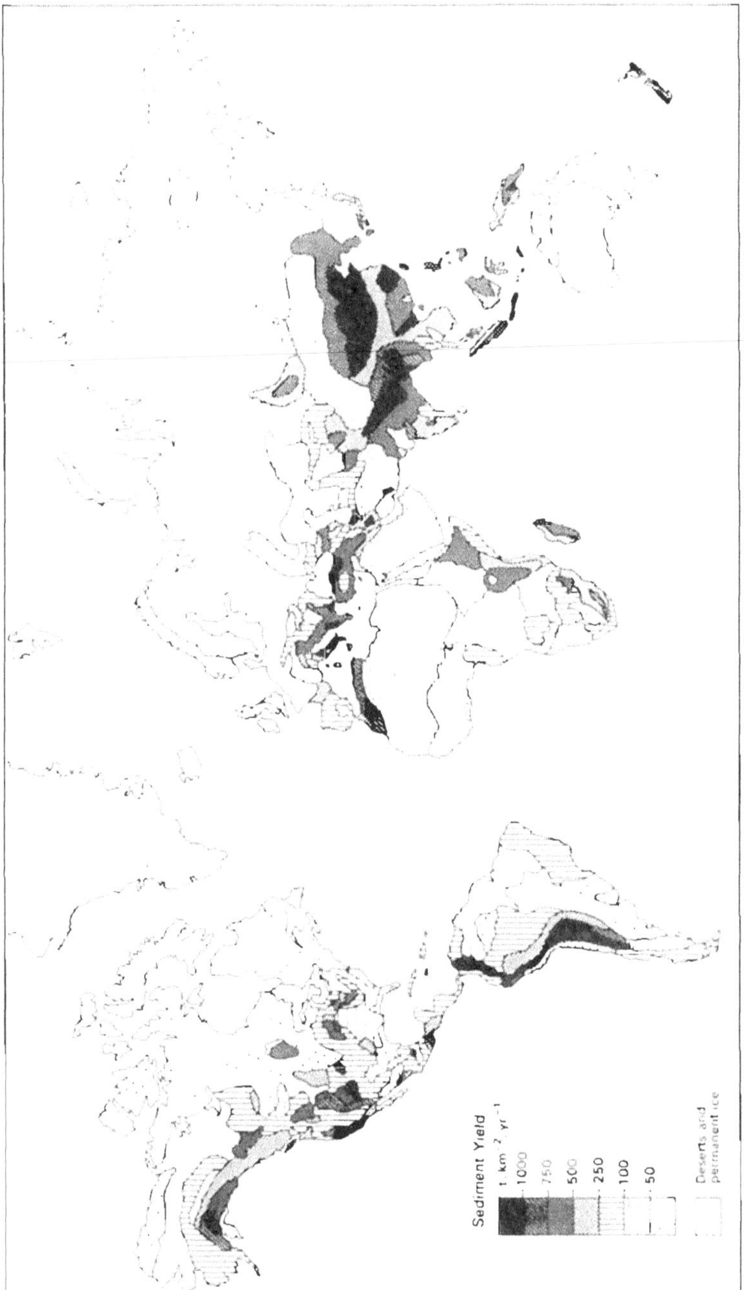

Figure 3.1 A tentative map of the global pattern of suspended sediment yield.

temporary erosion in the headwater areas. This situation is classically exemplified by the work of Trimble (*98, 99, 100*) in Georgia and in the Driftless Area of Wisconsin in the United States. Severe upland soil erosion occurred within these areas during the late 19th and early 20th centuries, but most of this sediment was deposited in the valley floors. Only about 5 percent of the eroded material reached the basin outlets. Soil conservation measures implemented in the 1930s and thereafter reduced upland erosion by about 25 percent but failed to reduce sediment yield significantly, because sediment stored in the valley deposits was remobilized (cf Figure 3.14). Meade described a similar situation for other rivers in the eastern United States (*62*). Another, but essentially opposite, example of this problem is to be found in the work of Graf (*41*) who has suggested that much of the reduction in the sediment loads of the Colorado River and its tributaries in the southwest United States which has occurred since the 1940s is due to a dramatic reversal in the behavior of the alluvial valleys and channels upstream, rather than reduction in erosion rates within the watershed. In this case erosional or throughput channel conditions were replaced by aggradation, causing increased losses to channel storage. Graf (*41*) indicated that the amount of sediment currently stored in the canyon floors of the Colorado and its tributaries is equal to about 200 times the present annual sediment yield.

A third problem in relating sediment yield downstream to soil erosion upstream stems from the fact that the sediment transported by a river represents material derived from a variety of sources in addition to upland soil erosion. These sources include channel and gully erosion and mass movements reaching the channel network. For example, in a study of the Ader Dutchi Massif region in Niger, Heusch (*50*) used plot studies to estimate local soil erosion rates within the 117 km^2 basin at about 8 t/ha/yr.. On the other hand, the sediment yield from the basin, estimated from reservoir surveys, was about 40 t/ha/yr. Heusch concluded that about 75 percent of the sediment reaching the basin outlet was derived from channel erosion and gullying. If the SDR associated with soil erosion from the slopes of the watershed was less than 100 percent, this value of 75 percent may itself be a substantial underestimate of the contribution derived from the additional sources.

The data in Table 3.2 effectively exemplify the magnitude of the problems involved in relating sediment yield data from river basins to rates of soil erosion in the watershed upstream. Table 3.2 compares measured sediment yields from a number of African river basins with the estimates of contemporary soil erosion rates within these basins obtained from the United Nation's Food and Agriculture Organization map of current rates of soil degradation by water erosion in Africa north of the equator (*36*). In all

Table 3.2 Comparisons of measured suspended sediment yields from several African drainage basins with estimated rates of contemporary soil loss by water erosion depicted on the FAO map of soil degradation (*36*).

River	Country	Basin Area (km^2)	Suspended Sediment Yield ($t/km^2/yr$)	FAO Estimate of Soil Loss ($t/km^2/yr$)
Watari	Nigeria	1,450	483a*	1,000–5,000
Bunsuru	Nigeria	5,900	438a	1,000–5,000
Senegal	Mali	157,400	14.6b	1,000–5,000
Faleme	Mali	15,000	40b	1,000–5,000
Hammam	Algeria	485	198c	1,000–5,000
Kebir Ouest	Algeria	1,130	92c	1,000–5,000
Mesanu	Ethiopia	150	1,680d	5,000–20,000

*Reported by (a) Oyebande (*77*), (b) Comité Inter Africain D'études hydrauliques, (c) Demmak (*21*), Virgo and Munro (*103*).

cases the estimates of soil erosion rates are about an order of magnitude greater than the reported sediment yields. This discrepancy could be explained in terms of the SDR being of the order of 10 percent in these drainage basins. This would be consistent with current views on the likely magnitude of such ratios in sizeable river basins. It also suggests that the sediment delivery problem probably represents the major difficulty in any attempt to relate downstream sediment yields to local erosion rates in the watershed upstream.

RELIABILITY OF EXISTING SEDIMENT YIELD DATA

In some locations, appropriate sediment yield data may be available from existing monitoring programs. However, before such data are used for further research or planning, careful consideration must be given to their reliability. A useful, although perhaps extreme, example of the potential magnitude of this problem is provided by recent work on the suspended sediment loads of New Zealand rivers by Griffiths (*42*) and Adams (*1*). Both workers used the same basic discharge and concentration data collected by the New Zealand Ministry of Works to estimate the mean annual suspended sediment load of the Cleddau River, which drains a 155 km^2 basin in the southwestern part of South Island. Both used rating curve procedures to calculate the loads, but their published loads of 13,300 $t/km^2/yr$ and 275 $t/km^2/yr$ differ by nearly two orders of magnitude.

In some situations, several different estimates of sediment yield may be available from existing sources as a result of the use of different load

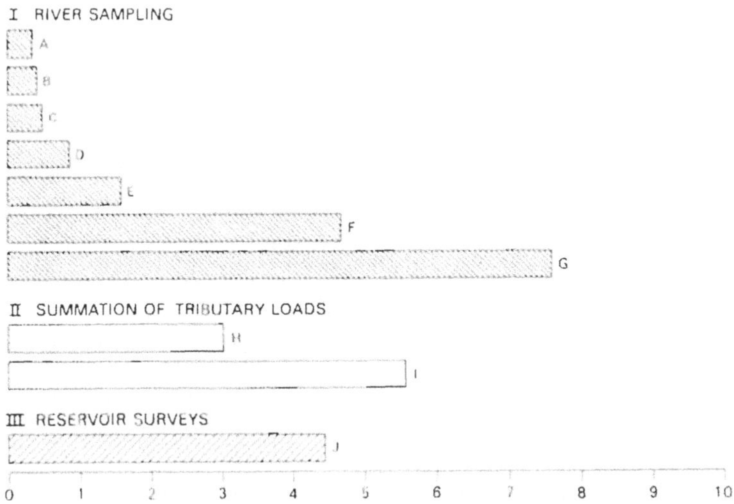

Figure 3.2 Estimates of the mean annual suspended sediment load of the Upper Tana River, at Kamburu, Kenya, produced by different workers. Based on Walling(*109*).

calculation procedures or different periods of record. For example, Walling (*109*) reports how several researchers have attempted to estimate the suspended sediment yield of the Upper Tana River in Kenya over a period of 20 years (Figure 3.2). These estimates range over an order of magnitude. Estimates A-E were derived using essentially the same basic flow and sediment concentration data. Estimates F and G were based on a new sampling program. Comparison of their load estimates for the main river with those for the upstream tributaries prompted other researchers to suggest that the former were underestimates. They produced revised estimates (H and I) by summing and extrapolating the estimates for the individual upstream tributaries. More recent surveys of sedimentation at Kamburu Reservoir coupled with an estimated trap efficiency for the reservoir of 75 percent, provide an alternative basis for estimating the sediment load. Current opinion would, in fact, indicate that the actual mean annual suspended sediment load of this river probably now exceeds 5 million t/yr, although it is likely to have increased markedly over the past 20 years as a result of land use change.

Walling and Webb (*115, 117*), Dickinson (*23*), and Allen and Petersen (*3*) have discussed factors involved in assessing the reliability of sediment yield data. These include the representativeness of the suspended sediment samples collected from the river channel and the efficiency of the laboratory techniques used to determine the concentration of individual samples. Many measurement programs take samples at or near the water surface. Use of such samples could underestimate the sediment load of a river, perhaps by as much as 25 percent. Conversely, from a study on the Washita River in Oklahoma (U.S.), Allen and Petersen (*3*) suggested that equal transit rate (ETR) sampling (ie., collection of a composite sample by depth integration at several equally spaced verticals across a stream) by operators with limited experience could lead to overestimation of the long-term load by 30 percent. Although national and international standards relating to sampling equipment and techniques as well as laboratory procedures have now been established (cf Table 3.3), sediment sampling still involves many problems and uncertainties. Careful attention must be given to the likely reliability of the methods employed in the monitoring programs used to obtain existing data (*120*).

Greater problems surround the methods used to assess long-term sediment loads and annual sediment yields from intermittent measurements of instantaneous suspended sediment load. These problems center around the marked and rapid fluctuations in suspended sediment concentration exhibited by most rivers. Accurate assessment of suspended sediment loads for specific periods necessitates detailed records of such fluctuations in sediment concentration that may be combined with the records of water discharge. Continuous records of stream discharge are available at most measuring stations, but an equivalent record of sediment concentration may be difficult to obtain by a program of manual sampling. On large rivers, it may be possible to collect sufficient manual samples to define meaningfully the record of sediment concentration during periods of fluctuating concentration, and procedures are available to establish the trend between individual samples (*81*). On smaller streams, however, concentrations fluctuate more rapidly during flood events, and operational constraints may limit the frequency at which manual samples may be collected. Attempts have been made to resolve these difficulties by developing equipment and instruments capable of automatic collection of data on sediment concentration. Many researchers have used automatic pump-sampling equipment (*114*). Continuous recording turbidity meters (*9, 34, 43*) and nuclear probes (*56, 82, 93*) have also been used successfully in a few cases. Difficulties may arise in relating the point values of concentration obtained with such equipment to the mean value for the cross-section, but this limitation is frequently of little significance when compared with the positive improvements in temporal resolution obtained.

More often, however, frequent manual samples or additional information from automatic samplers or recording equipment are not available, and the detailed record of fluctuations in sediment concentration cannot be

Table 3.3 Reports and manuals describing procedures for measuring suspended sediment loads.

(i) Allen, P.B. (1981) *Measurement and Prediction of Erosion and Sediment Yield.* ARM-S-15. U.S. Department of Agriculture, Washington D.C.

(ii) American Society of Civil Engineers (1975) *Sedimentation Engineering.* ASCE, New York, New York. Chapter III.

(iii) Guy, H.P. (1969) *Laboratory theory and methods for sediment analysis. Techniques of Water Resources Investigations of the U.S. Geological Survey,* Book 5, Chapter C1. U.S. Geological Survey, Reston, Virginia, U.S.

(iv) Guy, H.P. and Norman, V.W. (1970) *Field methods for measurement of fluvial sediment. Techniques of Water Resources Investigations of the U.S. Geological Survey,* Book 3, Chapter C2. U.S. Geological Survey, Reston, Virginia, U.S.

(v) International Standards Organisation (1977) *Method for Measurement of Suspended Sediment.* ISO Standard 4363.

(vi) International Standards Organisation (1977) *Requirements for Suspended Sediment Samplers.* ISO Standard 3716.

(vii) Long, Yuqian (1989) *Manual on Operational Methods for the Measurement of Sediment Transport.* Operational Hydrology Report No. 29, WMO-No. 686, World Meteorological Organization, Geneva, Switzerland.

(viii) U.S. Geological Survey (1978) *Sediment,* Chapter 3 in *National Handbook of Recommended Methods for Water Data Acquisition.* U.S. Geological Survey, Reston, Virginia, U.S.

(ix) Ward, P.R.B. (1984) *Measurement of sediment yields.* In R.F. Hadley and D.E. Walling (editors) *Erosion and Sediment Yield: Some Methods of Measurement and Modelling.* Geobooks, Norwich, England. pp. 37-70.

(x) World Meteorological Organization (1981) *Measurement of River Sediments.* Operational Hydrology Report No. 16, WMO-No. 561, World Meteorological Organization, Geneva, Switzerland.

(xi) World Meteorological Organization (1981) *Guide to Hydrological Practices.* WMO-No. 168, World Meteorological Organization, Geneva, Switzerland.

defined. In this case, researchers must use indirect load calculation procedures, which involve either interpolation or extrapolation of the available concentration data. In this context, interpolation procedures assume that the values of concentration or sediment discharge obtained from instantaneous samples represent a much longer period of time, for example, days or weeks; whereas sediment rating curve techniques may be viewed as the classic example of an extrapolation procedure. In the latter case a limited number of sediment concentration measurements are extrapolated over the period of interest by developing a relationship between concentration or sediment discharge and stream discharge and applying this relationship to the streamflow record (*15, 64, 106*). The streamflow record may be in the form of a flow-duration curve or a continuous series. Researchers in many studies have applied rating relationships developed from a short period of measurement to streamflow records covering a much longer period of time. Estimates of suspended sediment load obtained using these indirect load calculation procedures may involve considerable error, and the resulting data must be treated with caution.

Figure 3.3A shows the degree of scatter that may be associated with a suspended sediment concentration/discharge relationship. In this case the concentrations associated with a specific level of discharge range over several orders of magnitude. Although improved relationships between concentration and discharge may be obtained by subdividing the data set according to season and rising and falling stages of the hydrograph, any estimate of sediment concentration derived using such rating relationships will be associated with a very considerable standard error. Furthermore, it cannot be assumed that fluctuations in sediment concentration and discharge are synchronous. Frequently, sediment concentrations during a flood event will peak before the peak of water discharge, and concentrations estimated from a concentration/discharge relationship will fail to reproduce this behavior and may therefore result in a significant overestimation of the loads involved. Williams (*122*) provides a valuable discussion of the various forms of the sediment concentration/discharge relationship that have been reported in the literature. Because rating relationships are commonly fitted to logarithmic plots using linear least squares regression, they are frequently biased toward underestimation of sediment concentrations by the large number of samples collected under stable flow conditions generated by a regular sampling program. This problem can, however, be reduced by use of group average procedures for fitting the rating curve (*16, 39*). Further bias toward underestimation is introduced by the use of log transformed data in a linear least squares regression (*29*). Bias correction factors have been proposed to reduce this problem (*28, 30, 47*) but their value remains uncertain. Walling and Webb (*119*) report results from an analysis of detailed sediment load data for a river in Devon, United Kingdom, which indicate that application of these bias correction factors to rating curve calculations introduces little improvement in the accuracy of the resultant estimates of sediment yield, since they address only one of the several causes of error. Estimates of sediment yield obtained using rating curves can therefore generally be expected to underestimate the true sediment yield (*115*), possibly by as much as 80 percent. For some rivers, it cannot be assumed that the rating relationship will remain constant over a period of years and that the

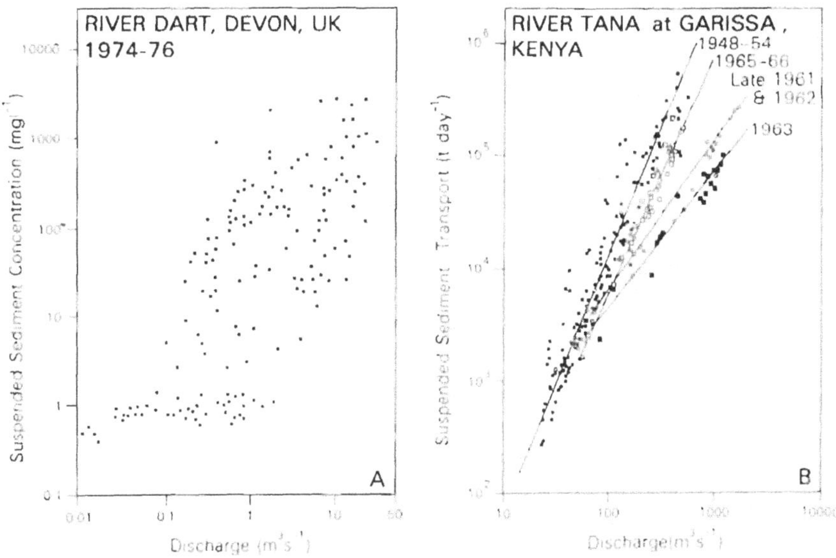

Figure 3.3 Uncertainties associated with suspended sediment rating curves. A demonstrates the degree of scatter that frequently will be associated with a suspended sediment concentration/discharge relationship. B illustrates inter-annual variation in the precise form of a suspended sediment load/discharge relationship, based on Dunne (*26*).

samples generated by a short period of monitoring can be used to develop a relationship applicable to the entire period. Land use change may cause non-stationary conditions. Dunne (*26*) reported an interesting example of inter-annual variation in the sediment discharge/water discharge relationship for the Tana River at Garissa, Kenya (Figure 3.3B). His data showed an essentially random pattern rather than any progressive shift. Long-term sediment yields estimated using a rating curve based on samples collected during only a limited part of the overall period should therefore be treated with caution.

As a final example of potential problems associated with infrequent sampling and use of extrapolation and interpolation techniques to estimate long-term sediment yields, Figure 3.4 compares the actual suspended sediment load of the River Creedy in Devon, England, for a seven-year period with nearly 100 load *estimates* for the same period. These estimates were obtained using typical (infrequent) manual sampling strategies and a selection of indirect load calculation procedures. In this case, a continuous-recording turbidity meter was used to provide a detailed record

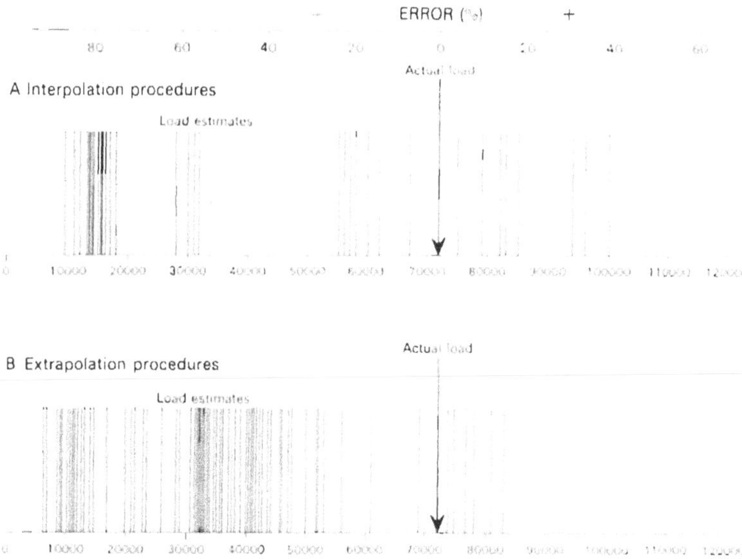

Figure 3.4 A comparison of suspended sediment load estimates for River Creedy in Devon, England, obtained using interpolation and extrapolation procedures with the actual load for the period 1972-1979.

of sediment concentration. Walling and Webb (*115*) used this record to calculate the actual sediment load and as the basis for deriving data sets representing the manual sampling strategies. The various estimates span a wide range, and under-estimation by as much as 60 percent is common.

Although published or tabulated sediment yield data may appear authoritative, it is essential that the user carefully considers their reliability and accuracy before using them for further analysis and interpretation. However, few guidelines for assessing reliability and accuracy are available currently. In many cases, assessment of the limitations of a particular measurement program may permit only very general conclusions about the margins of error that may be involved, but it is important to take account of such uncertainty.

DESIGNING SEDIMENT YIELD MONITORING PROGRAMS

Soil conservation research and planning may require the establishment of a sediment yield monitoring program to supplement or verify existing information or to provide data from a hitherto undocumented area or for specific experimental investigations. It is important that the aforementioned problems of data reliability be minimized. For example, if a program is designed to monitor the effects of conservation measures in reducing sediment yields, the sediment load data should be sufficiently accurate to produce meaningful conclusions. Further discussion of monitoring programs can focus on manual sampling strategies, use of automatic sampling devices and equipment for recording suspended sediment concentrations, and investigation of suspended sediment properties.

Manual sampling strategies

Existing manuals and technical reviews provide detailed and valuable discussions of currently available suspended sediment samplers, sampling techniques, and laboratory procedures for determining suspended sediment concentration. Some of these are listed in Table 3.3. Particular attention is given to the characteristics of good samplers and to techniques of sample collection. These publications should be consulted before a measurement program is initiated. In most circumstances, the sediment samplers developed by the Federal Interagency Sedimentation Project under the auspices of the U.S. Water Resources Council (Table 3.4) will prove appropriate and many of these samplers are commercially available. However, similar devices will have been developed in other countries and Figure 3.5 shows two examples of suspended sediment samplers used in China.

The manuals and reviews listed in Table 3.3 give much less attention to sampling frequency because they are more concerned with measurements of instantaneous sediment discharge than measurements of long-term sediment loads and yields. In view of the great uncertainty associated with any estimate of sediment yield based on infrequent sampling and the application of indirect load calculation procedures such as rating curves, manual sampling programs should, wherever possible, aim to document the continuous record of sediment concentration. This record can be combined with equivalent water discharge data to provide values of sediment load for the required time period. This will normally involve subdividing the records into short periods (e.g. 30 minutes) within which concentrations and water discharge can be assumed to be essentially constant, and calculating and summing the loads for these intervals.

In designing a sampling strategy to document the continuous record of sediment concentration, careful consideration must be given to the likely behavior of the river and the timing of significant concentration fluctuations. During periods of stable flow and low concentration, a low sampling frequency, daily or even weekly, will suffice because concentrations remain essentially stable. Any errors incurred are likely to be of limited significance to the final long-term load estimate. Most of the sediment

Table 3.4 Characteristics of U.S.-series depth-integrating and point-integrating samplers for collecting suspended sediment samples.

Name	Type of Sampler*	Method of Suspension	Mass (kg)	Overall Length (m)	Sample Container Size (ml)	Distance between Nozzle & Sampler Bottom (mm)	Remarks
US DH-48	DI	Rod	2.00	0.33	473	90	For wading
US DH-59	DI	Cable	10.2	0.42	473	114	For hand-line operation
US DH-75P	DI	Rod	0.4	0.26	500	83	For sampling only in subfreezing temeratures
US DH-75Q	DI	Rod	0.4	0.29	1,000	114	Similar to US DH 75P
US DH-76	DI	Cable	10.9	0.47	946	80	Similar to US DH 59
US D-49	DI	Cable	28.0	0.61	473	103	
US D-49AL	DI	Cable	18.0	0.61	473	103	Similar to US D-49
US D-74	DI	Cable	28.2	0.66	473 or 946	103	Similar to US D-49
US D-74AL	DI	Cable	11.4	0.66	473 or 946	111	Similar to US D-74
US P-50	PI	Cable	135.6	1.12	473 or 946	140	
US P-61-A1	PI	Cable	47.5	0.71	473 or 946	109	
US P-63	PI	Cable	90.4	0.86	473 or 946	150	
US P-72	PI	Cable	17.7	0.71	473 or 946	109	Similar to P-61-A1 but for hand-line operation

*Type: DI = depth-integrating. PI = point-integrating.

Figure 3.5 These two suspended sediment samplers employed by Chinese scientists on the headwaters of the Yangtze River provide examples of point-integrating (upper) and instantaneous (lower) samplers.

load will be moved during major flood events, and it is essential that these are sampled intensively to record fluctuations in suspended sediment concentration. In the case of a river that drains a large area, say greater than 1000 km^2, these fluctuations probably will be easy to follow with a sequence of manual samples, although it may be necessary to sample at frequent intervals, perhaps every one to two hours, on the rising stage of the flood. On smaller rivers, however, fluctuations in sediment concentration may be more rapid, and operational problems may preclude establishing an effective manual sampling program. Alternative automated equipment should be used in these circumstances.

Figure 3.6 illustrates some of the salient characteristics of suspended sediment transport in the River Creedy, which drains a 262-km^2 basin in Devon, England. The concentration and load-duration curves (Figure 3.6A, B) indicate that most of the overall sediment load is transported during only 5 percent of the time, when suspended sediment concentrations exceed about 100 mg/l. Clearly, sediment sampling should focus on this time period, although the relatively even distribution of major sediment transporting events through the year (Figure 3.6C) poses problems for undertaking event-based sampling. These problems would be reduced significantly where a river exhibits a marked seasonal regime, with major events being limited to one particular time of year, for example, snowmelt or seasonal rains. Figure 3.6C demonstrates the impracticality of using extrapolation procedures based on concentration/discharge relationships to reduce the need for frequent sampling. The pattern of variation of sediment concentration during this sequence of storm events shows that there is no well-defined relationship between sediment concentration and discharge.

Automatic sampling equipment and continuous monitoring

Where suspended sediment concentrations in a river fluctuate rapidly, automatic sampling equipment and devices for continuous monitoring of sediment concentration may be required to augment or replace manual sampling. Single-stage samplers provide a simple means of obtaining samples automatically at unattended sites. These samplers consist of individual sample bottles with intake nozzles and exhaust ports mounted in a vertical sequence (*44*). The concentrations obtained from such samplers should be viewed as only approximate because (a) samples are collected at a fixed point in the vertical, usually at or near the channel margin, and (b) there may be additional movement of sediment into the bottle during the period that the bottle is submerged. The resulting data are further limited by the fact that samples are obtained at predetermined water levels, rather than

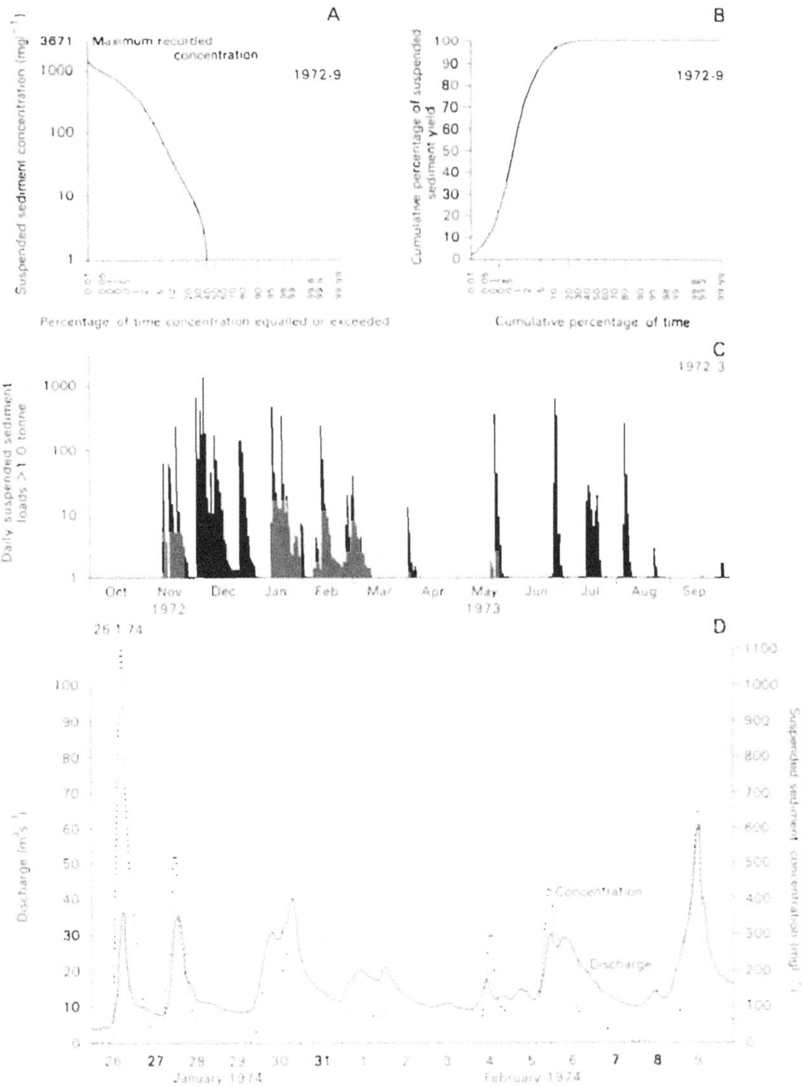

Figure 3.6 Characteristics of the suspended sediment transport regime on the River Creedy, Devon, England.

on a time basis, and are commonly restricted to rising stage conditions. During a multi-peaked event, samples may be available only from the initial hydrograph peak. However, there are devices that operate during falling stages.

Automatic pumping samplers afford a more effective means of sample collection (Figure 3.7). These samplers can be programmed to collect samples at predetermined time intervals or increments of stage or discharge. They can be distinguished further according to the two basic principles used for withdrawing samples from the flow: a pump or the creation of a vacuum in the sample bottle or a primary chamber. Automatic pumping samplers commonly include 36 or more bottles and in most cases they may be battery operated (*83, 108, 120*). Because the sample is obtained from a fixed inlet and is not collected isokinetically, these samplers are suited more to small rivers and to situations where the suspended load is comprised predominantly of silt and clay particles. Results obtained should be calibrated against manual samples collected from the overall cross section. Although pumping samplers are usually programmed to operate at predetermined time intervals or stage increments, in order to document the variation of suspended sediment concentration during flood events, Thomas (*94, 95*) has described an alternative automated sampling strategy and load calculation technique aimed at providing accurate estimates of long-term loads. Known as the SALT sampling procedure, sampling is controlled by a microprocessor and emphasizes those time periods when high sediment loads are being transported. Because sampling is probability-based, the variance and confidence limits of the final load estimate may be calculated using standard statistical procedures.

The availability of reliable equipment for continuous monitoring of suspended sediment concentrations would clearly be a major benefit in any sediment yield monitoring program, and considerable effort has been devoted to developing such instruments. To date, there is no entirely satisfactory and generally applicable instrument, but several have been used with encouraging results. These include two major types: optical turbidity meters and nuclear gauges. Optical instruments (*9, 34, 42, 101, 120*) measure the turbidity of water by the attenuation of transmitted light (absorptiometers), or by monitoring the intensity of the light scattered by the suspended sediment particles (nephelometers). Figure 3.8 shows an example of a photoelectric turbidity meter of the absorptiometer type. In both cases, values of optical turbidity, generally defined in terms of formazin standards, must be calibrated against direct measurements of suspended sediment concentration, because the particle size distribution and the organic matter content, density, and mineralogy of the sediment particles

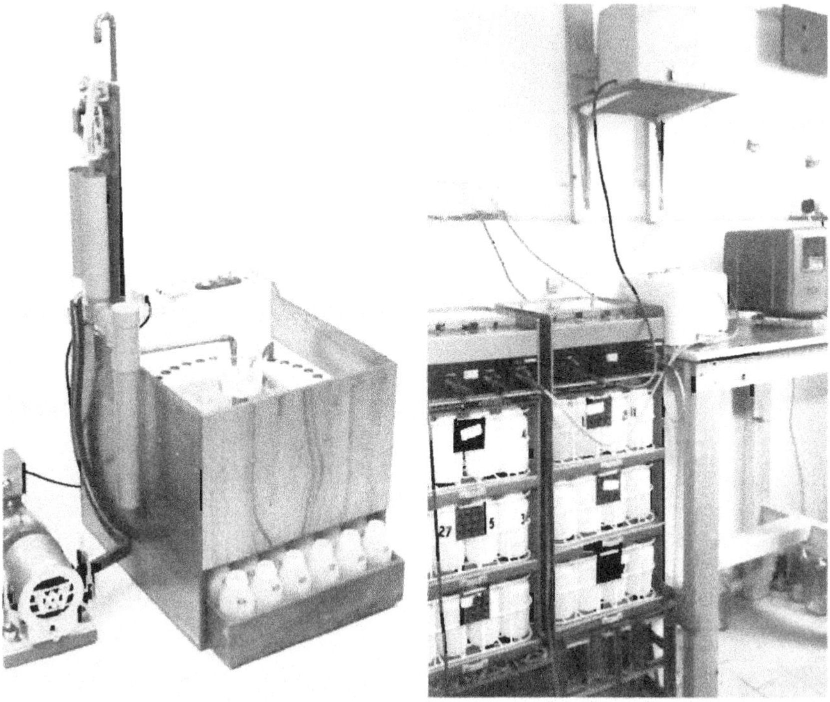

Figure 3.7 These are typical examples of pumping samplers used for automatic collection of river water samples. Both employ peristaltic pumps to deliver water to the sampler, which incorporates a rotating nozzle capable of filling the set of 36 (left) or 48 (right) bottles.

influences the relationship between turbidity and sediment concentration (*37*). With some instruments, fouling of cell windows and ambient light effects may also affect the calibration adversely. In some cases the sensor may be mounted directly in the river (e.g. Figure 3.8). In others, the river water is circulated by a pump to the measuring cell housed on the river bank.

Reports regarding the effectiveness of optical turbidity meters for continuous monitoring of suspended sediment concentrations are somewhat conflicting. In some studies, researchers apparently have used turbidity meters very successfully. In others, it has proved difficult, if not impossible, to develop a meaningful calibration between turbidity and sediment concentration. However, the balance of evidence suggests that such instruments frequently provide an effective means of monitoring suspended sedi-

Figure 3.8 A photoelectric turbidity monitor and probe (inset) can be mounted directly in a river, as at this monitoring station on the Sagana River in Kenya, to provide a continuous record of wash load concentrations. The probe is mounted inside the perforated tube and may be readily raised for cleaning. (Photographs of Sagana River monitoring station provided by Hydraulics Research Ltd., Wallingford, UK.)

ment concentrations, although they are limited to the range of concentration below 10,000 mg/l. Furthermore, they are applicable primarily to the clay and silt fraction of the suspended load. Fish (*32*) demonstrated that optical turbidity meters are virtually insensitive to sand-sized material, unless the associated concentrations are high.

Nuclear sediment gauges are based on similar principles but measure the absorption or scattering of gamma radiation rather than light (*6, 56, 82, 93*). They are suitable for concentrations above a minimum threshold of about 500 mg/l. These gauges commonly use probes that are either fixed within or lowered into the channel. However, the instruments are more sophisticated than optical turbidity meters and have been restricted mostly to specialized experimental measurements, rather than routine monitoring. Other devices such as ultrasonic sensors (*31, 33*) and vibrating U-tube densimeters (*87, 89*) have been used, but these devices are again essentially experimental.

Manual sampling and use of equipment for continuous monitoring of suspended sediment concentrations may frequently be complementary. For example, continuous-monitoring devices could provide a useful means of interpolating fluctuations in sediment concentrations between manual samples for which accurate values of sediment concentration have been obtained. Bolton (*8*) suggested that in many instances it may be convenient to distinguish the fine-grained washload and the coarser, suspended bed material components of the total suspended load, using optical turbidity meters to monitor the former and manual sampling to measure the latter. He concluded that it may be possible to obtain a clearly defined rating relationship between suspended bed material concentrations and flow so that the record for this component can be extrapolated from limited sampling.

Study of sediment properties

Traditionally, measurements of suspended sediment transport by rivers have emphasized documentation of the magnitude of the loads transported. There is now increased awareness of the role of fine sediment in the transport of nutrients and contaminants and in non-point pollution from agricultural areas (*2, 40, 69, 88, 102*). This has focussed attention on the physical and chemical properties of the sediment transported, both as a component of river water quality and in terms of the relationship between the properties of sediment and source material and, therefore, the selectivity of erosion and conveyance processes. The former may have important implications for the off-farm impacts of soil erosion, but the latter is perhaps more relevant to soil erosion and conservation research, because erosion by water may preferentially mobilize and transport specific fractions and constituents of the soil that are of particular importance in maintaining fertility. It may therefore be appropriate to consider the particle size characteristics and the organic matter, nutrient, and mineral content of transported sediment and associated enrichment ratios (*112*). Taking the example of particle size, Figure 3.9 demonstrates the considerable range of particle size distributions associated with suspended sediment from rivers in different areas of the world. Walling and Moorehead (*113*) have shown how this variability reflects a number of controls, including the character of the catchment soils, and the selectivity of the erosion and conveyance system.

In addition, Figure 3.10 illustrates the potential contrasts in the particle size of suspended sediment and source material. It compares the proportion of silt, clay, and sand in suspended sediment samples collected by the U.S. Geological Survey from four small drainage basins with equivalent information on the soils, based on their textural classification. Looking also at other properties, Table 3.5 provides an example of the potential enrichment of transported sediment in a number of constituents, relative to source material. This is based on the work of Novotny and associates (*70*) in a 21-km^2 basin draining a predominantly agricultural area in the headwaters of the Menomonee River in Wisconsin (U.S.). Information on the properties of the suspended sediment transported from a drainage basin also can provide valuable information on the major types of sediment source involved. If various potential sediment sources can be 'fingerprinted' by diagnostic properties, comparisons with the suspended sediment may indicate the dominant source of that sediment (*51, 73, 105*).

In some cases, laboratories can analyze properties of the small samples of particulate material recovered from manual or automatic samplers. However, where a wide range of analysis is intended or where a particular laboratory procedure requires a considerable amount of sediment, researchers must collect bulk samples of suspended sediment. Experience suggests that continuous flow centrifuges are most useful in recovering sediment from bulk samples of river water. Ongley and Blachford (*74*) reported the use of a specialized sampling/recovery system in which water is pumped directly from the river into the centrifuge. They found a good recovery performance for particles down to 0.45 μm. Similar equipment is discussed by Burrus and associates (*13*) and by Ongley and Thomas (*75*).

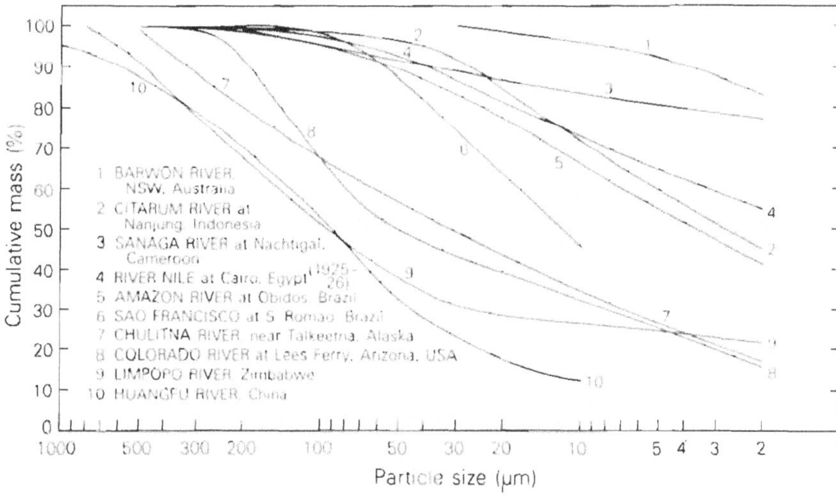

Figure 3.9 The range of particle size distributions associated with suspended sediment transported by a selection of world rivers [based on Walling and Moorehead (*113*)].

Table 3.5 Enrichment of clay and pollutants in suspended sediment transported by a headwater tributary of the Menomonee River, Wisconsin, US [based on Novotny et al. (*70*)].

Constituent	Soil	Suspended Sediment	Enrichment Ratio
Clay (%)	27	91	3.4
Total P	810	1700	2.1
Pb	19	39	2.0
Zn	69	280	4.0
Cu	25	45	1.8
Cr	29	56	1.9
Ni	17	45	2.6

All chemical elements expressed in parts per million.

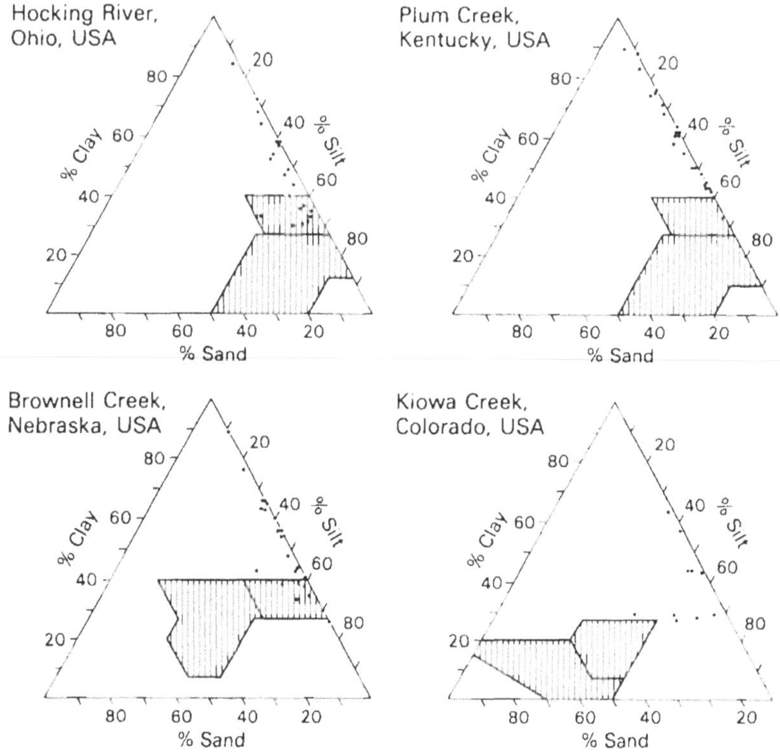

Figure 3.10 Comparison of the particle size composition of suspended sediment and soils in four U.S. drainage basins. Textural classes of the dominant soils are denoted by shaded zones on the trilinear plots. Based on data from Flint (*35*), Anttila (*5*), and Mundorff (*66,67*).

ADDITIONAL SOURCES OF INFORMATION ON SEDIMENT YIELDS

Reservoir surveys

Reservoir sedimentation surveys may also be used to assess the sediment yield from drainage basins that discharge into reservoirs (*17*). Such surveys have several potential advantages over river measurements. For example, a single survey will provide, in many cases, a meaningful retrospective estimate of sediment yield averaged over a period of several years. Provided the trap efficiency of the reservoir can be estimated and the volumes and densities of deposited sediment assessed accurately, there should be less uncertainty associated with the resultant sediment yield values than with some direct measurements. Furthermore, once the ranges and benchmarks are established, resurveys can be undertaken relatively easily to update the sediment yield record. In addition to an estimate of the trap efficiency of the reservoir, information on the history of the reservoir, including the date of impoundment and whether any sediment has been removed, should be available. Sediment yields calculated from reservoir survey data, however, will include both bedload and suspended load and will necessarily lack the temporal resolution of river monitoring data.

Procedures for reservoir sedimentation surveys have been developed for both small and large impoundments and some of the relevant manuals and reviews which should be consulted for details are listed in Table 3.6. Sonic depth recorders are commonly used to measure water depth along predetermined ranges, and sediment depth can be calculated from a previous survey of the original reservoir bottom or from soundings of the sediment depth. In view of the difficulties of sounding sediment depth, it is advisable to undertake an initial survey of the reservoir bottom before or immediately after impoundment in order to avoid the need for such measurements. Values of volume-weight or density of deposited sediments can be obtained by collecting undisturbed volumetric samples of the deposits or by in-situ measurements using a gamma probe (*59, 60*). Use of the latter is, however, restricted to saturated sediments. To determine the mean annual sediment yield from the contributing drainage basin, the weight of deposited sediment must be adjusted for reservoir trap efficiency. Heinemann (*48*) discussed some of the problems involved in obtaining accurate estimates of trap efficiency. Clearly, there is a need for further research on this topic, because most existing estimation procedures are based on data assembled from reservoirs in the United States. In the absence of more detailed procedures, many researchers use the curves developed by Brune (*12*), which relate trap efficiency to the capacity/inflow ratio of the reservoir and an approximate index of sediment particle size.

Although reservoir surveys are aimed primarily at providing estimates of long-term sediment yields, recent work by Laronne (*52*) and Hereford (*49*) indicates that in some situations detailed analysis of the sediment deposits may be capable of providing information for shorter timescales. Working on reservoirs in semi-arid areas in Israel and Utah (U.S.), where it was possible to gain access to the sediment deposits when the reservoir was empty, these authors were able to identify sedimentation units or "event couplets" in the sediment deposits which represented the deposition associated with individual flood events. Detailed three dimensional analysis of these units could provide data on the sediment yields of individual flood events and permit the reconstruction of event-based sediment yields.

Table 3.6 Reports and manuals describing procedures for undertaking reservoir sedimentation surveys.

(i) American Society of Civil Engineers (1975) *Reservoir deposits.* In *Sedimentation Engineering.* American Society of Civil Engineers, New York, New York, USA. pp. 349-382.
(ii) Long Yuqian (1989) *Measurement of erosion and deposition in reservoirs and river reaches.* In *Manual on Operational Methods for the Measurement of Sediment Transport, World Meteorological Organization Operational Hydrology Report.* No. 29, WMO-No. 686, World Meteorological Organization, Geneva, Switzerland, pp. 69-86.
(iii) Rausch, D.L. and H.G. Heinemann (1976) *Reservoir sedimentation survey methods.* In *Hydrological Techniques for Upstream Conservation.* Conservation Guide No. 2. Food and Agriculture Organization, United Nations, Rome, Italy. pp. 29-43.
(iv) Rausch, D.L. and H.G. Heinemann (1984) *Measurement of reservoir sedimentation.* In R.F. Hadley and D.E. Walling (editors) *Erosion and Sediment Yield: Some Methods of Measurement and Modelling.* Geobooks, Norwich, England. pp. 179-200.
(v) UNESCO (1985) *Methods of Computing Sedimentation in Lakes and Reservoirs.* Technical Document in Hydrology, Unesco, Paris, France.

Lake sediment studies

The lack of long-term sediment transport records hampers the study of long-term patterns of sediment yield from a drainage basin. The availability of records covering, for example, the last 100 years would be of considerable value in attempting to understand temporal variations in erosion and sediment yield in response to changes in catchment land use. Record lengths increase as time proceeds, but opportunities to document the effects of particular changes in land use may not recur. Furthermore, there are many circumstances in which it would be useful to have reliable information on the sediment yields occurring in the past under undisturbed natural conditions, as a means of defining the baseline against which to assess subsequent changes. In this context, researchers should focus more attention on the potential of lake sedimentation studies for illuminating the environmental history of the upstream drainage basin (*17, 71, 76*).

Lake sediments are capable of providing a wide range of general information regarding past environmental conditions in the watershed. Thomas (*96*), for example, provides a valuable review of the wide range of circumstantial evidence concerning past erosion rates and human activity that can be derived from lake sediment cores. However, recent studies by Dearing and his co-workers (*18, 19, 20*) and by Souch and Slaymaker (*91*) represent a major advance in that they have used lake sediment cores to reconstruct the *actual sediment yields* involved. There are four major requirements for such reconstruction. First, there must be a suitable lake which has trapped sediment eroded from the drainage basin over a considerable period of time. Ideally the lake should have a high trap efficiency which has been essentially stable over the intended period of reconstruction. Second, dating methods capable of providing an absolute chronology for the sedimentation represented by the sediment cores must be available. Third, and most important, a method of intercore correlation is required, to enable the volumes of sediment stored between synchronous levels in the lake deposit to be calculated. Fourth, it must be possible to distinguish the allochthonous inputs of sediment eroded from within the surrounding drainage basin from autochthonous sediment originating from within the lake itself. Radiometric dating techniques involving lead-210, cesium-137 and carbon-14 (*14, 72*) and paleomagnetic dating methods (*97*) provide a reliable means of obtaining absolute chronologies for sediment cores extending back for several thousand years. Continuous measurements of magnetic susceptibility profiles on retrieved sediment cores (*7*), coupled with visual analysis of downcore stratigraphic changes can be used to establish the necessary intercore correlations.

Figures 11 and 12 provide an illustration of the potential of lake sediment investigations for reconstructing records of sediment yield over both the relatively recent past and over a period of several thousand years, based on the work of Dearing and his co-workers (*18, 19, 20*). In the case of Seeswood Pool, a small reservoir in the English Midlands constructed around 1765, the reconstructed record of sediment yield (Figure 3.11A) highlights the substantial increases that have occurred over the past 60 years as a result of the intensification of farming activity. The reconstructed record for Llyn Peris, a lake in North Wales, United Kingdom, (Figure 3.11B) shows an almost order of magnitude increase in sediment yields over the past 150 years, which can be related to the erosional effects of mining, quarrying, overgrazing, and recent construction activity in this upland catchment. In this case maximum and minimum values of sediment yield are shown for the individual periods, in order to take into account some of the uncertainties associated with individual estimates. The example from Lake Havgardssjon (Figure 3.12), relates to a small lake in the hummocky ground moraine landscape of southern Skane, Sweden. In this study a grid of 47 cores was used to reconstruct the sediment yields for the period since 1550, and evidence from a single 4 m core was used to provide a more tentative extension of the record back to 3050 BC.

In addition to providing evidence of past sediment yields, lake sediment investigations also can be coupled with recent work on sediment properties to provide information on changing sediment sources, if downcore changes in sediment properties can be linked to sediment influx from dif-

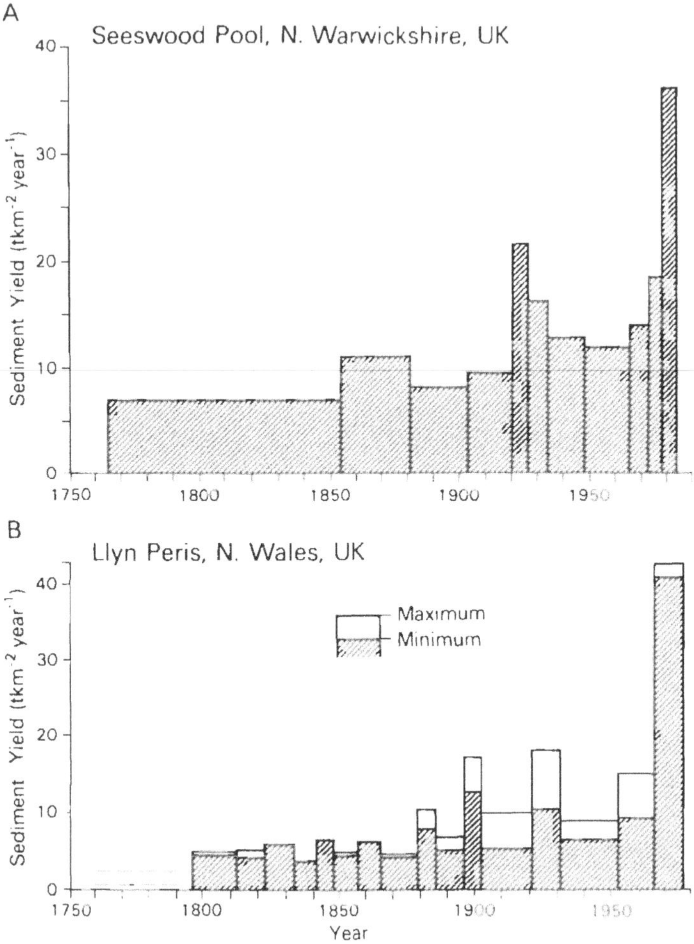

Figure 3.11 Examples of lake sediment-based recontructions of sediment yield for two drainage basins in Britain. (A) is based on Dearing and Foster (*19*) and (B) on Dearing and associates (*18*).

ferent sources or to a change in the relative importance of the sources involved. In the case of the Lake Havgardssjon study, downcore changes in the magnetic susceptibility of sediment from the 4 m core (Figure 3.12) were interpreted in terms of changes in sediment grain size and source. High susceptibility values were linked to the influx of coarser sediment, much of which would have been derived from accelerated channel erosion. A recent study on the Burrinjuck Reservoir in eastern New South

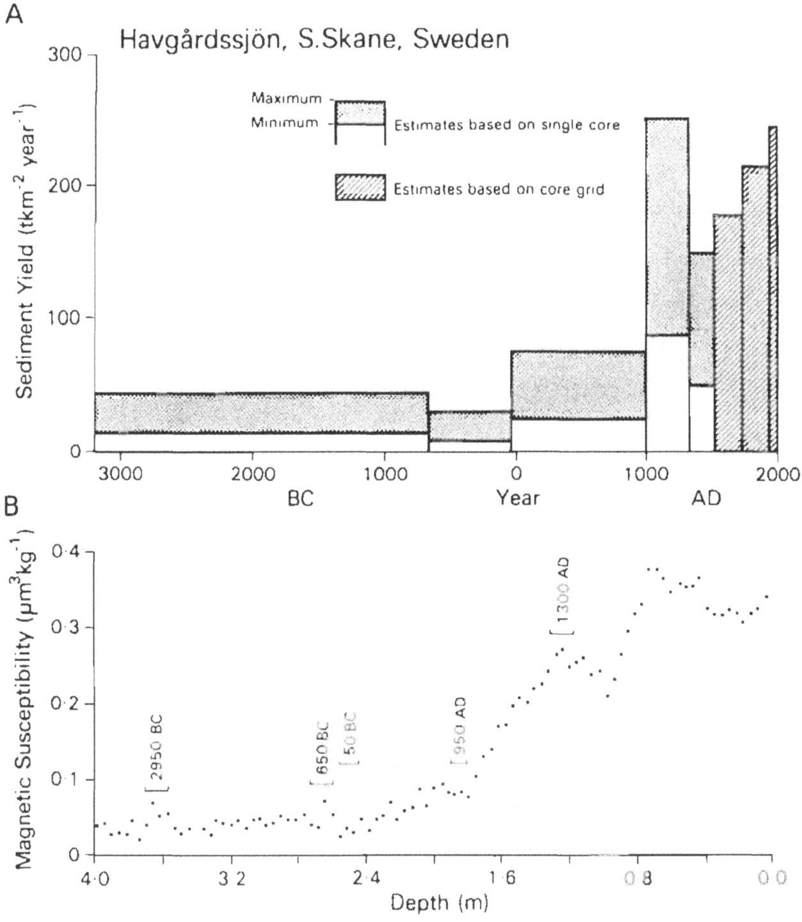

Figure 3.12 A lake sediment-based reconstruction of the sediment yield of the Lake Havgardssjon basin, Sweden, from 3050 BC to the present (A) and a magnetic susceptibility profile of the lake sediment with approximate sediment ages (B). Based on Dearing and associates (*20*).

Wales, Australia, reported by Wasson and his co-workers (*121*) used measurements of the unsupported lead-210 and the radium-226 content of the sediment cores to distinguish temporal trends in the relative importance of topsoil and subsoil as sediment sources.

LINKING DOWNSTREAM SEDIMENT YIELD TO UPSTREAM EROSION

As emphasized previously, there are several problems involved in linking downstream sediment yields to upstream rates of soil erosion. This task therefore faces many uncertainties. The traditional approach in theory involves the application of a SDR to calculate the gross erosion within the drainage basin, that is,

Gross erosion = sediment yield/SDR

In practice, however, few rigorous guidelines exist for predicting the SDR for a specific basin (*107*).

The magnitude of the SDR for a particular basin will be influenced by a wide range of geomorphological and environmental factors, including the nature, extent, and location of the sediment sources; relief and slope characteristics; the drainage pattern and channel conditions; vegetation cover; land use; and soil texture. Several studies have attempted to produce empirical prediction equations for this variable. Basin area has been isolated frequently as a dominant control (*4*). The U.S. Soil Conservation Service has developed a generalized relationship between SDR and drainage basin area (Figure 3.13A), which has been widely applied in the central and eastern United States (*86*). The inverse relationship between SDR and area has been explained in terms of decreasing slope and channel gradients and increasing opportunities for deposition associated with increasing basin size. Other attempts to develop prediction equations have included such variables as relief ratio and main channel slope (Table 3.7).

The failure to produce a generally applicable prediction equation for SDR is due partly to the complexity of sediment delivery processes and their interaction with catchment characteristics and partly to a lack of definitive assessments of the dependent variable itself. Assessments that have been undertaken are themselves based primarily on a comparison of measured sediment yield with an estimate of gross erosion. The latter is generally derived from an estimate of sheet erosion based on a soil loss equation, corrected to take account of additional contributions from channel and gully erosion. These estimates of gross erosion, therefore, are open to considerable uncertainty. The various relationships developed for particular areas presented in Figure 3.13B emphasize the considerable diversity of values associated with basins of a given size. The delivery characteristics of basins in the southeastern Piedmont in the United States contrast markedly with those of watersheds in the loess region of China, where delivery ratios close to 100 percent characterize a wide range of basin size.

Table 3.7 Examples of proposed relationships between sediment delivery ratio and catchment characteristics.

Region	Equation*	Author
Kansas, U.S.	$\text{Log DR} = 2.962 + 0.869 \log R - 0.854 \text{ Log L}$	(57)
Southeastern U.S.	$\text{Log DR} = 4.5 - 0.23 \log 10 A - 0.510 \times \text{colog R/L} - 2.786 \log BR$	(86)
Brushy Creek, Texas, U.S.	$DR = 0.627 \text{ SLP}^{0.403}$	(124)
Texas, U.S.	$DR = 1.366 \times 10^{-11} A^{-0.100} R/L^{0.363} CN^{5.444}$	(123)
Pigeon Roost Creek, Mississippi, U.S.	$DR = 0.488 - 0.006 A + 0.010 RO$	(68)
Dali River Basin, Shaanxi, China	$DR = 1.29 + 1.37 \ln Rc - 0.025 \ln A$	(65)

*DR = sediment delivery ration; R = basin relief; L = basin length; A = basin area; R/L = relief/length ratio; BR = bifurcation ratio; SLP = percent slope of main stem channel; CN = SCS curve number; RO = annual runoff; Rc = gully density. Note: units vary between equations.

Some consistency in the slope of the delivery ratio-area relationship is apparent for several regions in the United States, lending support to the view that an exponent of -0.125 is typical (*4*). But even here a wide range of intercept values exists.

A more rigorous approach to defining and investigating the sediment delivery characteristics of a drainage basin is provided by the sediment budget concept originally advocated by Dietrich and Dunne (*24*) and developed by Lehre (*53, 54*), Swanson and associates (*92*), and by Phillips (*79*). In this essentially conceptual approach or framework, the various sediment sources within a basin are defined, and the sediment mobilized from these sources is routed to and through the channel system by considering the various sinks.

Figure 3.14 provides a diagrammatic representation of such budgets for three drainage basins where investigators have attempted to define tentatively the source-yield linkages. In all cases the proportion of the eroded sediment delivered to the basin outlet is relatively small, ranging from 53 percent to 5.5 percent, but substantial differences exist between the catchments represented in the precise form of the budget and in the location and importance of the various sinks.

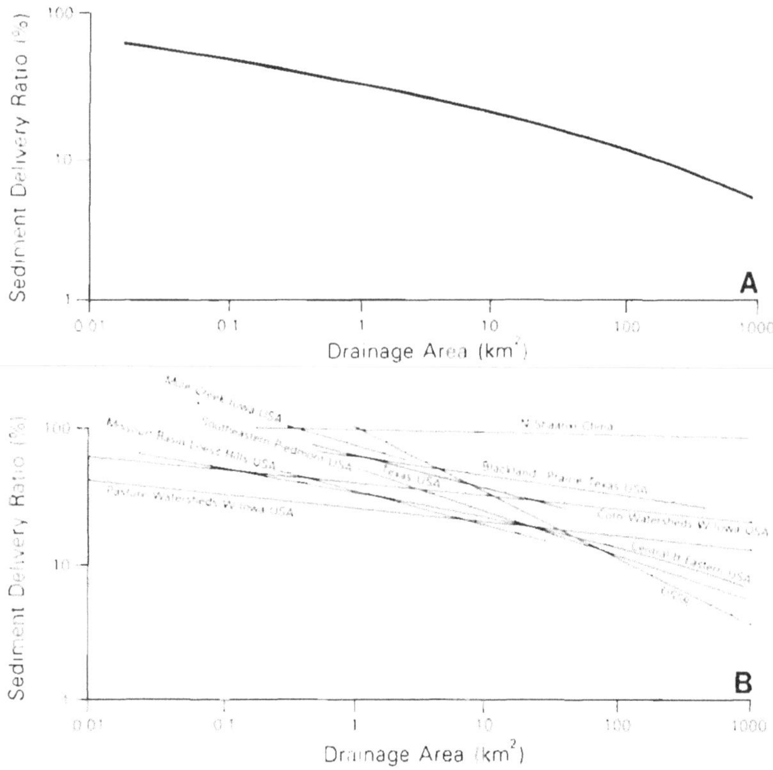

Figure 3.13 (A) The relationship between sediment delivery ratio and drainage basin area developed by the U.S. Soil Conservation Service for the central and eastern United States and (B) similar relationships proposed by other workers for different areas. B is based on Roehl (*86*), Sokolovskii (*90*), Piest and associates (*80*), Renfro (*84*), ASCE (*4*), Williams (*123*), and Mou and Meng (*65*).

As yet, the availability of techniques for quantifying the various sources and sinks involved in a sediment budget is rather limited and the effort needed to assemble the information necessary to construct reliable budgets of the type illustrated in Figure 3.14 could be very great. Development of such techniques represents an important research need. The use of sediment properties to fingerprint sediment sources (*51, 73, 78, 105, 112*) and the use of cesium-137 measurements to investigate the amount and spatial distribution of deposited sediment within a drainage basin (*11, 55, 58, 61, 110, 111*) must, however, be seen as important advances toward this goal.

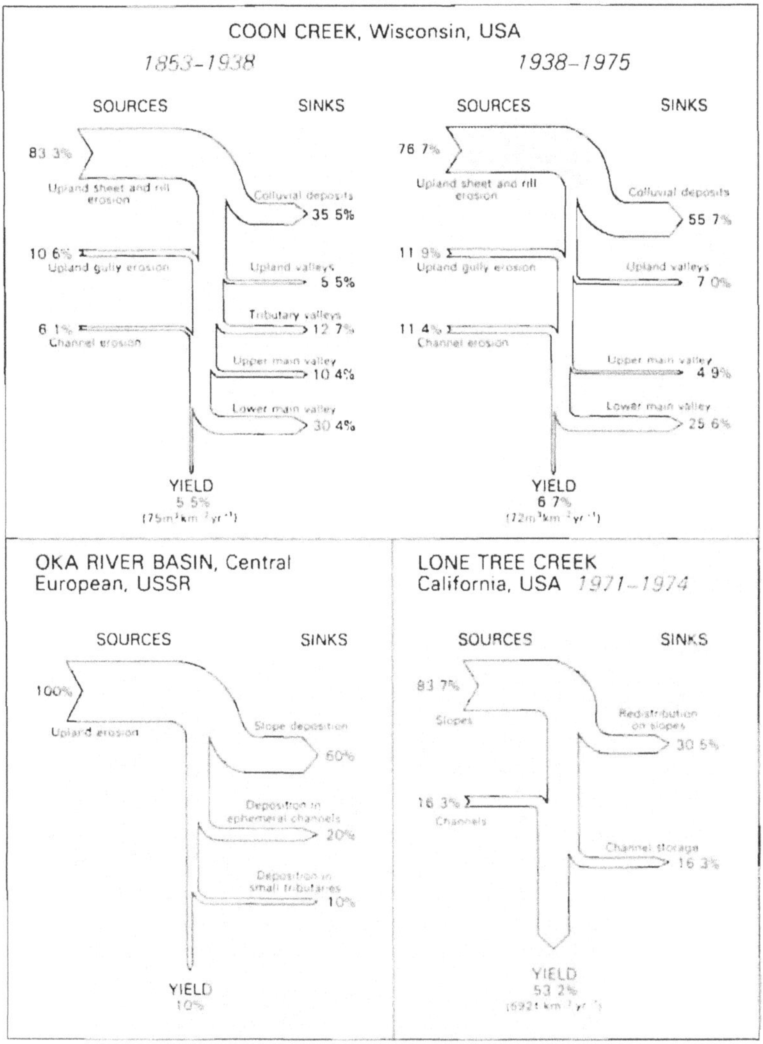

Figure 3.14 Tentative sediment budgets for Coon Creek, Wisconsin (360 km^2), 1853–1938 and 1938–1975; Lone Tree Creek, California (1.74 km^2); and the Oka River, USSR. Based on data presented by Trimble (*100*), Lehre (*54*), and Zaslavsky (*125*).

REFERENCES

1. Adams, J. 1980. *High sediment yields from major rivers of the western Southern Alps, New Zealand.* Nature 287: 88-89.
2. Allan, R.J. 1986. *The role of particulate matter in the fate of contaminants in aquatic ecosystems.* Scientific Series No. 142, Inland Waters Directorate, Environment Canada, Ottawa, Canada.
3. Allen, P.B., and D.V. Petersen. 1981. *A study of the variability of suspended sediment measurements.* In: *Erosion and Sediment Transport Measurement, Proceedings, Florence Symposium, June 1981.* Publication No. 133. International Association of Hydrological Sciences, Wallingford, England. pp. 203-211.
4. American Society of Civil Engineers. 1975. *Sedimentation Engineering.* New York. 745 pp.
5. Anttila, P.W. 1970. *Sedimentation in Plum Creek subwatershed No. 4, Shelby County, North-Central Kentucky.* Water Supply Paper 1798-G. U.S. Geological Survey, Reston, Virginia.
6. Berke, B., and L. Rakoczi. 1981. *Latest achievements in the development of nuclear suspended sediment gauges.* In: *Erosion and Sediment Transport Measurement, Proceedings, Florence Symposium, June 1981.* Publication No. 133. International Association of Hydrological Sciences, Wallingford, England. pp. 91-96.
7. Bloemendal, J., F. Oldfield, and R. Thompson. 1979. *Magnetic measurements used to assess sediment influx at Llyn Goddionduon.* Nature 280: 50-53.
8. Bolton, P. 1983. *Sediment discharge measurement and calculation.* Technical Note OD/TN2. Hydraulics Research, Wallingford, England.
9. Brabben, T. 1981. *Use of turbidity monitors to assess sediment yield in East Java, Indonesia.* In: *Erosion and Sediment Transport Measurement, Proceedings, Florence Symposium, June 1981.* Publication No. 133. International Association of Hydrological Sciences, Wallingford, England. pp. 105-113.
10. Branski, J. 1975. *Ocena denudacji dorzecze wisley na podatawie wynikow ponmiarow rumowiska unoszonego.* Prace Instytutu Meteorologii i Gospodarki Wodnej 6: 1-58.
11. Brown, R.B., G.F. Kling, and N.B. Cutshall. 1981. *Agricultural erosion indicated by ^{137}Cs redistribution: II. Estimates of erosion rates.* Soil Science Society of America Journal 45: 1191-1197.
12. Brune, G.M. 1953. *Trap efficiency of reservoirs.* Transactions, American Geophysical Union 34: 407-418.
13. Burrus, D., R.L. Thomas, J. Dominik, and J.P. Vernet. 1989. *Recovery and concentration of suspended solids in the Upper Rhone River by continuous flow centrifugation.* Hydrological Processes 3: 65-74.
14. Campbell, B.L. 1983. *Applications of environmental caesium-137 for the determination of sedimentation rates in reservoirs and lakes and related catchment studies in developing countries.* In: *Radioisotopes in Sediment Studies, Technical Document 298.* International Atomic Energy Agency, Vienna, Austria.

15. Campbell, F.B., and H.A. Bauder. 1940. *A rating-curve method for determining silt-discharge of streams*. Transactions, American Geophysical Union 21: 603-607.
16. Colby, B.R. 1956. *Relationship of sediment discharge to streamflow*. U.S. Geological Survey Open File Report, U.S. Geological Survey, Reston, Virginia.
17. Davis, M.B. 1976. *Erosion rates and land use history in southern Michigan*. Environmental Conservation 3: 139-148.
18. Dearing, J.A., J.K. Elner, and C.M. Happey-Wood. 1981. *Recent sediment flux and erosional processes in a Welsh upland lake catchment based on magnetic susceptibility measurements*. Quaternary Research 16: 356-372.
19. Dearing, J.A., and I.D.L. Foster. 1987. *Limnic sediments used to reconstruct sediment yields and sources in the English Midlands since 1765*. In: V. Gardiner (ed.) International Geomorphology 1986. John Wiley and Sons, Chichester, England. pp. 853-868.
20. Dearing, J.A., H. Hakansson, B. Liedberg-Jonsson, A. Persson, S. Skansjo, D. Widholm, and F. El Daoushy. 1987. *Lake sediments used to quantify the erosion response to land use change in southern Sweden*. Oikos 50: 60-78.
21. Demmak, A. 1982. *Contribution à l'étude de l'érosion et des transports solides en Algerie Septentrionale*. Thèse de Docteur-Ingenieur, Université de Paris, Paris, France.
22. Dendy, F.E., and G.C. Bolton. 1976. *Sediment yield-runoff drainage area relationships in the United States*. Journal of Soil and Water Conservation 32: 264-266.
23. Dickinson, W.T. 1981. *Accuracy and precision of suspended sediment loads*. In: *Erosion and Sediment Transport Measurement, Proceedings, Florence Symposium, June 1981*. Publication No. 133. International Association of Hydrological Sciences, Wallingford, England. pp. 195-202.
24. Dietrich, W.B., and T. Dunne. 1978. *Sediment budget for a small catchment in mountainous terrain*. Zeitschrift fur Geomorphologie, Supplementband 29: 191-206.
25. Douglas, I. 1973. *Rates of denudation in selected small catchments in eastern Australia*. Occasional Papers in Geography No. 21. University of Hull, Hull, England.
26. Dunne, T. 1977. *Evaluation of erosion conditions and trends*. In: *Guidelines for Watershed Management*. Conservation Guide No. 1. Food and Agriculture Organization, United Nations, Rome, Italy. pp. 53-83.
27. Emmett, W.W. 1984. *Measurement of bedload in rivers*. In: R.F. Hadley and D.E. Walling (eds.) *Erosion and Sediment Yield: Some Methods of Measurement and Modelling*. Geobooks, Norwich, England. pp. 91-109.
28. Farr, I.S., and R.T. Clarke. 1984. *Reliability of suspended load estimates in chalk streams*. Arch. Hydrobiol. 102: 1-19.
29. Ferguson, R.I. 1986. *River loads underestimated by rating curves*. Water Resources Research 22: 74-76.

30. Ferguson, R.I. 1987. *Accuracy and precision of methods for estimating river loads.* Earth Surface Processes and Landforms 12: 95-104.
31. Ficken, J.H. 1986. *Instruments for measuring suspended sediment.* In: *Proceedings of the Fourth Federal Interagency Sedimentation Conference 1986.* Subcommittee on Sedimentation of the Interagency Advisory Committee on Water Data. pp. 1-40--1-50.
32. Fish, I.L. 1983. *Partech turbidity monitors. Calibration with silt and the effects of sand.* Technical Note OD/TNR. Hydraulics Research, Wallingford, England.
33. Flammer, G.H. 1962. *Ultrasonic measurement of suspended sediment.* Bulletin 1141-A. U.S. Geological Survey, Reston, Virginia.
34. Fleming, G. 1969. *Suspended solids monitoring: A comparison between three instruments.* Water and Water Engineering 72: 377-383.
35. Flint, R.F. 1972. *Fluvial sediment in Hocking River subwatershed 1 (North Branch Hunters Run) Ohio.* Water Supply Paper 1798-I. U.S. Geological Survey, Reston, Virginia.
36. Food and Agriculture Organization, United Nations. 1979. *A Provisional Methodology for Soil Degradation Assessment.* Rome, Italy.
37. Gippel, C.J. 1989. *The use of turbidimeters in suspended sediment research.* Hydrobiologia 176/177: 465-480.
38. Glymph, L.M. 1954. *Studies of sediment yields from watersheds.* Publication No. 36. International Association of Hydrological Sciences, Wallingford, England. pp. 173-191.
39. Glysson, G.D. 1987. *Suspended-sediment transport curves.* U.S. Geological Survey Open-File Report 87-218. U.S. Geological Survey, Reston, Virginia.
40. Golterman, H.L., P.G. Sly, and R.L. Thomas. 1983. *Study of the Relationship Between Water Quality and Sediment Transport.* Technical Papers in Hydrology No. 26. UNESCO, Paris, France.
41. Graf, W.L. 1985. *Channel processes and sediment yield in arid region drainage basins.* In: T. Spencer (ed.) International Geomorphology 1985. Abstracts First International Conference Geomorphology, University of Manchester, Manchester, England. p. 223.
42. Griffiths, G.A. 1979. *High sediment yields from major rivers of the western Southern Alps, New Zealand.* Nature 282: 61-63.
43. Grobler, D.C., and A. van B. Weaver. 1981. *Continuous measurement of suspended sediment in rivers by means of a double beam turbidity meter.* In: *Erosion and Sediment Transport Measurement, Proceedings, Florence Symposium, June 1981.* Publication No. 138. International Association of Hydrological Sciences, Wallingford, England. pp. 97-103.
44. Guy, H.P., and V.W. Norman. 1970. *Field methods for measurement of fluvial sediment.* In: *Techniques of Water Resources Investigations of the U.S. Geological Survey* (Book 3, Chapter C2). U.S. Geological Survey, Reston, Virginia.
45. Hadley, R.F., and L.M. Shown. 1976. *Relations of erosion to sediment yield.* In: *Proceedings, Third Federal Inter-Agency Sedimentation Conference.* U.S. Water Resources Council, Washington, D.C. 1-132—1-139.

46. Hadley, R.F., and D.E. Walling (eds.). 1984. *Erosion and Sediment Yield: Some Methods of Measurement and Modelling*. Geobooks, Norwich, England.
47. Hansen, D., and D. Bray. 1987. *Generation of annual suspended sediment loads for the Kennebacasis using sediment rating curves*. In: *Proceedings 8th Canadian Hydrotech. Conference, Quebec, May 1987*. Canadian Society of Civil Engineers.
48. Heinemann, H.G. 1984. *Reservoir trap efficiency*. In: R.F. Hadley and D.E. Walling (eds.) *Erosion and Sediment Yield: Some Methods of Measurement and Modelling*. Geobooks, Norwich, England. pp. 201-218.
49. Hereford, R. 1987. *Sediment-yield history of a small basin in southern Utah, 1937-1976: Implications for land management and geomorphology*. Geology 15: 954-957.
50. Heusch, B. 1980. *Erosion in the Ader Dutchi Massif (Niger)*. In: M. De Boodt and D. Gabriels (eds.) *Assessment of Erosion*. John Wiley and Sons, Chichester, England. pp. 521-529.
51. Klages, M.G., and Y.P. Hsieh. 1975. *Suspended solids carried by the Gallatin River of southwestern Montana: II. Using mineralogy for inferring sources*. Journal of Environmental Quality 4: 68-73.
52. Laronne, J.B. 1987. *Rhythmic couplets: sedimentology and prediction of reservoir design periods in semiarid areas*. In: L. Berkofsky and M.G. Wurtele (eds.) *Progress in Desert Research*. Rowman and Littlefield Publishers, New Jersey. pp. 229-244.
53. Lehre, A.K. 1981. *Sediment budget of a small coast range drainage basin*. In: *Erosion and Sediment Transport in Pacific Rim Steeplands*. Publication No. 132. International Association of Hydrological Sciences, Wallingford, England. pp. 123-139.
54. Lehre, A.K. 1982. *Sediment budget in a small coast range drainage basin in north-central California*. In: F.J. Swanson, R.J. Janda, T. Dunne, and D.N. Swanston (eds.) *Sediment Budgets and Routing in Forested Drainage Basins*. General Technical Report PNW-141. Forest Service, U.S. Department of Agriculture, Portland, Oregon.
55. Loughran, R.J., B.L Campbell, and G.L. Elliott. 1982. *The identification and quantification of sediment sources using ^{137}Cs*. In D.E. Walling (ed.) *Recent Developments in the Explanation and Prediction of Erosion and Sediment Yield*. Publication No. 137. International Association of Hydrological Sciences, Wallingford, England. pp. 361-369.
56. Lu Zhi, Yuren Lui, Leling Sun, Xianglin Xu, Yujing Yang, and Lingqui Kong. 1981. *The development of nuclear sediment concentration gauges for use on the Yellow River*. In: *Erosion and Sediment Transport Measurement, Proceedings, Florence Symposium*. Publication No. 133. International Association of Hydrological Sciences, Wallingford, England. pp. 83-90.
57. Maner, S.B. 1958. *Factors affecting sediment delivery rates in the Red Hills physiographic area*. Transactions, American Geophysical Union. 39: 669-675.
58. McCallan, M.E., B.M. O'Leary, and C.W. Rose. 1980. *Redistribution of cesium-137 by erosion and deposition on Australian soil*. Australian Journal of Soil Research 18: 119-128.

59. McHenry, J.R. 1971. *Sediment-measurement techniques: reservoir deposits.* American Society of Civil Engineers Proceedings. 97: 1253-1257.
60. McHenry, J.R., and F.E. Dendy. 1964. *Measurement of sediment density by attenuation of transmitted gamma rays.* Soil Science Society of America Proceedings 28: 817-822.
61. McHenry, J.R., and J.C. Ritchie. 1977. *Estimating field erosion losses from fall out cesium-137 measurements.* In: *Erosion and Solid Matter Transport in Inland Waters, Proceedings, Paris Symposium, July 1977.* Publication 122. International Association of Hydrological Sciences, Wallingford, England. pp. 26-33.
62. Meade, R.H. 1982. *Sources, sinks and storage of river sediments in the Atlantic drainage of the United States.* Journal of Geology 90: 235-252.
63. Meade, R.H. 1988. *Movement and storage of sediment in river systems.* In: A. Lerman and M. Meybeck (eds.) *Physical and Chemical Weathering in Geochemical Cycles.* Kluwer, Dordrecht, The Netherlands. pp. 165-179.
64. Miller, C.R. 1951. *Analysis of flow duration sediment rating curve method of computing sediment yield.* U.S. Bureau of Reclamation. Washington, D.C.
65. Mou, J., and Q. Meng. 1980. *Sediment delivery ratio as used in the computation of the watershed sediment yield.* Beijing, China.
66. Mundorff, J.C. 1964. *Fluvial sediment in Kiowa Creek basin, Colorado.* Water Supply Paper 1798-A. U.S. Geological Survey, Reston, Virginia.
67. Mundorff, J.C. 1966. *Sedimentation in Brownell Creek, Subwatershed No. 1, Nebraska.* Water Supply Paper 1798-C. U.S. Geological Survey, Reston, Virginia.
68. Mutchler, C.K. and A.J. Bowie. 1976. *Effect of land use on sediment delivery ratios. In: Proceedings, Third Federal Inter-Agency Sedimentation Conference.* U.S. Water Resources Council, Washington D.C. 1-11—1-21.
69. Novotny, V., and G. Chesters. 1981. *Handbook of Nonpoint Pollution: Sources and Management.* Van Nostrand Reinhold, New York.
70. Novotny, V., G.V. Simsiman, and G. Chesters. 1986. *Delivery of pollutants from nonpoint sources.* In: *Drainage Basin Sediment Delivery, Proceedings, Albuquerque Symposium, August 1986.* Publication No. 159, International Association of Hydrological Sciences, Wallingford, England. pp. 133-140.
71. Oldfield, F. 1977. *Lakes and their drainage basins as units of sediment-based ecological study.* Progress in Physical Geography 1: 460-504.
72. Oldfield, F., and P.G. Appleby. 1984. *Empirical testing of ^{210}Pb-dating models for lake sediments.* In: E.Y. Haworth and J.W.G. Lund (eds.) *Lake Sediments and Environmental History.* University of Leicester Press, Leicester, England.
73. Oldfield, F., T.A. Rummery, R. Thompson, and D.E. Walling. 1979. *Identification of suspended sediment sources by means of magnetic measurements: Some preliminary results.* Water Resources Research 15: 211-218.
74. Ongley, E.D., and D.P. Blachford. 1982. *Application of continuous-flow centrifugation to contaminant analysis of suspended sediment in fluvial streams.* Environmental Technology Letters 3: 219.

75. Ongley, E.D., and R.L. Thomas, 1989. *Dewatering suspended solids by continuous flow centrifugation: Practical considerations*. Hydrological Processes 3: 255-260.
76. O'Sullivan, P.E. 1979. *The ecosystem watershed concept in the environmental sciences—a review*. International Journal of Environmental Studies 13: 273-281.
77. Oyebande, L. 1981. *Sediment transport and river basin management in Nigeria*. In: R. Lal and E.W. Russell (eds.) *Tropical Agricultural Hydrology*. John Wiley and Sons, Chichester, England. pp. 201-225.
78. Peart, M.R., and D.E. Walling. 1988. *Techniques for establishing suspended sediment sources in two drainage basins in Devon, UK: a comparative assessment*. In: *Sediment Budgets, Proceedings of the Porto Alegre Symposium, December 1988*. Publication No. 174. International Association of Hydrological Sciences, Wallingford, England. pp. 269-279.
79. Phillips, J.D. 1986. *The utility of the sediment budget concept in sediment pollution control*. Professional Geographer 38: 246-252.
80. Piest, R.F., L.A. Kramer, and H.G. Heinemann. 1975. *Sediment movement from loessial watersheds*. In: *Present and Prospective Technology for Predicting Sediment Yields and Sources*. ARS-S-40. Agricultural Research Service, U.S. Department of Agriculture, Washington, D.C. pp. 130-141.
81. Porterfield, G. 1972. *Computation of fluvial-sediment discharge*. In: *Techniques of Water Resources Investigations of the U.S. Geological Survey* (Book 3, Chapter C3), U.S. Geological Survey, Reston, Virginia.
82. Rakoczi, L. 1976. *Possibilities and limitation of nuclear suspended sediment gauging*. In: *Modern Developments in Hydrometry*. Publication No. 427. World Meteorological Organization, Geneva, Switzerland. pp. 138-150.
83. Renard, K.G., J.R. Simanton, and C.E. Fancher. 1986. *Small watershed automatic water quality sampler*. In: *Proceedings of the Fourth Federal Interagency Sedimentation Conference 1986*. Subcommittee on Sedimentation of the Interagency Advisory Committee on Water Data. pp. 1-51—1-58.
84. Renfro, G.W. 1975. *Use of erosion equations and sediment delivery ratios for predicting sediment yield*. In: *Present and Prospective Technology for Predicting Sediment Yields and Sources*. ARS-S-40. Agricultural Research Service, U.S. Department of Agriculture, Washington, D.C. pp. 33-45.
85. Robinson, A.R. 1977. *Relationships between soil erosion and sediment delivery*. In: *Erosion and Solid Matter Transport in Inland Waters, Proceedings, Paris Symposium, July 1977*. Publication No. 122. International Association of Hydrological Sciences, Wallingford, England. pp. 159-167.
86. Roehl, J.E. 1962. *Sediment source areas, delivery ratios and influencing morphological factors*. Publication 59. International Association of Hydrological Sciences, Wallingford, England. pp. 202-213.
87. Seely, E.H. 1982. *The Goodwin Creek Research Catchment: Part IV. Field Data Acquisition*. In: *Hydrological Research Basins and their Use in Water Resources Planning, Proceedings, Berne Symposium, September 1982*. Landeshydrologie, Berne, Switzerland. pp. 173-182.

88. Shear, H., and A.E.P. Watson (eds.). 1977. *The Fluvial Transport of Sediment-Associated Nutrients and Contaminants.* International Joint Commission, Windsor, Ontario.
89. Skinner, J V., and J.P. Beverage. 1986. *Continuous measurement of suspended-sediment concentration.* In: *Proceedings of the Fourth Federal Interagency Sedimentation Conference 1986.* Subcommittee on Sedimentation of the Interagency Advisory Committee on Water Data. pp. 1-29—1-39.
90. Sokolovskii, D.L. 1968. *Rechnoi Stok. Osnovy Teorii i Metodiki Paschetov.* Gidrometeorologicheskoe Izdatel'stvo, Leningrad, Soviet Union.
91. Souch, C., and O. Slaymaker. 1986. *Temporal variability of sediment yield using accumulations in small ponds.* Physical Geography 7: 140-153.
92. Swanson, F.J., R.J. Janda, T. Dunne, and D.N. Swanston. 1982. *Sediment budgets and routing in forested drainage basins.* General Technical Report PNW-141. Forest Service. U.S. Department of Agriculture, Portland, Oregon.
93. Tazioli, G.S. 1981. *Nuclear techniques for measuring sediment transport in natural streams – examples from instrumented basins.* In: *Erosion and Sediment Transport Measurement, Proceedings, Florence Symposium, June 1981.* Publication No. 133. International Association of Hydrological Sciences, Wallingford, England. pp. 63-81.
94. Thomas, R.B. 1985. *Estimating total suspended sediment yield with probability sampling.* Water Resources Research 21: 1381-1388.
95. Thomas, R.B. 1986. *Calibrating SALT: a sampling scheme to improve estimates of suspended sediment yield.* In: *Monitoring to Detect Changes in Water Quality Series, Proceedings, Budapest Symposium, July 1986.* Publication No. 157. International Association of Hydrological Sciences, Wallingford, England. pp. 79-88.
96. Thomas, R.L. 1988. *Lake sediments as indicators of changes in land erosion rates.* In: A. Lerman and M. Meybeck (eds.) *Physical and Chemical Weathering in Geochemical Cycles.* Kluwer, Dordrecht, The Netherlands. pp. 143-164.
97. Thompson, R., and G.M. Turner. 1979. British geomagnetic master curves 10,000-0 yr for dating European sediments. Geophysical Research Letters. 6: 249-252.
98. Trimble, S.W. 1974. *Man-induced soil erosion on the Southern Piedmont.* Soil and Water Conservation Society, Ankeny, Iowa.
99. Trimble, S.W. 1976. *Sedimentation in Coon Creek Valley, Wisconsin. In: Proceedings, Third Federal Inter-Agency Sedimentation Conference.* U.S. Water Resources Council, Washington, D.C. pp. 5-100—5-112.
100. Trimble, S.W. 1981. *Changes in sediment storage in the Coon Creek Basin, Driftless Area, Wisconsin.* 1853 to 1975. Science 214: 181-183.
101. Truhlar, J.F. 1978. *Determining suspended sediment loads from turbidity records.* Hydrological Sciences Bulletin 23: 409-417.
102. U.S. Department of Agriculture. 1976. *Control of Water Pollution from Cropland. Volume II. An overview.* Washington, D.C.
103. Virgo, K.J., and R.N. Munro. 1978. *Soil and erosion features of the Central Plateau region of Tigrai, Ethiopia.* Geoderma 20: 131-157.

104. Wade, J.C., and E.O. Heady. 1978. *Measurement of sediment control impacts on agriculture*. Water Resources Research 14: 1-8.
105. Wall, C.J., and L.P. Wilding. 1976. *Mineralogy and related parameters of fluvial suspended sediments in northwestern Ohio*. Journal of Environmental Quality 5: 168-173.
106. Walling, D.E. 1977. *Limitations of the rating curve technique for estimating suspended sediment loads, with particular reference to British rivers*. In: *Erosion and Solid Matter Transport in Inland Waters, Proceedings, Paris Symposium, July 1977*. Publication No. 122. International Association of Hydrological Sciences, Wallingford, England. pp. 34-48.
107. Walling, D.E. 1983. *The sediment delivery problem*. Journal of Hydrology 65: 209-237.
108. Walling, D.E. 1984. *Dissolved loads and their measurement*. In: R.F. Hadley and D.E. Walling (eds.) *Erosion and Sediment Yield: Some Methods of Measurement and Modelling*. Geobooks, Norwich, England. pp. 111-177.
109. Walling, D.E. 1984. *The sediment yields of African rivers*. In: *Challenges in African Hydrology and Water Resources, Proceedings of the Harare Symposium, July 1984*. Publication No. 144. International Association of Hydrological Sciences, Wallingford, England. pp. 265-283.
110. Walling, D.E., and S.B. Bradley. 1988. *The use of caesium-137 measurements to investigate sediment delivery from cultivated areas in Devon, UK*. In: *Sediment Budgets, Proceedings, Porto Alegre Symposium, December 1988*. Publication No. 174. International Association of Hydrological Sciences, Wallingford, England. pp. 325-335.
111. Walling, D.E., and S.B. Bradley. 1989. *Rates and patterns of contemporary floodplain sedimentation: A case study of the River Culm, Devon, UK*. Geojournal 19: 53-62.
112. Walling, D.E., and P. Kane. 1984. *Suspended sediment properties and their geomorphological significance*. In: T.P. Burt and D.E. Walling (eds.) *Catchment Experiments in Fluvial Geomorphology*. Geobooks, Norwich, England. pp. 311-334.
113. Walling, D.E., and P.W. Moorehead. 1989. *The particle size characteristics of fluvial suspended sediment: an overview*. Hydrobiologia 176/177: 125-149.
114. Walling, D.E., and A. Teed. 1971. *A simple pumping sampler for research into suspended sediment transport in small catchments*. Journal of Hydrology 13: 325-337.
115. Walling, D.E., and B.W. Webb. 1981. *The reliability of suspended sediment load data*. In: *Erosion and Sediment Transport Measurement, Proceedings, Florence Symposium, June 1981*. Publication No. 133. International Association of Hydrological Sciences, Wallingford, England. pp. 177-194.
116. Walling, D.E., and B.W. Webb. 1983. *Patterns of sediment yield*. In: K.J. Gregory (ed.) *Background to Palaeohydrology*. John Wiley and Sons, Chichester, England. pp. 69-100.
117. Walling, D.E., and B.W. Webb. 1985. *Estimating the discharge of contaminants to coastal waters by rivers: Some cautionary comments*. Marine Pollution Bulletin 16: 488-492.

118. Walling, D.E., and B.W. Webb. 1987. *Material transport by the world's rivers.* In: *Water for the Future: Hydrology in Perspective, Proceedings of the Rome Symposium, April 1987.* Publication No. 164. International Association of Hydrological Sciences, Wallingford, England. pp. 313-329.
119. Walling, D.E., and B.W. Webb. 1988. *The reliability of rating curve estimates of suspended sediment yield: some further comments.* In: *Sediment Budgets, Proceedings of the Porto Alegre Symposium, December 1988.* Publication No. 174. International Association of Hydrological Sciences, Wallingford, England. pp. 337-350.
120. Ward, P.R.B. 1984. *Measurement of sediment yields.* In: R.F. Hadley and D.E. Walling (eds.) *Erosion and Sediment Yield: Some Methods of Measurement and Modelling.* Geobooks, Norwich, England. pp. 67-70.
121. Wasson, R.J., R.L. Clark, P.M. Nanninga, and J. Waters. 1987. *^{210}Pb as a chronometer and tracer, Burrinjuck Reservoir, Australia.* Earth Surface Processes and Landforms 12: 399-414.
122. Williams, G.P. 1989. *Sediment concentration versus water discharge during single hydrologic events.* Journal of Hydrology 111: 89-106.
123. Williams, J.R. 1977. *Sediment delivery ratios determined with sediment and runoff models.* In: *Erosion and Solid Matter Transport in Inland Waters, Proceedings, Paris Symposium, July 1977.* Publication No. 122. International Association of Hydrological Sciences, Wallingford, England. pp. 168-179.
124. Williams, J.R., and H.D. Berndt. 1972. *Sediment yield computed with universal equation.* Proceedings, American Society of Civil Engineers, Journal of Hydraulics Division 98(HY2): 2087-2098.
125. Zaslavsky, M.N. 1979. *Erozia Pochv (soil erosion).* Mysl. Publishing House, Moscow, Soviet Union.

Chapter 4

RAINFALL SIMULATORS for SOIL EROSION RESEARCH

L. D. Meyer

Rainfall simulators are research tools designed to apply water in a form similar to natural rainstorms. They are useful for many types of soil erosion and hydrologic experiments. However, rainstorm characteristics must be simulated properly, runoff/erosion data analyzed carefully, and results interpreted judiciously to obtain reliable information for the conditions to which the simulated rainstorms are applied.

The major advantages of rainfall-simulator research are fourfold: it is more rapid, more efficient, more controlled, and more adaptable than natural rainfall research. Researchers can measure hydrologic or soil erosion characteristics of newly developed cropping and management practices in a relatively short time. Simulated storms can be applied for selected durations on selected treatment conditions, and measurements from a few such storms often can indicate conclusively at least relative differences for those treatments. Plot preparation prior to application of such storms usually takes much less time than plot maintenance for studies that depend on natural rainfall. Frequently, parts of farm fields that have the desired soil characteristics and cropping management histories suffice as research areas. Plots and equipment can be inspected immediately prior to and during data collection. Researchers can make measurements and observations during simulated storms that are difficult or impossible during natural rainstorms. In addition to their use for field studies, rainfall simulators are readily adaptable for highly controlled laboratory research of basic infiltration, runoff, and erosion processes.

Rainfall simulator research also has major limitations and disadvantages. The cost and time required to construct a suitable rainfall simulator with related equipment and the personnel needed to operate an effective simulated rainfall research program generally are the greatest hurdles. The difficulty of simulating natural rainfall characteristics and of properly interpreting data obtained with rainfall simulators, especially if they fail to fully achieve such characteristics, also impedes such research. Other major limitations are the relatively small area to which rain can be applied by most rainfall simulators and the necessary compromise in the characteristics of rain produced by large area rainfall simulators. These and other limitations, discussed in greater detail later, need to be recognized and resolved before embarking on a simulated rainfall research effort.

Rainfall simulators do not eliminate the need for natural rainfall experiments. For example, studies that rely on natural rainfall are essential for obtaining data on long-term average erosion or infiltration amounts and on variability and extremes resulting from climatic differences. However, to be of maximum value, research results must be available as soon as possible after their need is recognized. Conclusive results from field-plot studies that rely on natural rainfall often require many years to obtain a representative sampling of rainstorms that includes typical combinations of critical storms and critical treatment conditions. Management practices of interest often change appreciably or may even be abandoned during this period. Furthermore, many needed studies never would be initiated if they had to rely on natural rainfall because the results would not be available soon enough. For such situations, rainfall simulators may be an essential component in productive research programs.

CHARACTERISTICS OF RAINFALL SIMULATORS

The ideal rainfall simulator would be inexpensive to build and operate, would simulate rainfall perfectly, would be simple to move, and could be used whenever and wherever needed. Most researchers realize that such a utopian rainfall simulator is impossible to acquire. Thus, different rainfall simulators embody different characteristics to meet different research goals. Perhaps the most important characteristics of natural rainfall that need to be closely simulated for soil and water management research are raindrop size distribution, raindrop impact velocity, and appropriate rainstorm intensities. These three characteristics are key factors in soil detachment, soil surface sealing, and resulting runoff. These and other desirable characteristics for rainfall simulators to be used in erosion and hydrologic studies include the following:

- *Drop size distribution near that of natural rainstorms.* Natural rainfall consists of a wide distribution of drop sizes that range from near zero to about 7 mm in diameter. The median drop diameter is between 1 and 3 mm for erosive rainstorms and tends to increase with rain intensity (*9*).

- *Drop impact velocities near those of natural raindrops.* Raindrop fall velocities vary from near zero for mist-sized drops to more than 9 m/sec for the largest sizes. A common-sized raindrop of 2 mm falls at a velocity of 6 to 7 m/sec (*6*).
- *Intensities in the range of storms for which results are of interest.* Intensities of natural rainfall vary from near zero to several millimeters per minute. Generally, very low intensities are not of major interest for erosion and hydrologic studies, and very high intensities are so rare that they may be of limited interest. Intensities between 0.2 and 2 mm/min are usually of greatest importance.
- *Research area of sufficient size to represent the treatments and conditions being evaluated.* Rainfall simulators should be capable of applying rainfall to plots that are large enough for a realistic test of treatment characteristics. Square-meter and smaller plots may be sufficient for studying raindrop impact (interrill) erosion, but much longer plots are necessary for evaluating scour and transport by runoff.
- *Drop characteristics and intensity of application fairly uniform over the study area.*
- *Raindrop application nearly continuous throughout the study area.*
- *Angle of impact not greatly different from vertical for most drops.*
- *Capability of applying the same simulated rainstorm(s) repeatedly.*
- *Satisfactory rainstorm characteristics when used during common field conditions, such as high temperatures and moderate winds.*
- *Portability for movement from one research site to another.*

Developers of rainfall simulators should strive for these characteristics, especially those of greatest importance for their specific uses, but they should not expect to achieve all characteristics perfectly. Certain compromises are necessary when simulating rainstorms that are scientifically acceptable in comparison with natural rainstorms. Furthermore, researchers should avoid becoming so involved in developing and improving rainfall simulators that little time is left for their use. The goal of rainfall simulator research should be the collection of accurate and useful data, not perfection of a rainfall simulator.

TYPES OF RAINFALL SIMULATORS

During the last 50 years, researchers have used a broad range of techniques and equipment for simulating rainfall. These techniques and equipment have ranged from walking up and down the slope with common sprinkling cans to elaborate, pushbutton-operated electronic and hydraulic machines (*3, 7, 12, 21, 24*). The drop formers that produce the simulated raindrops are the key components of rainfall simulators. Once the drop-producing method is chosen, suitable mechanical and electronic components can be developed for operating them.

The major methods used to produce simulated raindrops for erosion and hydrologic research can be grouped into two rather broad categories: those involving nozzles from which water is forced at a significant velocity by pressure and those where drops form and fall from a tip starting at essentially zero velocity (*3, 21, 24*). Nozzles produce a wide range of drop sizes as do rainstorms, but the large nozzle orifices that are necessary to obtain large drops usually require that the nozzle spray only intermittently to reduce application rates to typical rain intensities. Tips produce only one drop size or a very limited range of sizes, so they are used mostly for fundamental studies when a carefully controlled drop size is important.

Early rainfall simulators were rather crude in comparison with today's standards because little information was available on rainfall characteristics. In particular, scientists had not recognized the importance of raindrop impact on soil detachment. Thus, the primary concern was to apply water uniformly over the research area with little regard to drop characteristics. Subsequent research that evaluated characteristics of rainstorms has provided an understanding of appropriate goals for the design of rainfall simulators. Most designers of equipment developed in recent decades have carefully considered available raindrop size and velocity data as a basis for more realistic simulation of rainstorms. However, drop size distribution data are not available for some types of climate and seasonal variations, so the capability of available rainfall simulators to reproduce rainstorms for those conditions is unknown.

Many of the rainfall simulators currently used for soil conservation research are described in the proceedings of the 1979 rainfall simulator workshop (*24*). A summary of the drop sizes, rain intensities, plot sizes, and literature references for each of them is given on pages 122-130 of that report. Some rainfall simulators can apply rainstorms to runoff-plot sized areas. Others are suitable for only very small field plots or laboratory studies. Figures 4.1 and 4.2 show some of the more widely used designs.

Several rainfall simulators have been designed for use on field plots that are similar in size to those used for natural rainfall studies of runoff and erosion. The rainulator (*16*) was the first simulator designed to apply rainstorms with drop characteristics near those of natural rainfall on several runoff plots simultaneously. Simulated rain is applied by downward-spraying nozzles that are moved laterally across the plots and border areas. Spray application is intermittent, and only a few intensities can be simulated. The equipment is cumbersome to move from site to site.

The rotating-boom rainfall simulator (*23*) uses the same nozzles as the rainulator, but they are located along arms extending from a central vertical shaft that rotates slowly. Only two intensities are possible, plot size is limited, and application is intermittent. The entire rainfall simulator is mounted on a trailer and can be readily moved from one site to another.

The programmable rainfall simulator (*4*) and Kentucky rainfall simulator (*18*) use rapidly oscillating nozzles that reduce spray intermittency to short periods. They can produce a wide range of intensities. Both are electronically and mechanically complex. The Kentucky machine is on wheels for more rapid movement to adjacent plots.

Sprinkler irrigation equipment and the Colorado State University RREF rainfall simulator (*8*) can be used on even larger runoff plots and small watersheds. These types of simulators are less successful in achieving natural rainfall characteristics, especially drop size distribution and uniformity of application.

Some rainfall simulators have been designed primarily for research on small field plots of about 1 m^2 and for laboratory studies. The Purdue sprinkling infiltrometer (*1*), a similar, more recent design (*11*), and the rotating-disk rainfall simulator (*19, 5*) use stationary nozzles, but the latter gives a better simulation of raindrop-impact energy. The interrill rainfall simulator (*14*) uses a rapidly oscillating nozzle to produce a wide range of intensities at energies very near natural rainfall. This design has also been adapted for use on longer plots by adding more nozzles (*15*). Other designs use yarn or capillary tubing of various materials to form drops, mostly for laboratory studies. Drop sizes are moderate to large, and all are about the same diameter for any specific design. Impact velocity is considerably less than terminal for fall heights less than 5 to 10 m. Designs that are being used for erosion research include those by Blackburn, Bubenzer, Gifford, and Römkens (*24*).

Recent modifications of stationary-nozzle rainfall simulators to improve simulated rainfall characteristics include air injected into the water-supply conduit (*22*), solenoids to reduce intensities (*17*), and further refinements in rotating-disk components (*2*).

Further details about rainfall simulators can be obtained from referenced publications and from the researchers who developed or now use such equipment. Many of the original designs have been modified to improve performance or adapt the equipment for other uses.

Any researcher who is considering the development of a rainfall simulator and/or use of simulated rainfall for erosion or hydrologic research should carefully consider the following check-list suggested by Bubenzer (*3*):

Figure 4.1 Examples of rainfall simulators used on large runoff plots: above, rainulator (top) and programmable simulator. Facing page: Kentucky simulator (top) and rotating-boom simulator.

USDA
RAIN SIMULATOR

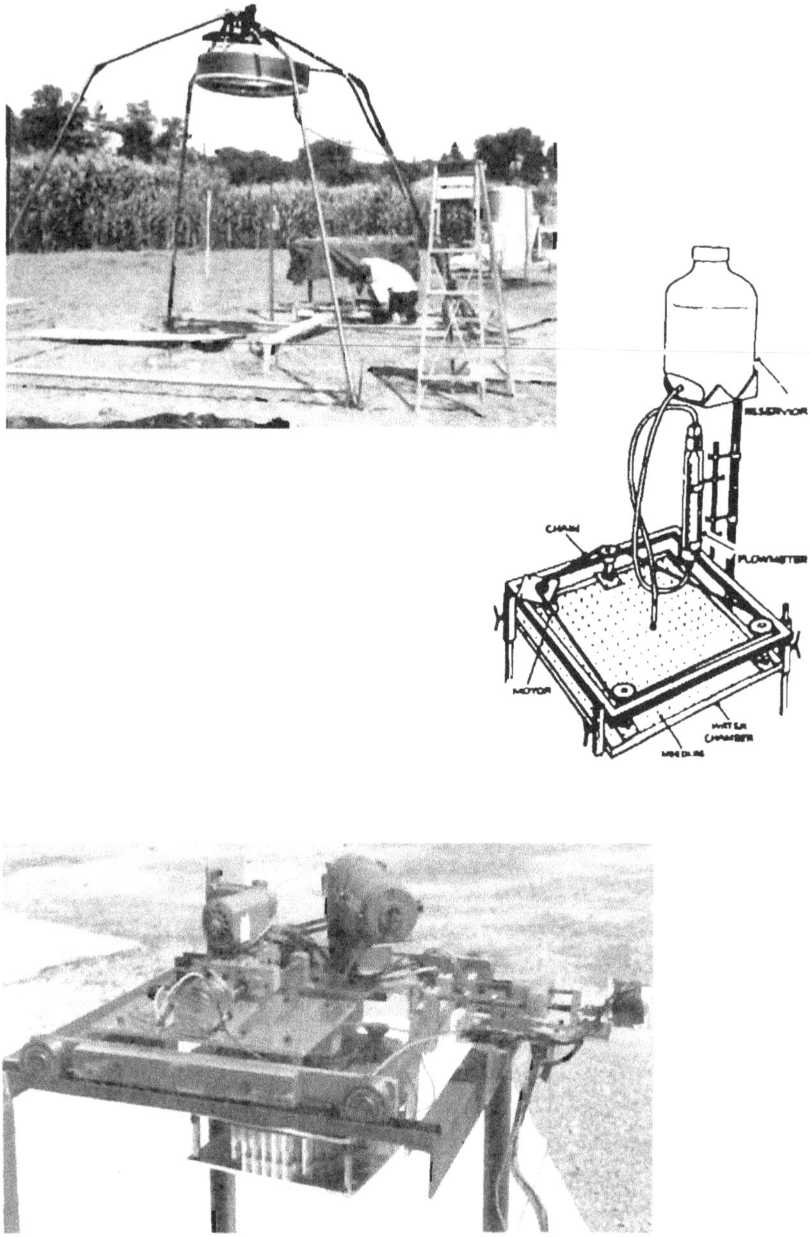

Figure 4.2 Examples of rainfall simulators used on small field plots: above, rotating-disk simulator (left top) modulator-type infiltrometer (right, center), and variable-intensity laboratory simulator (left bottom). Facing page: rill/row simulator (top) and interrill simulator.

- Define the research objectives clearly.
- Determine if the research objectives can reasonably be met using simulated rainfall.
- Analyze the hydrologic process under study to determine those components which will require accurate simulation.
- Consider plot and soil conditions that will interact with the hydrologic process.
- Determine the short- and long-term economic and labor resources available.

If these considerations suggest that rainfall simulator research is feasible, the researcher should thoroughly study the available literature to see if one or more of the existing rainfall simulators might be suitable. Researchers who are currently using potentially suitable equipment should be contacted to obtain their opinions concerning that simulator for the specific use envisioned and other information concerning its use. Research with rainfall simulators involves many problems and pitfalls, and most researchers are glad to help others avoid problems that they have encountered.

Once specific types of equipment have been selected for further consideration, an effort should be made to visit research locations that are using such equipment to see it in operation and learn some of the principles and details first-hand. There is no substitute for a personal encounter with the equipment, its operation, and those operating it. The time and money required can reduce costs, minimize frustrations, and save many hours of effort. Even when suitable equipment or designs are not available, the researcher could benefit from discussions with researchers who are experienced in simulated rainfall research before embarking on a rainfall simulator development program.

PERSONNEL REQUIREMENTS AND RELATED EQUIPMENT

Much more is involved in conducting productive simulated rainfall research than simply the availability of a rainfall simulator and the expertise to operate it. A researcher should ensure that adequate personnel, supplementary equipment, and sufficient time and facilities to process samples and data are available before constructing or otherwise acquiring a rainfall simulator.

Personnel

The number of people and necessary skills required to conduct simulated rainfall research vary, depending on the size of the rainfall simulator, the number of samples to be taken, and the analyses to be made during the experiments. Most major rainfall simulator field programs need a crew of at least three or four, both for assembly and disassembly and to conduct research activities during the tests. Most of their time and effort is required to prepare plots, install runoff measuring and sampling equipment, assemble and dissemble the rainfall simulator, and complete other measurements. Application of simulated rainstorms takes only a relatively small percentage of the total time required for operation, especially for large rainfall simulators. In addition to field work, considerable time and effort are needed to analyze the great quantity of samples and data normally collected for simulator experiments. Frequently, two to four months of fieldwork produce enough samples and data to keep researchers busy for the remainder of the year.

Pumps, hoses, and water supply

Rainfall simulators often require a large quantity of water, depending upon the area covered and the amount and duration of the rainfall to be simulated. For small rainfall simulators, it may be possible to haul enough water to the site. Large rainfall simulators usually require a water source near the research area. The water supply must be of a suitable quality, both chemically and physically, because variations in properties such as organic content and electrolyte concentrations can affect infiltration and erosion rates. Wells, clear ponds or lakes, or streams that will be clear at the time of the tests are usually acceptable water supplies for large rainfall simulators. Because the water usually must be pumped from the source to the rainfall simulator, pumps and hoses of sufficient size are necessary to deliver water at the required pressure and quantity. Storage tanks or lined pits near the experiment site may be useful for storing water that can only be obtained slowly from the source or to provide water at a uniform temperature.

Runoff equipment

Equipment is needed to measure runoff rates and to sample runoff for sediment concentration. Flumes and samplers similar to those used for natural-rainfall plots (*20*) can be adapted for use with rainfall simulators. However, they must be modified to make them easy to transport and install from site to site. On moderate-sized and larger plots, periodic "grab" samples may provide the necessary information. On very small plots, all runoff and soil loss may be collected in a succession of containers, rather than measuring the runoff and sampling a portion of it. Automated samplers usually are unnecessary, because some of the crew can also collect runoff samples during the simulated rainstorms.

Plot borders

The sides and tops of rainfall simulator plots must be accurately delineated by a barrier or border, just as for plots tested under natural rainfall. Strips of metal are frequently used for this purpose. The metal should be driven deep enough into the ground to avoid major subsurface water movement and should extend above the ground far enough that water from outside the plot does not enter and vice versa. Galvanized steel sheets with corrugations that parallel the narrow dimension make suitable borders for large plots, because they combine strength with reasonable mass. At the lower end of a plot, collectors are needed to collect the runoff for measurement and sampling.

Rain application measurement

The amount and rate of rain application are important for interpreting the resulting data. Rainfall simulators either need to be calibrated accurately for the conditions on which they are used, or the applied rain needs to be measured during the simulated rainstorms. For larger plots, suspended troughs running diagonally across the plot are effective in sampling the variability that occurs at different locations. Small aluminum channels leading to covered containers at their lower end work well for this purpose. An alternative is use of numerous small wedge gauges that are well distributed over the research plots.

Other equipment

Various other types of ancillary equipment may be necessary to conduct simulated rainfall experiments, depending upon the situation. Movement from site to site may require a truck or other means of transportation. An enclosed trailer equipped as a laboratory is useful for analyses during field studies. Numerous containers are required to collect runoff samples, and sediment analysis facilities need to be available to analyze the samples. A portable generator may be required to supply electrical power. Wind screens may be necessary for conducting experiments during moderate winds. Clocks or stopwatches are needed for timing samples, and a camera should be available at all times to document plot conditions. Other equipment may be needed to measure surface roughness, ground cover, plot slope, soil strength, bulk density, and soil moisture. These types of items are important components of a successful rainfall simulator research program.

SIMULATED RAINFALL RESEARCH PROCEDURES

In conducting simulated rainfall research, the researcher must make many decisions about conditions to be simulated and procedures to be followed. These vary from situation to situation, depending upon the purpose of the experiment and the equipment being used.

Rainfall intensity variation

Rainfall intensities vary widely in both time and space during most natural rainstorms. A rainfall simulator that can vary the intensity during simulated rainstorms may be desirable, but this is difficult to achieve with many types of equipment. Variation of intensities from drip-type equipment is difficult but sometimes can be achieved by varying the pressure or water flow to the tips. For nozzle-type equipment, changing the effective area covered per nozzle or the time between successive spray applications can vary the intensity. Even if variations in rainfall intensity during simulated rainstorms can be produced, selection of appropriate combinations of intensities and durations is difficult. A researcher should seriously consider if such effort can be justified, if a specific variable-intensity storm can be repeated identically for each treatment, and if the resulting data can be interpreted meaningfully. In many cases, a constant-intensity storm or a series of constant-intensity storms that each have different intensities is satisfactory.

Sequence of simulated rainstorms

Through the hydrologic year, natural rainstorms occur on a wide range of soil moisture and cover conditions. Seldom can simulated rainfall experiments be conducted for all conditions of interest. The researcher must choose the conditions that will best provide the desired information. Rainstorms sometimes occur on relatively dry soil and at other times on wet to very wet soil. A sequence of simulated storms that evaluates erosion first on relatively dry conditions and, thereafter, on wetter soil conditions usually provides informative data. Furthermore, the initial soil moisture at different research sites seldom can be established at the same level, but the wetness of all plots will be similar and therefore comparable after the initial storm.

If a researcher is studying different rainfall intensities, the obvious preference is to apply each intensity to each of the moisture conditions of interest. However, this often is not feasible because it is time consuming and requires many plots. An alternative is to apply a series of storms at different intensities on the same research plots, following the studies at different antecedent moistures.

Agricultural crops and other land use situations often change greatly through the year. Both plant growth and tillage affect the soil condition and cover that exist at different times throughout the year and, consequently, affect the erosion and hydrologic response. Where such changes are significant, the researcher should attempt to make tests at different times of the year so the trend for major cropping conditions can be determined throughout the year.

Length of simulated rainstorms

The duration of a rainfall simulator test is usually less critical than other decisions. If rainfall intensity-frequency-duration data are available for the area, they may be considered in selecting the rainstorm duration. Another important consideration, of course, is the water supply available for the experiments. Most tests should be long enough that runoff is well established and infiltration is somewhat constant before rainfall is stopped or its intensity is changed. Generally, this means that the rainstorm at the initial driest condition will be the longest, with succeeding storms somewhat shorter. Because timed runoff and erosion samples are generally collected periodically throughout the simulated rainstorms, the researcher can use the data to determine results from shorter storms than those actually applied.

Comparing simulated rainfall with natural rainfall

Ever since the characteristics of natural rainfall were determined several decades ago, researchers have sought a parameter to indicate how closely the simulated rainfall attained the important characteristics of natural rainfall. The most widely used parameter has been the kinetic energy of rain at impact. Basic physics suggests that kinetic energy, or the similar momentum parameter, should be an important parameter. However, the area over which this energy or momentum is dissipated at impact may also be important. For example, eight drops 2 mm in diameter equal the mass of one 4-mm drop, but the horizontal cross section of eight 2-mm drops is twice that of a 4-mm drop. The kinetic energy or momentum of the 2-mm drops, although only slightly less due to slower terminal velocity, will be dissipated over twice the area. Thus, the erosiveness of a large drop may be much greater than for smaller drops with the same total kinetic energy. Kinetic energy alone probably is not an adequate parameter for comparison. The conclusions in my 1965 report on possible parameters still seem appropriate: "until some parameter is proved to be adequate for comparison, this analysis suggests (a) that both the drop-size distribution and drop-fall velocity of natural rainfall should be simulated as closely as possible and (b) that an appreciable sacrifice of either for the other is unwise.

One of the parameters may be chosen as a guide, but its influence should be secondary to a comparison with actual raindrop characteristics" (*13*).

Suitable studies with rainfall simulators

Rainfall simulators are well suited for field studies that compare different soil, cropping, and cover conditions. The resulting data give relative values rather than providing absolute rates of erosion. To obtain realistic estimates of annual erosion rates, results from long-term studies under natural rainfall are usually required for at least one of the treatments. The ratio of natural to simulated erosion for that treatment may be used with the other conditions to estimate their annual rates. However, recent research (*10*) has indicated that plot data from rainfall simulator studies also can be used to obtain good erosion predictions for larger areas, provided the data used for parameter derivation cover erosion processes of relevance to the field situations of interest.

Appropriate studies with rainfall simulators include those designed to evaluate different crop and soil-surface covers, crop residue and land use management, types and methods of tillage, steepness and length of slope, relative erodibility of different soils, type and sequence of crops in rotations, critical periods in the annual crop cycle, and fundamental mechanics of erosion and runoff. However, treatments that have only minor differences are poorly suited for evaluation by either simulated or natural rainfall because considerable uncontrollable experimental and natural variation usually is associated with such research, making detection of small differences impossible. This is especially true for the heterogeneous conditions encountered during most field studies. Therefore, studies with simulated rain should be limited primarily to tests of conditions that may be expected to produce data with appreciable differences.

Studies of the effects of fertility, rotations, or long-term tillage on erosion rates require plots that have been maintained for several years prior to testing. Other studies, such as those comparing a single characteristic of different crops, can often be conducted in a single season. Studies of surface-residue rates, cover of a single crop, slope differences, and critical periods often can be made on suitable farm fields. The relative erodibilities of different soils can be evaluated by applying identical simulated storms to similarly prepared plots at selected sites with similar land-use histories, thereby reducing the effects of climatic differences among locations that are present in the K factor of the universal soil loss equation (*25*).

To test the effects of treatments for slopes that are longer than the rainfall simulator, inflow water may be added at the top of the rainfall simulator plots to simulate runoff from upslope. The resulting data pertain only to the bottom section of the longer slopes, *not* to the entire length. Thus, results from such simulated slope lengths must be interpreted properly to avoid incorrect conclusions.

Of course, there are limitations to the types of research that can be conducted with rainfall simulators. Studies that require a wide variety of rainfall intensities or impact energies may not be possible with certain rainfall simulators. Experiments cannot be conducted when water in the lines and fittings will freeze or when crops are at a height that seriously distorts the application and energy patterns from those of natural rainfall. Often, the width and length desired for the research areas are greater than the rainfall simulator can accommodate. Plots with contoured rows generally cannot be evaluated adequately with plot-sized rainfall simulators, because the plot borders create unnatural barriers that affect the normal water flow; yet, plots with up-and-down-slope rows are not directly applicable to most farm conditions. Nevertheless, treatments for many types of studies can be applied on up-and-down-slope rows to determine the differences among treatments. Such results then should be adjusted for the effects of contouring or other water management practices.

Designing rainfall simulator experiments

Rainfall simulator experiments must be carefully designed statistically to ensure that the results provide the most possible information. Proper randomization of treatments and appropriate experimental techniques are important. Generally, all treatments should be tested in duplicate at least, although additional replicates often are desirable. Practical considerations may make compromises necessary in the number of treatments versus the number of replications, especially for field studies or when information on a range of conditions is more important than great accuracy. The researcher should remember that the goal of the research is useful information about the treatments being studied, not an impressive array of statistics.

Suitable sites

Researchers should consider the following factors in selecting suitable field sites for rainfall simulator tests:

- Past land-use history of research area.
- Soil type and variability on research area.
- Past erosion on the research plot area.
- Plot size, shape, slope, and uniformity.
- Availability of water supply.

- Suitability of water supply.
- Availability of and access to the research area.
- Availability of equipment for applying treatments.
- Disposal of runoff from the plot area.

The first four items should be determined accurately and recorded for each study site. Other important site information should also be documented.

Research plot conditions

Characteristics of the research plot often affect the suitability of the simulated rainfall. Where the soil surface of all treatments is well covered by plants, mulches, or other surface covers, the impact energy of the simulated rainfall may be of minor importance as far as erosion and infiltration are concerned. Where tall crops such as mature corn are growing, the height of the drop-former or nozzle and the type of nozzle movement may affect the suitability of the resulting simulated rain. Where the surface of the plot has a very steep slope and the rainfall simulator is designed to apply all simulated rain from the same elevation, the impact energy and intensity distribution may vary excessively from the upper end to the lower end of a long plot.

Rainfall simulators have been used on plots ranging in size from small circles to greater than a hectare. On areas up to several square meters, elaborate rainfall simulation methods, such as capillary-tubing drop-formers, are feasible. However, they seldom are usable on areas of tens or hundreds of square meters. For such areas, nozzles that each cover a considerable area are more suitable.

For conditions where raindrop impact significantly affects the erosion or infiltration rate, the impact characteristics of rainfall must be simulated adequately. Interrill erosion cannot be evaluated properly unless raindrop impact characteristics are similar to those of appropriate rainstorms. Infiltration rates cannot be evaluated properly on soils that seal from raindrop impact unless the simulated rain acts on these soils in the same manner as natural rainfall.

Results from small areas that are suitable for evaluations of interrill erosion and localized infiltration cannot be extrapolated directly to larger areas where flowing runoff would cause rill erosion or where other characteristics may affect infiltration rate. For such conditions, rainfall simulators must apply rainfall to large enough areas so that runoff can accumulate or so that land irregularities affecting infiltration can be evaluated properly.

On soils or topographies where gully erosion or subsurface flow may be major considerations, research areas may need to be quite large. For such studies, it may be most important to apply water relatively uniformly over a very large area, even if it means sacrifice in the drop impact characteristics. Agricultural irrigation equipment then may be the only feasible approach and may be quite appropriate, particularly if good soil-surface cover is present.

Comparative versus quantitatively accurate results

Some research seeks to evaluate erosion or infiltration quantitatively. Other research is concerned primarily with comparisons among several treatments. For quantitatively accurate results, major rainfall characteristics such as drop size distribution, impact velocity, and rain intensity must be nearly identical to those of rainfall. Where comparison of several conditions is the primary goal, studies at one or a limited number of intensities and some compromise in drop characteristics may be acceptable. However, researchers should clearly distinguish for themselves and others the quantitative accuracy of the results.

Laboratory research

Rainfall simulators also are useful for many laboratory studies. Some field rainfall simulators are adaptable for laboratory use. Other rainfall simulators have been designed specifically for laboratory research. The researcher must weigh the advantages of conducting research in the laboratory under controlled conditions against the disadvantages of the simulated, often quite unnatural, conditions necessitated by the laboratory environment. Laboratory studies are not affected by weather and may be conducted throughout the year.

Adjustment for minor differences in application intensity

Identical rates of application are generally impossible when using rainfall simulators at different times under different conditions at different locations. Usually the variation in rain intensity is only a few percent, but adjustment of the resulting runoff and soil loss to a common (design) application intensity is desirable.

For studies where high-intensity rainfall is applied at a constant rate, researchers can assume that a small variation between the actual application intensity and the design intensity will have little, if any, effect on the infiltration rate. Therefore the adjusted runoff may be computed as:

$$\text{Adjusted runoff} = \text{measured runoff} + (\text{design intensity} - \text{measured intensity}) \times (\text{duration of applied rain})$$

A researcher needs to know the effect of intensity on erosion to adjust soil loss determinations. Several analyses and studies have suggested that erosion is approximately proportional to the intensity squared. Unless a different relationship is known to exist for the specific situation, soil loss may be adjusted for small intensity differences as

$$\text{Adjusted soil loss} = \left(\frac{\text{design intensity}}{\text{measured intensity}}\right)^2 \times \text{measured soil loss}$$

Interpretation of rainfall simulator data

The primary result from studies using simulated rainfall is a relative comparison of the treatments during an intense, simulated rainstorm or series of rainstorms. Considerable data interpretation is necessary to obtain indications of average annual values, such as those used for the USLE (*25*). The researcher must carefully consider how the relationships among soil erosion and the various independent variables change with time. Extrapolation to field conditions can be made only with careful analysis. However, with proper interpretation, results from rainfall simulator storms potentially can provide very useful information in a much more convenient manner than by relying totally on natural rainfall.

RAINFALL SIMULATORS: ASSET OR BURDEN?

Rainfall simulators can be great assets to soil erosion and hydrologic research if they are designed and used properly and if the resulting data are interpreted judiciously. However, simulated rainfall is not a magic method for satisfying all erosion and hydrologic research needs. Researchers must weigh carefully both the benefits to be derived from the use of rainfall simulators and the problems that the use of rainfall simulators may cause. Often, simulated rainfall is the only feasible means for conducting needed research. Many research studies could never be considered if they could not be conducted using rainstorms applied by rainfall simulators.

Many decisions need to be made and many questions resolved in considering a research program involving rainfall simulators. But the greatest question to be faced is: Can the conditions of interest be adequately evaluated using simulated rainfall? Unless a researcher can answer that question positively, with considerable assurance of success, a rainfall simulator can be more of a burden than an asset to a research program. On the other hand, where rainfall simulators are applicable, they can be marvelous research tools in hydrologic and erosion research. The researcher has the responsibility to objectively resolve the above issues and to use appropriate research equipment and techniques that will produce reliable data to help answer important research needs.

REFERENCES

1. Bertrand, A.R., and J.F. Parr. 1961. *Design and operation of the Purdue sprinkling infiltrometer.* Research Bulletin No. 723. Purdue University, 16 pp.
2. Bubenzer, C.D., M. Molnau, and D.K. McCool. 1985. *Low intensity rainfall with a rotating disk simulator.* Transactions, American Society of Agricultural Engineers 28: 1230-1232.
3. Bubenzer, C.D. 1980. *An overview of rainfall simulators.* ASAE Paper No. 80-2033, American Society of Agricultural Engineers, St. Joseph, Michigan.
4. Foster, G.R., W.H. Neibling, and R.A. Nattermann. 1982. *A programmable rainfall simulator.* ASAE Paper No. 82-2570, American Society of Agricultural Engineers, St. Joseph, Michigan.
5. Grierson, I.T., and J.M. Oades. 1977. *Rainfall simulator for field studies of runoff and erosion.* Journal of Agricultural Engineering Research 22: 37-44.
6. Gunn, R., and G.D. Kinzer. 1949. *Terminal velocity of water droplets in stagnant air.* Journal of Meterology 6: 243-248.
7. Hall, M.J. 1970. *A critique of methods of simulating rainfall.* Water Resources Research 6: 1104-1114.
8. Holland, M.E. 1969. *Design and testing of a rainfall system.* Colorado State University Experiment Station CER 69-70. MEH 21. Ft. Collins, Colorado.
9. Laws, J.O., and D.A. Parsons. 1943. *Relation of raindrop size to intensity.* Transactions, American Geophysical Union 24: 452-460.
10. Loch, R.J., D.M. Silburn, and D.M. Freebairn. 1989. *Evaluation of the CREAMS model. II. Use of rainulator data to derive soil erodibility parameters and prediction of field soil losses using derived parameters.* Australian Journal of Soil Research 27: 563-576.
11. Luk, S.H., A.D. Abrahams, and A.J. Parsons. 1986. *A simple rainfall simulator and trickle system for hydro-geomorphological experiments.* Physical Geology 7: 344-356.
12. Meyer, L.D. 1958. *An investigation of methods for simulating rainfall on standard runoff plots and a study of the drop size, velocity, and kinetic energy of selected spray nozzles.* USDA-ARS Special Report No. 81. U.S. Department of Agriculture. 43 pp.
13. Meyer, L.D. 1965. *Simulation of rainfall for soil erosion research.* Transactions, American Society of Agricultural Engineers 8: 63-65.
14. Meyer, L.D., and D.L. McCune. 1958. *Rainfall simulator for runoff plots.* Agricultural Engineering 39: 644-648.
15. Meyer, L.D., and W.C. Harmon. 1979. *Multiple-intensity rainfall simulator for erosion research on row sideslopes.* Transactions, American Society of Agricultural Engineers 22: 100-103.
16. Meyer, L.D., and W.C. Harmon. 1985. *Sediment losses from cropland furrows of different gradients.* Transactions, American Society of Agricultural Engineers 28: 448-453.
17. Miller, W.P. 1987. *A solenoid-operated, variable intensity rainfall simulator.* Soil Science Society of America Journal 51: 832-834.

18. Moore, I.D., M.C. Hirschi, and B.J. Barfield. 1983. *Kentucky rainfall simulator*. Transactions, American Society of Agricultural Engineers 26: 1085-1089.
19. Morin, J., D. Goldberg, and I. Seginer. 1967. *A rainfall simulator with a rotating disk*. Transactions, American Society of Agricultural Engineers 10: 74-77.
20. Mutchler, C.K., C.E. Murphree, and K.C. McGregor. 1994. *Laboratory and field plots for erosion studies*. (An accompanying article in this manual).
21. Mutchler, C.K., and L.F. Hermsmeier. 1965. *A review of rainfall simulators*. Transactions, American Society of Agricultural Engineers 8: 63-65.
22. Shelton, C.H., R.D. von Bernuth, and S.P. Rajbhandari. 1985. *A continuous-application rainfall simulator*. Transactions, American Society of Agricultural Engineers 28: 1115-1119.
23. Swanson, N.P. 1965. *Rotating boom rainfall simulator*. Transactions, American Society of Agricultural Engineers 8: 71-72.
24. United States Department of Agriculture. 1979. *Proceedings of the rainfall simulator workshop*. Tucson, Arizona (E.L. Neff, Chairman) ARM-W-10, 185 pp.
25. Wischmeier, W.H. and D.D. Smith. 1978. *Predicting rainfall erosion losses - a guide to conservation planning*. USDA-SEA Agriculture Handbook 537. U.S. Department of Agriculture. 58 pp.

Chapter 5

The REVISED UNIVERSAL SOIL LOSS EQUATION

K. G. Renard, J. M. Laflen,
G. R. Foster, and D. K. McCool

The Universal Soil Loss Equation (USLE) (*6*), has been the workhorse of erosion prediction and conservation planning technology in the U.S. and even worldwide. In 1985, at a meeting of U.S. Department of Agriculture (USDA) and other erosion researchers, it was decided that the USLE should be revised to incorporate additional research and technology developed after the 1978 USLE handbook. Work on the revision was initiated seriously late in 1987, and resulted in a new handbook and technology called RUSLE, the Revised USLE.

RUSLE maintains the basic structure of the USLE, namely

$$A = R\ K\ L\ S\ C\ P \qquad [1]$$

where A is the computed soil loss, R is the rainfall-runoff erosivity factor, K is the soil erodibility factor, L is the slope length factor, S is the slope steepness factor, C is the cover-management factor, and P is the supporting practices factor. This empirically-based equation, derived from a large mass of field data, computes sheet and rill erosion using values representing the four major factors affecting erosion. These factors are climate erosivity represented by R, soil erodibility represented by K, topography represented by LS, and land use and management represented by CP.

Whereas the basic USLE structure has been retained, the algorithms used to calculate the individual factors have been changed significantly in RUSLE. Perhaps most important has been the computerization of the technology to assist with individual factor determinations.

R-FACTOR

The rainfall-runoff erosivity term in RUSLE is calculated as the product of storm kinetic energy times the maximum 30-minute storm depth and summed for all storms in a year. The R-factor represents the input that drives the sheet and rill erosion processes. Thus differences in R values represent differences in erosivity of the climate. Of the RUSLE factors, R is the one most exactly computed from input data.

In the western U.S. (Figures 5.1, 5.2, and 5.3), new R values have been calculated using over 1000 point values, a significant addition to the information available in *Agriculture Handbook 537* (6). Whereas the old R isoerodent maps for the west had maximum point values of about 50 units (hundreds of foot · tonforce · inch/acre · hour · year), new values are as large as 350 units along the Pacific coastal areas (Figure 5.3). Some changes are also involved in the eastern states (east of the 105th meridian) (Figure 5.4). Another change in the R-factor is to reduce R values where flat slopes occur in regions of long intense rainstorms. Ponded water on the soil reduces the erosivity of the rain. Finally, an R equivalent approach is being used in the frozen soil areas of the Pacific Northwest to reflect the combined effect of rain or snowmelt on frozen or partly-thawed soil.

Part of the R-factor calculation involves a seasonal distribution to permit weighting of the soil erodibility value, K, and the cover-management factor, C. To facilitate these calculations, climate data files have been developed (called a city code) for climatically homogeneous areas. These computer files include information such as the frost-free duration in days, monthly precipitation and temperature, and 15-day distributions of R. Typical values are included in the computer program for at least one station in each of 120 climatic regions (Figure 5.5) of the contiguous 48 states plus numerous stations in Hawaii. The computer program includes a provision for addition of new climate data sets using data provided by the user. This also allows easy use of RUSLE in other regions of the world.

K-FACTOR

The K-factor is a measure of the inherent erodibility of a given soil under the standard condition of the unit USLE plot maintained in continuous fallow. Values for K typically range from about 0.10 to 0.45 (U.S. customary units), with high-sand and high-clay content soils having the lower values and high-silt content soils having the higher values. Users have little difficulty choosing a K-factor value because the Soil Conservation Service (SCS) has identified K values for all major soil mapping units

Figure 5.1 Western U.S. isoerodent map.

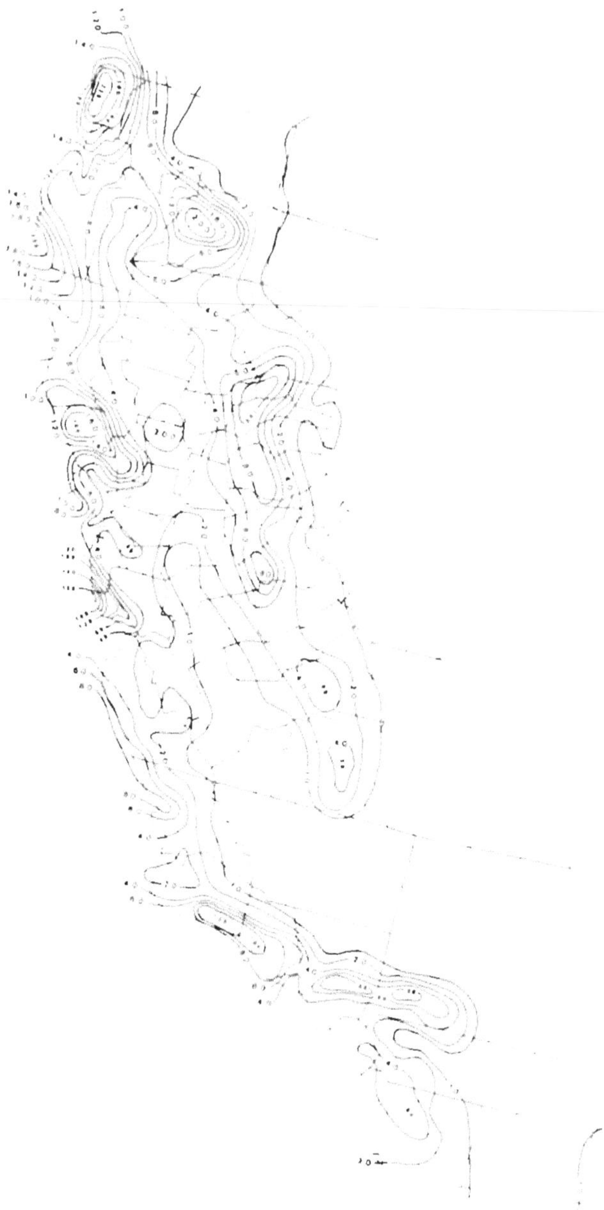

Figure 5.2 California isoerodent map.

Figure 5.3 Isoerodent map of Oregon and Washington.

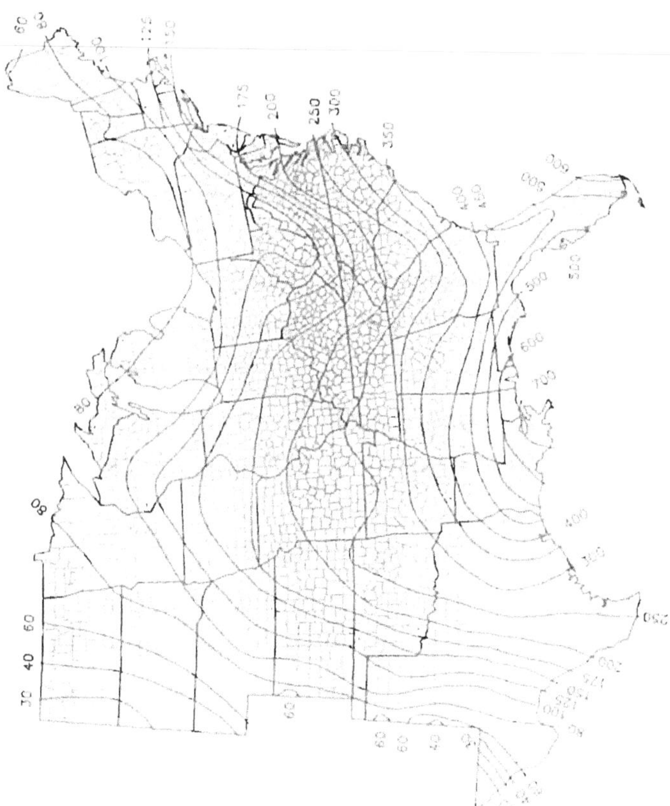

Figure 5.4 Isoerodent map of eastern United States.

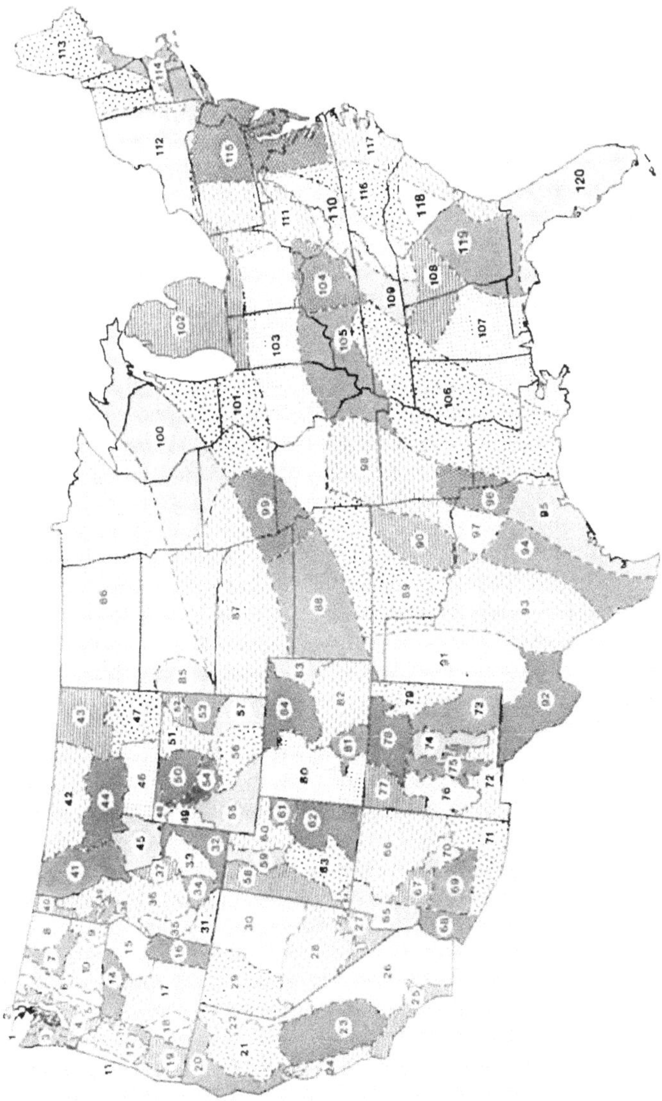

Figure 5.5 EI distribution zones for the conterminious U.S.

in the United States. However, the site-specific K value, and its seasonal variation, can be quite different from the K value given in soil survey information.

The soil erodibility nomograph is a popular tool for estimating K values, but it does not apply to some soils. Updating the K-factor for RUSLE involved developing guides so the user could identify soils where the nomograph does not apply and estimate K using alternative methods. Erodibility data from around the world have been reviewed, and an equation has been developed that gives an estimate of K as a function of an "average" diameter of the soil particles. K-values for volcanic soils of Hawaii are also estimated with an alternative algorithm to the erodibility nomograph.

RUSLE also varies K seasonally, a major change over the USLE procedure. Experimental data show that K is not constant but varies with season, being highest in early spring and lowest in mid-fall or when the soil is frozen. The seasonal variability is addressed by weighing the instantaneous estimate of K in proportion to EI (the percent of annual R) for twice monthly intervals. Instantaneous estimates of K are made from equations relating K to the frost-free period and the annual R-factor.

An additional change incorporated in RUSLE is to account for rock fragments on and in the soil, a common occurrence on western rangelands. Rock fragments on the soil surface are treated like mulch in the C-factor, while K is adjusted for rock in the soil profile to account for effects on runoff.

An example output from the computer program which weights the instantaneous K values in proportion to the distribution of twice monthly EI values is given in Table 5.1. For this illustration, the nomograph K was given as 0.28 (customary U.S. units) and the average weighted K was 0.25. The maximum K was estimated to be 0.707 with a minimum of 0.104. Note that the corresponding observed K values were somewhat lower. In other instances the difference between the nomograph and weighed K go in the opposite directions from this example. K_{max} and K_{min} are determined with regression equations relating K_i to the annual R factor and the frost free period.

L- AND S-FACTOR

More questions and concerns are expressed over the L-factor than any of the other USLE factors. One reason is that the choice of a slope length involves judgment; different users choose different slope lengths for similar situations. RUSLE includes improved guides for choosing slope length values to give greater consistency among users.

Table 5.1 Measured and calculated monthly K values for a Barnes loam soil near Morris, Minnesota (R=100, K_{nom}=0.28).

Date	% of annual EI	t_i	T_{av}(°F)	K_i^*(cal)	%(EI_i)K_i	K_i^*(obs.)
01/01-03/15	0	37	15	0.104	0	0
03/16-03/31	1	82	31	0.612	0.612	
04/01-04/15	1	98	40	0.707	0.707	0.404
04/16-04/30	1	113	40	0.672	0.672	
05/01-05/15	3	128	57	0.548	1.644	0.393
05/16-05/31	5	143	57	0.447	2.235	
06/01-06/15	12	159	66	0.360	4.317	0.229
06/16-06/30	13	174	66	0.293	3.814	
07/01-07/15	13	189	72	0.239	3.11	0.328
07/16-07/31	14	204	72	0.195	2.733	
08/01-08/15	14	220	71	0.157	2.199	0.205
08/16-08/31	13	235	71	0.128	1.665	
09/01-09/15	5	251	60	0.105	0.525	0.042
09/16-09/30	3	266	60	0.120	0.360	
10/01-10/15	1	281	50	0.137	0.137	0.065
10/16-10/31	1	296	50	0.157	0.157	
11/01-11/15	0	312	34	0.182	0	0
11/16-12/31	0	342	20	0.104	0	
Total	**100**				**24.888**	

$K_{av} = \Sigma(EI_i)K_i/100 = 0.25$

$t_{max} = 110$ $t_{min} = 250$

$K_{max}/K_{nom} = 2.5$ $K_{max} = 0.700$

$K_{max}/K_{min} = 6.7$ $K_{min} = 0.104$

*K_i values are given in tons/acre per erosion index unit. For SI-units of t·ha·h/ha·MJ·mm, mulitply K_i values by 0.1317.

t_i = Julian day for date interval used in calculation (interval midpoint)

K_i = soil erodibility for day t_i

K_{nom} = nomograph soil erodibility

The attention given to the L-factor is not always warranted because soil loss is less sensitive to slope length than to any other USLE factor. For typical slope conditions, a 10 percent error in slope length measurement results in a 5 percent error in computed soil loss.

RUSLE uses four separate slope length relationships (Tables 5.2, 5.3, 5.4, and 5.5). Three are functions of slope steepness as in the USLE, and

Table 5.2 Values for topographic factor, LS, for low ratio of rill to interrill erosion, such as for rangeland and other consolidated soil conditions with cover (applicable to thawing soil where both interrill and rill erosion are significant).

	Slope length in feet																
Slope	<3	6	9	12	15	25	50	75	100	150	200	250	300	400	600	800	1000
%																	
0.2	0.05	0.05	.0.05	0.05	0.05	0.05	0.05	0.05	0.05	0.05	0.05	0.05	0.05	0.05	0.05	0.05	0.05
0.5	0.08	0.08	0.08	0.08	0.08	0.08	0.08	0.08	0.09	0.09	0.09	0.09	0.09	0.09	0.09	0.09	0.09
1.0	0.12	0.12	0.12	0.12	0.12	0.13	0.13	0.14	0.14	0.15	0.15	0.15	0.15	0.16	0.16	0.17	0.17
2.0	0.20	0.20	0.20	0.20	0.20	0.21	0.23	0.25	0.26	0.27	0.28	0.29	0.30	0.31	0.33	0.34	0.35
3.0	0.26	0.26	0.26	0.26	0.26	0.29	0.33	0.36	0.38	0.40	0.43	0.44	0.46	0.48	0.52	0.55	0.57
4.0	0.33	0.33	0.33	0.33	0.33	0.36	0.43	0.46	0.50	0.54	0.58	0.61	0.63	0.67	0.74	0.78	0.82
5.0	0.38	0.38	0.38	0.38	0.38	0.44	0.52	0.57	0.62	0.68	0.73	0.78	0.81	0.87	0.97	1.04	1.10
6.0	0.44	0.44	0.44	0.44	0.44	0.50	0.61	0.68	0.74	0.83	0.90	0.95	1.00	1.08	1.21	1.31	1.40
8.0	0.54	0.54	0.54	0.54	0.54	0.64	0.79	0.90	0.99	1.12	1.23	1.32	1.40	1.53	1.74	1.91	2.05
10.0	0.60	0.63	0.65	0.66	0.68	0.81	1.03	1.19	1.31	1.51	1.67	1.80	1.92	2.13	2.45	2.71	2.93
12.0	0.61	0.70	0.75	0.80	0.83	1.01	1.31	1.52	1.69	1.97	2.20	2.39	2.56	2.85	3.32	3.70	4.02
14.0	0.63	0.76	0.85	0.92	0.98	1.20	1.58	1.85	2.08	2.44	2.73	2.99	3.21	3.60	4.23	4.74	5.18
16.0	0.65	0.82	0.94	1.04	1.12	1.38	1.85	2.18	2.46	2.91	3.28	3.60	3.88	4.37	5.17	5.82	6.39
20.0	0.68	0.93	1.11	1.26	1.39	1.74	2.37	2.84	3.22	3.85	4.38	4.83	5.24	5.95	7.13	8.10	8.94
25.0	0.73	1.05	1.30	1.51	1.70	2.17	3.00	3.63	4.16	5.03	5.76	6.39	6.96	7.97	9.65	11.04	12.26
30.0	0.77	1.16	1.48	1.75	2.00	2.57	3.60	4.40	5.06	6.18	7.11	7.94	8.68	9.99	12.19	14.04	15.66
40.0	0.85	1.36	1.79	2.17	2.53	3.30	4.73	5.84	6.78	8.37	9.71	10.91	11.99	13.92	17.19	19.96	22.41
50.0	0.91	1.52	2.06	2.54	3.00	3.95	5.74	7.14	8.33	10.37	12.11	13.65	15.06	17.59	21.88	25.55	28.82
60.0	0.97	1.67	2.29	2.86	3.41	4.52	6.63	8.29	9.72	12.16	14.26	16.13	17.84	20.92	26.17	30.68	34.71

Table 5.3 Values for topographic factor, LS, for moderate ratio of rill to interrill erosion, such as for row-cropped agricultural and other moderately consolidated soil conditions with little to moderate cover (not applicable to thawing soil).

	Slope length in feet																
Slope	<3	6	9	12	15	25	50	75	100	150	200	250	300	400	600	800	1000
%																	
0.2	0.05	0.05	0.05	0.05	0.05	0.05	0.05	0.05	0.05	0.05	0.05	0.05	0.05	0.05	0.06	0.06	0.06
0.5	0.07	0.07	0.07	0.07	0.07	0.08	0.08	0.08	0.09	0.09	0.09	0.09	0.09	0.10	0.10	0.10	0.10
1.0	0.11	0.11	0.11	0.11	0.11	0.12	0.13	0.14	0.14	0.15	0.16	0.17	0.17	0.18	0.19	0.20	0.20
2.0	0.17	0.17	0.17	0.17	0.17	0.19	0.22	0.25	0.27	0.29	0.31	0.33	0.35	0.37	0.41	0.44	0.47
3.0	0.22	0.22	0.22	0.22	0.22	0.25	0.32	0.36	0.39	0.44	0.48	0.52	0.55	0.60	0.68	0.75	0.80
4.0	0.26	0.26	0.26	0.26	0.26	0.31	0.40	0.47	0.52	0.60	0.67	0.72	0.77	0.86	0.99	1.10	1.19
5.0	0.30	0.30	0.30	0.30	0.30	0.37	0.49	0.58	0.65	0.76	0.85	0.93	1.01	1.13	1.33	1.49	1.63
6.0	0.34	0.34	0.34	0.34	0.34	0.43	0.58	0.69	0.78	0.93	1.05	1.16	1.25	1.42	1.69	1.91	2.11
8.0	0.42	0.42	0.42	0.42	0.42	0.53	0.74	0.91	1.04	1.26	1.45	1.62	1.77	2.03	2.47	2.83	3.15
10.0	0.46	0.48	0.50	0.51	0.52	0.67	0.97	1.19	1.38	1.71	1.98	2.22	2.44	2.84	3.50	4.06	4.56
12.0	0.47	0.53	0.58	0.61	0.64	0.84	1.23	1.53	1.79	2.23	2.61	2.95	3.26	3.81	4.75	5.56	6.28
14.0	0.48	0.58	0.65	0.70	0.75	1.00	1.48	1.86	2.19	2.76	3.25	3.69	4.09	4.82	6.07	7.15	8.11
16.0	0.49	0.63	0.72	0.79	0.85	1.15	1.73	2.20	2.60	3.30	3.90	4.45	4.95	5.86	7.43	8.79	10.02
20.0	0.52	0.71	0.85	0.96	1.06	1.45	2.22	2.85	3.40	4.36	5.21	5.97	6.68	7.97	10.23	12.20	13.99
25.0	0.56	0.80	1.00	1.16	1.30	1.81	2.82	3.65	4.39	5.69	6.83	7.88	8.86	10.65	13.80	16.58	19.13
30.0	0.59	0.89	1.13	1.34	1.53	2.15	3.39	4.42	5.34	6.98	8.43	9.76	11.01	13.30	17.37	20.99	24.31
40.0	0.65	1.05	1.38	1.68	1.95	2.77	4.45	5.87	7.14	9.43	11.47	13.37	15.14	18.43	24.32	29.60	34.48
50.0	0.71	1.18	1.59	1.97	2.32	3.32	5.40	7.17	8.78	11.66	14.26	16.67	18.94	23.17	30.78	37.65	44.02
60.0	0.76	1.30	1.78	2.23	2.65	3.81	6.24	8.33	10.23	13.65	16.76	19.64	22.36	27.45	36.63	44.96	52.70

Table 5.4 Values for topographic factor, LS, for high ratio of rill to interrill erosion, such as for freshly prepared construction and other highly disturbed soil conditions with little or no cover (not applicable to thawing soil).

	Slope length in feet															
Slope	<3	6	9	12	15	25	50	75	100	150	200	250	300	400	600	800	1000
%																	
0.2	0.05	0.05	0.05	0.05	0.05	0.05	0.05	0.05	0.05	0.05	0.06	0.06	0.06	0.06	0.06	0.06	0.06
0.5	0.07	0.07	0.07	0.07	0.07	0.07	0.08	0.08	0.09	0.09	0.10	0.10	0.10	0.11	0.12	0.12	0.13
1.0	0.09	0.09	0.09	0.09	0.09	0.10	0.13	0.14	0.15	0.17	0.18	0.19	0.20	0.22	0.24	0.26	0.27
2.0	0.13	0.13	0.13	0.13	0.13	0.16	0.21	0.25	0.28	0.33	0.37	0.40	0.43	0.48	0.56	0.63	0.69
3.0	0.17	0.17	0.17	0.17	0.17	0.21	0.30	0.36	0.41	0.50	0.57	0.64	0.69	0.80	0.96	1.10	1.23
4.0	0.20	0.20	0.20	0.20	0.20	0.26	0.38	0.47	0.55	0.68	0.79	0.89	0.98	1.14	1.42	1.65	1.86
5.0	0.23	0.23	0.23	0.23	0.23	0.31	0.46	0.58	0.68	0.86	1.02	1.16	1.28	1.51	1.91	2.25	2.55
6.0	0.26	0.26	0.26	0.26	0.26	0.36	0.54	0.69	0.82	1.05	1.25	1.43	1.60	1.90	2.43	2.89	3.30
8.0	0.32	0.32	0.32	0.32	0.32	0.45	0.70	0.91	1.10	1.43	1.72	1.99	2.24	2.70	3.52	4.24	4.91
10.0	0.35	0.37	0.38	0.39	0.40	0.57	0.91	1.20	1.46	1.92	2.34	2.72	3.09	3.75	4.95	6.03	7.02
12.0	0.36	0.41	0.45	0.47	0.49	0.71	1.15	1.54	1.88	2.51	3.07	3.60	4.09	5.01	6.67	8.17	9.57
14.0	0.38	0.45	0.51	0.55	0.58	0.85	1.40	1.87	2.31	3.09	3.81	4.48	5.11	6.30	8.45	10.40	12.23
16.0	0.39	0.49	0.56	0.62	0.67	0.98	1.64	2.21	2.73	3.68	4.56	5.37	6.15	7.60	10.26	12.69	14.96
20.0	0.41	0.56	0.67	0.76	0.84	1.24	2.10	2.86	3.57	4.85	6.04	7.16	8.23	10.24	13.94	17.35	20.57
25.0	0.45	0.64	0.80	0.93	1.04	1.56	2.67	3.67	4.59	6.30	7.88	9.38	10.81	13.53	18.57	23.24	27.66
30.0	0.48	0.72	0.91	1.08	1.24	1.86	3.22	4.44	5.58	7.70	9.67	11.55	13.35	16.77	23.14	29.07	34.71
40.0	0.53	0.85	1.13	1.37	1.59	2.41	4.24	5.89	7.44	10.35	13.07	15.67	18.17	22.95	31.89	40.29	48.29
50.0	0.58	0.97	1.31	1.62	1.91	2.91	5.16	7.20	9.13	12.75	16.16	19.42	22.57	28.60	39.95	50.63	60.84
60.0	0.63	1.07	1.47	1.84	2.19	3.36	5.97	8.37	10.63	14.89	18.92	22.78	26.51	33.67	47.18	59.93	72.15

Table 5.5 Values for topographic factor, LS, for thawing soils where most of the erosion is caused by surface flow.

	Slope length in feet												
Slope	15	25	50	75	100	150	200	250	300	400	600	800	1000
%													
0.2	0.02	0.03	0.04	0.05	0.06	0.07	0.09	0.10	0.10	0.12	0.15	0.17	0.19
0.5	0.04	0.05	0.07	0.09	0.10	0.12	0.14	0.16	0.17	0.20	0.24	0.28	0.31
1.0	0.06	0.08	0.11	0.14	0.16	0.20	0.23	0.26	0.28	0.32	0.40	0.46	0.51
2.0	0.11	0.14	0.20	0.25	0.29	0.35	0.41	0.46	0.50	0.58	0.71	0.82	0.91
3.0	0.16	0.21	0.29	0.36	0.42	0.51	0.59	0.66	0.72	0.83	1.02	1.17	1.31
4.0	0.21	0.27	0.38	0.47	0.54	0.66	0.77	0.86	0.94	1.08	1.33	1.53	1.71
5.0	0.26	0.33	0.47	0.58	0.67	0.82	0.94	1.06	1.16	1.34	1.64	1.89	2.11
6.0	0.31	0.40	0.56	0.69	0.79	0.97	1.12	1.26	1.38	1.59	1.95	2.25	2.51
8.0	0.41	0.52	0.74	0.91	1.05	1.28	1.48	1.65	1.81	2.09	2.56	2.96	3.31
10.0	0.48	0.62	0.88	1.08	1.25	1.53	1.77	1.98	2.16	2.50	3.06	3.54	3.95
12.0	0.54	0.70	0.98	1.21	1.39	1.71	1.97	2.20	2.41	2.78	3.41	3.94	4.40
14.0	0.59	0.76	1.08	1.32	1.53	1.87	2.16	2.41	2.64	3.05	3.74	4.31	4.82
16.0	0.64	0.82	1.17	1.43	1.65	2.02	2.33	2.61	2.86	3.30	4.04	4.67	5.22
20.0	0.73	0.94	1.33	1.63	1.88	2.30	2.66	2.97	3.25	3.76	4.60	5.31	5.94
25.0	0.83	1.07	1.51	1.85	2.13	2.61	3.02	3.37	3.69	4.27	5.23	6.03	6.75
30.0	0.91	1.18	1.67	2.05	2.36	2.89	3.34	3.73	4.09	4.72	5.78	6.68	7.47
40.0	1.07	1.38	1.95	2.39	2.75	3.37	3.90	4.36	4.77	5.51	6.75	7.79	8.71
50.0	1.19	1.54	2.18	2.67	3.08	3.77	4.35	4.87	5.33	6.16	7.54	8.71	9.74
60.0	1.30	1.67	2.37	2.90	3.35	4.10	4.74	5.30	5.80	6.70	8.20	9.47	10.59

of the susceptibility of the soil to rill erosion relative to interrill erosion. A slope length relationship has been developed specifically for the frozen soil area of the Pacific Northwest of the U.S. (*3, 4*). A guide in the computer program and users manual helps identify the appropriate relationship for the particular field conditions.

Soil loss is much more sensitive to changes in slope steepness than to changes in slope length. In the USLE, a 10 percent error in slope steepness gives about a 20 percent error in computed soil loss. Thus, special attention should be given to obtaining good estimates of slope steepness. RUSLE has a more nearly linear slope steepness relationship than the USLE. Computed soil loss for slopes less than 20 percent are similar in USLE and RUSLE. However, on steep slopes, RUSLE computed soil loss is reduced up to about half that with USLE. Experimental data and field observations, especially on rangelands, do not support the USLE quadratic relationship when extended to steep slopes. RUSLE also provides a slope steepness relationship for short slopes subject primarily to interrill erosion and a steepness relationship for the frozen soil area of the Pacific Northwest.

In most practical applications, a slope segment previously estimated as a single plane or uniform slope can be a poor representation of the topography. In RUSLE and its computer program, complex slopes can be readily represented to provide a better approximation of the topographic effect. For example, a 300-ft slope having three equal segments at 5 percent, 15 percent, and 10 percent slope has an LS value of 2.9 for a moderate rill to interrill ratio. The same slope described as 300 feet at 10 percent has a LS values of 2.44 or a 20 percent decrease.

C-FACTOR

The C-factor is perhaps the most important USLE/RUSLE factor because it represents conditions that can most easily be managed to reduce erosion. Values for C can vary from near zero for a very well protected soil to 1.5 for a finely tilled, ridged surface that produces much runoff and leaves the soil highly susceptible to rill erosion.

Values for C are a weighted average of soil loss ratios that represent the soil loss for a given condition at a given time, to that of the unit plot (a unit plot is one maintained in clean-tilled fallow). Thus, soil loss ratios vary during the year as soil and cover conditions change. To compute C, soil loss ratios (SLR) are weighted according to the distribution of erosivity during a year (i.e., from the information in the city code and climate data). In RUSLE, a subfactor method is used to compute SLRs as a function of four subfactors (*2*) given as

$$C = PLU \cdot CC \cdot SC \cdot SR \quad [2]$$

where PLU is prior land use, CC is crop canopy, SC is surface or ground cover (including erosion pavement) and SR is the surface roughness. In the Pacific Northwest area where fallow is used to replenish soil moisture, an additional term is included reflecting the moisture status of the profile.

For cropland, CC and SC and the associated below-ground biomass are calculated from a crop and operation file. A residue decomposition calculation (*1*) is used to make estimates of the amount of residue cover on the soil surface and buried residue in the soil profile. The crop and operation file gives the amount of below ground living biomass, canopy cover, and canopy fall height by 15-day periods (see Figures 5.6 and 5.7). On rangeland, the user inputs ground cover, canopy cover; below-ground biomass is then estimated from above-ground biomass using ratios that are specific to different ecological zones (*5*). Surface roughness values are also specified by the user from a list of typical values for different rangeland cover conditions.

Ground (surface) cover is the term of the subfactors having the greatest effect on erosion. The inclusion of erosion pavements results in large changes in the value of the subfactor. Figure 5.8 illustrates the sensitivity of the elements considered in the subfactors on the final C-factor.

Grazed effects on rangelands, pasture, and meadows are reflected in the grazing-induced changes in canopy height, surface cover, and root biomass. Finally, surface cover, as used in the USLE, reflected vegetation and litter; in RUSLE, surface cover is given as 1.0 minus the amount of bare soil which reflects the addition of litter in the form of rock and stone to the conventional vegetative litter.

In *Agriculture Handbook 537*, SLRs were given in several tables for differing crops and crop growth periods. In many instances, the SLRs were developed for row spacing, crop varieties, and practices no longer used and in other instances, the SLRs were just not available. In the RUSLE approach, with a farming operation (Figure 5.6) and crop file (Figure 5.7) for which some fundamental data are available, SLRs can be calculated and weighted at 15-day intervals to obtain a crop or annual cover-management factor. Thus it is possible with the computer program to assemble SLRs for most agronomic crops used in modern agriculture.

P-FACTOR

Of the USLE/RUSLE factors, values for the P-factor are the least reliable. The P-factor mainly represents how surface conditions affect flow paths and flow hydraulics. For example, with contouring, runoff flows

Field operation: moldboard (5-7 inches)

Effect #1: 2—% dist.: 100 roughness: 1.3 % left: 30 depth: 6
Effect #2: 8
Effect #3: 1
Effect #4: 1
Effect #5: 1

enter effects in the order in which they occur

Operation effect list:

1. no effect
2. the soil surface is disturbed
3. crop residue is added to the surface
4. other residue is added to the surface
5. crop residue is removed from the field
6. the crop is harvested
7. the crop is planted
8. the crop is killed
9. call in a new crop growth file

Figure 5.6 Field operation example for a moldboard plow.

crop name: corn: ______________ category: 1
weight per harvest unit: 56 residue-to-yield ratio: 1
carbon: nitrogen ration: 100 stem constant u/R: 0.00120
residue at 30% cover (#/A): 1250 harvest units: bushels

Days After Planting	*Root Mass (pounds/acre/inches)*	*Canopy Cover (%)*	*Fall Height (feet)*
15	0	0	0
30	23	10	.5
45	45	50	1.2
60	68	80	2.5
75	136	100	5
90	136	100	6
105	136	100	6
120	136	100	6

Figure 5.7 Crop file for corn as might be grown in Indiana.

around the slope in channels formed by tillage. The grade and flow velocities could be much lower than in up-and-down hill flow paths. There are many interacting variables that determine the effect of contouring. Size of storm, antecedent soil water, and tillage type to name a few of these variables, interact in such a way that a contouring factor may vary widely from storm to storm and field to field; these interactions have made it

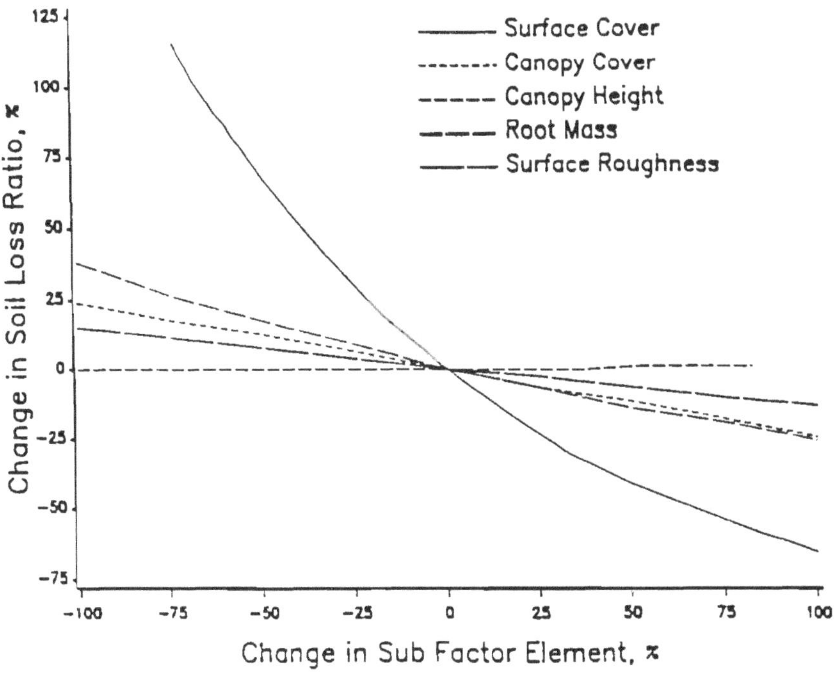

Figure 5.8 Sensitivity of soil loss ratio to subfactor elements.

difficult to document in the limited number of field studies dealing with contouring. Likewise, identifying these subtle characteristics in the field is difficult when applying RUSLE. Thus, P-factor values represent broad, general effects of such practices as contouring.

RUSLE P-factors are treated as the product of subfactors computed based on practices applied to the landscape. In RUSLE extensive data (both field and model) have been analyzed to reevaluate the effect of contouring. The results have been interpreted to give factor values for contouring as a function of ridge height, furrow grade, and climatic erosivity. New P-factor values for the effect of terracing account for grade along the terrace while a broader array of stripcropping conditions are considered in RUSLE than in USLE.

Finally, P-factors in RUSLE have been developed to reflect conservation practices on rangelands. The practices require estimates of surface roughness and runoff reduction.

COMPARISON OF RUSLE AND USLE

To illustrate some of the differences between RUSLE and USLE soil loss estimates, calculations were made for a continuous corn field with conventional tillage near Indianapolis, Indiana, and for a rangeland field near Tombstone, Arizona. Table 5.6 summarizes the comparison and illustrates differences in individual parameter values.

For these illustrations, the changes in R-values are relatively insignificant. K-factor changes using the time varying factor in RUSLE led to a smaller K value in Indianapolis and a larger value in Tombstone, a trend observed frequently in our experience to date. Breaking a 300-ft-long slope at 8 percent into three segments (top of slope to the bottom) of 100 ft at 6 percent, 150 ft at 10 percent and 50 ft at 6 percent (the same total elevation change) produced greatly different LS values.

On the Indianapolis location, the 1.72 LS value in the USLE increased to 1.94 in RUSLE, whereas the LS value for the RUSLE rangeland location decreased to 1.52, from 1.72, indicating the reduced rill to interrill erosion ratio on rangeland over that for cropland in addition to the slope segmenting.

The C-factor values in both instances were lower for the RUSLE estimates when compared to the values from *Agriculture Handbook 537*. In still other instances, the C-factor values have been observed higher from RUSLE than from the USLE.

The estimated soil loss for these two illustrations are both less with RUSLE than with USLE estimates from *Agriculture Handbook 537*. This should not be considered to be the case for all locations.

Of greatest significance is that C-factor values can be estimated with RUSLE for crops where SLRs were not available, i.e., there was no data in Table 5 and 10 of *Agriculture Handbook 537* to cover the particular crop and operation. Given that the user can obtain data for developing a crop file to cover the specific conditions encountered in his climatic conditions (data to describe at intervals after planting the root mass in the upper four in., canopy cover, fall height, carbon/nitrogen ratio, residue to yield ratio and characteristics of the residue stem), SLRs with which to calculate a C-factor can be made for any crop. Furthermore, new tillage implements can be added to the operations file to cover an infinite range of activities with which to simulate their effect on soil loss.

Table 5.6 Summary of RUSLE and USLE soil loss estimates for two locations.

	Location			
	Continuous corn/Indianapolis, IN		Rangeland/Tombstone, AZ	
Factor	RUSLE	USLE	RUSLE	USLE
R	180	175	65	70
*K	0.32	0.37	0.32	0.26
†LS	1.94	1.72	1.52	1.72
C	‡0.236	§0.252	††0.014	#0.038
P	1.0	1.0	1.0	1.0
A	26.4	28.1	0.44	1.19

* Used a Miami silt loam in Indianapolis and a Stronghold gravelly sandy loam in Tombstone.

† Used a 300 ft slope length at 8 percent in USLE and a three segment 100 ft at 6 percent, 150 ft at 10 percent, and 50 ft at 6 percent in RUSLE.

‡ Used continuous corn with 8-in. deep moldboard plowing on 4/10, tandem disk on 4/15, row planter on 4/20, row cultivator on 5/15, and harvest on 10/13 with 120 bu/A yield.

§ Used continuous corn with SLRs from line 1 of Table 5 in Agriculture Handbook 537.

†† Used 0.8 roughness, 60 percent ground cover, 25 percent canopy cover, 4000 lb/A root biomass in the upper 4 in., and a canopy height of 1 ft.

Used Table 10 of Agriculture Handbook 537 with grass, 25 percent cover, and 60 percent ground cover.

SUMMARY

RUSLE retains the basic structure of the USLE with soil loss estimated for a slope profile as the product of the terms reflecting the climatic erosivity represented by R, soil erodibility represented by K, topography represented by LS, and land use, and management and conservation support practices represented by CP. Incorporation of the technology in a computer program facilitates the calculations.

New climatic erosivity values have been calculated which reflect the variability of R in the mountainous conditions of the western U.S. Corrections have been introduced in high R-value areas with flat slopes. It is recognized that in these areas, ponded water on the soil reduces the erosivity of the rain.

The soil erodibility term K is varied seasonally being highest in spring and lowest in late fall synonymous with soil freezing. Although the K values are generally obtained from the erodibility nomograph, alternative methods are used for volcanic soils (such as dominate in Hawaii) and an alternative proposed for use as a function of average soil particle diameters.

The topographic factors L and S are evaluated using four separate slope length relationships. Three are a function of slope steepness and the susceptibility of the soil to rill erosion relative to interrill erosion. A specific slope length relationship has been developed for the frozen soil areas of the Pacific Northwest where rilling and freeze-thaw phenomenon dominate. In RUSLE, and its computer programs, complex slopes can be readily represented to provide a better approximation of the topographic effect.

Soil loss ratios are calculated with time dependent functions that address subfactors reflecting prior land use, crop canopy, surface or ground cover, and surface roughness. For time invariant conditions encountered on range and pasture land, the user inputs ground cover, canopy cover, and below-ground biomass. Grazing effects are reflected in the grazing induced changes in canopy height, surface cover, and root biomass.

The conservation practice value represents broad, general effects of such practices as contouring, strip cropping, subsurface drainage, and terracing. P-factors have also been developed to reflect conservation practices on rangeland and require the user to estimate surface roughness and runoff reductions.

REFERENCES

1. Gregory, J.M., T.R. McCarty, F. Ghidey, and E.E. Alberts. 1985. *Derivation and evaluation of a residue decay equation.* Transactions, American Society of Agricultural Engineers 28: 98-101, 105.
2. Laflen, J.M., G.R. Foster, and C.A. Onstad. 1985. *Simulation of individual-storm soil loss for modeling the impact of soil erosion on crop productivity.* In: S.A. El-Swaify, W.C. Moldenhauer, and A. Lo, (eds.) *Soil Erosion and Conservation.* Soil and Water Conservation Society, Ankeny, Iowa. pp. 285-295.
3. McCool, D.K., L.C. Brown, G.R. Foster, C.K. Mutchler, and L.D. Meyer. 1987. *Revised slope steepness factor for the universal soil loss equation.* Transactions, American Society of Agricultural Engineers 30: 1387-1396.
4. McCool, D.K., G.R. Foster, C.K. Mutchler, and L.D. Meyer. 1989. *Revised slope length factor for the universal soil loss equation.* Transactions, American Society of Agricultural Engineers 32(5): 1571-1576.
5. Weltz, M.A., K.G. Renard, and J.R. Simanton. 1987. *Revised universal soil loss equation for Western rangelands.* In: *U.S./Mexico Symposium on Strategies for Classification and Management of Native Vegetation for Food Production in Arid Zones*, U.S. Forest Service General Technical Report RM-150, pp. 104-111.
6. Wischmeier, W.H., and D.D. Smith. 1978. *Predicting rainfall erosion losses.* USDA Agriculture Handbook 537, U.S. Department of Agriculture.

Chapter 6

MODELING SOIL EROSION

M. A. Nearing, L. J. Lane, and V. L. Lopes

Soil loss is defined in erosion literature as the amount of soil lost in a specified time period over an area of land which has experienced net soil loss. Soil loss is expressed in units of mass per unit area, such as t/ha or kg/m^2, and may be for a single storm event, an average value for a number of years, or for any other specified time period. Soil loss is of interest primarily in terms of on-site effects of erosion such as loss of crop productivity. Sediment yield is defined as the amount of sediment which leaves a specified area of land in a given time period. Sediment yield refers to a mass of sediment which crosses a boundary, such as the edge of a field or outlet of a watershed, and may be expressed in units of total mass (kg), mass per unit width of the boundary (kg/m), or mass per unit area (kg/m^2). Sediment yield is important in terms of off-site effects of erosion such as siltation in ditches, streams, and reservoirs. Sediment is also a primary carrier of agricultural chemicals which can pollute streams and lakes. In most cases not all soil lost on a field becomes sediment yield, as some of the soil particles are deposited on the field before leaving the field boundary. In other words, most fields have some areas which experience net soil loss over time and some areas which experience net deposition over time. The difference between the spatially integrated net soil loss and the spatially integrated net deposition is what leaves the field, herein termed sediment yield.

Modeling soil erosion is the process of mathematically describing soil particle detachment, transport, and deposition on land surfaces. There are at least three reasons for modeling erosion: (a) erosion models can be used as predictive tools for assessing soil loss for conservation planning, project planning, soil erosion inventories, and for regulation; (b) physically-based mathematical models can predict where and when erosion is occurring, thus helping the conservation planner target efforts to reduce erosion; (c) models can be used as tools for understanding erosion processes and their interactions and for setting research priorities.

There are basically three types of erosion models (*21*): empirical, conceptual, and physically-based. Empirical models are based primarily on observation and are usually statistical in nature. Empirical models are based on inductive logic, and generally are applicable only to those conditions for which the parameters have been calibrated. The Universal Soil Loss Equation (USLE) is the empirical erosion model which has been used most widely for predicting soil erosion. The greatest criticism of the USLE has been its ineffectiveness in applications outside the range of conditions for which it was developed. Adaptation of the USLE to a new environment requires a major investment of resources and time to develop the database required to drive the model. The primary focus of the empirical models has been in predicting average soil loss, although some extensions to sediment yield estimates have been developed (*33, 37*).

Conceptual models lie somewhere between physically-based models and empirical models, and are based on spatially lumped forms of water and sediment continuity equations (*21*). The focus of the conceptual models has been to predict sediment yields, primarily using the concept of the unit hydrograph.

Physically-based models are intended to represent the essential mechanisms controlling erosion. The power of physically-based models is that they represent a synthesis of the individual components which affect erosion, including the complex interactions between various factors and their spatial and temporal variabilities. The result is synergistic, the model as a whole represents more than the sum of the individual pieces. The research scientist can use the physically-based erosion models to help identify which parts of the system are the most important to the overall erosion process, and therefore should be given attention in research and development of erosion prediction and control technology. The conservation planner can use a physically-based model as an interactive conservation design tool, targeting critical seasons or months in which major erosion events occur as well as critical positions on the hillslopes where the greatest soil loss takes place. The planner can also quickly suggest and evaluate new conservation strategies for individual fields.

The focus of this chapter is on development of physically-based erosion models. Chapter 5 discusses current developments in the more empirically-based Universal Soil Loss Equation. We will discuss only models for erosion by water; erosion by wind is discussed in Chapter 11. Also, in this chapter we draw much upon experiences related to the development of the USDA-Water Erosion Prediction Project (WEPP) erosion models (*22*). The objective of WEPP is to develop new generation erosion prediction technology for use by the conservation planner at the field level. The technology is based on fundamentals of erosion and hydrologic sciences and is computer driven. The two WEPP models referred to herein as examples of physically-based models are the WEPP hillslope profile model and the WEPP watershed model.

MECHANICS OF EROSION BY WATER

Erosion encompasses detachment, transport, and deposition of soil particles by the erosive forces of raindrops and surface flow of water. It is common in erosion models to divide, conceptually, the process of erosion on hillslopes into that related to rill flow mechanisms and interrill mechanisms. Within this framework rill flow may act to detach soil particles whenever the hydraulic shear stress in the rill is sufficient to overcome the binding forces between individual particles in the soil mass. The flow in the rill also acts as the transporting agent to carry detached soil, i.e. sediment, from both rill and interrill areas. Detachment on interrill areas is primarily induced by raindrop impact, since flow depths on interrill areas are by definition of negligible erosive power. Transport on interrill areas is primarily by broad shallow surface flow. The energy required to move (entrain) sediment is much less than that required to detach in-situ soil particles. Net transport by raindrop splash is very small, with only a small net downslope movement due to the slope effect (*39*).

Most erosion models rely upon the concept of transport capacity, which is defined as the maximum amount of sediment that a flow can carry without net deposition occurring. Several transport capacity equations have been developed for transport of sediment in large channels and adapted for use in upland erosion models. Choice of the "best" sediment transport equation is subjective, and opinions vary as to the most appropriate equation to use (*19*). Instead of which equation to use, a more important issue in modeling fundamental erosion processes is what transport capacity means and how it is used in the model. Transport capacity is basically a balance between entrainment and deposition rates of the sediment in the flow. The description of the entrainment process does not include a factor for cohesive soil forces, but considers only gravity forces of the sediment which must be overcome for the particle to be lifted into the flow. The implicit assumption, then, for erosion of cohesive soils is that cohesive forces are negligible once the soil has been initially detached from the in-situ soil mass.

Another implicit assumption when using a sediment transport equation to describe erosion is that deposition is a continual process. When we refer to "detachment" in describing soil loss we mean the process of removing in-situ soil particles from the bulk soil mass. The term "net detachment" refers to a balance between detachment, entrainment of previously detached particles, and deposition for the case when net movement of particles is from the soil surface into the flow. Some recent erosion models have avoided the explicit use of an existing sediment transport equation entirely. Hairsine (*15*) calculates rates for five simultaneous processes: detachment of soil by flow, entrainment of sediment by flow, detachment of soil by rainfall, entrainment of sediment by rainfall, and deposition of sediment. In that model net detachment is the case where the sum of detachment and entrainment rates exceeds the rate of deposition. Lopes and Lane (*23*) take a similar approach except that their model does not differentiate between entrainment of sediment and detachment of soil.

Interrill processes

Interrill areas are defined as the areas wherein erosion is dominated by detachment by raindrop impacts and transport by very shallow sheet flows. As mentioned above, net transport by splash is negligible, since splash has only a small net downslope movement. Obviously, however, raindrop impact on shallow flows may enhance the capacity of the flow to transport sediment (*16*). The three other factors which influence transport on interrill areas are runoff rate, interrill slope gradient, and surface roughness.

Many erosion models, including the WEPP model (*10, 31*), do not treat the individual process of detachment and transport on interrill areas, but only estimate sediment yield from interrill areas to rills as some function of rainfall intensity. The equation used in the WEPP erosion models is

$$D_i = K_i \, I^2 \qquad [1]$$

where D_i is rate of interrill sediment delivery to rills, K_i is an interrill erodibility parameter, and I is average rainfall intensity integrated over the duration of rainfall excess. Equation 1 is an empirical relationship based on extensive rainfall simulation studies on a number of different soils (*26*). Interrill models which lump processes to estimate only sediment yield are simple, stable, and give reasonable results. Kirkby (*17*) argues that sediment yields from interrill areas are relatively low compared to rill erosion rates for cases where soil losses are high, which is true in general. Interrill erosion may dominate, however, in certain cases, such as on rangelands and no-till situations or where slope angles are low and slope lengths are short (*32*).

Models which treat the individual processes of detachment and transport on interrill areas may have an advantage over models which predict only sediment yield from interrill areas if the individual processes can be described simply and effectively, and if the parameters within the model can be experimentally determined. Gilley et al. (*12*) proposed a model which explicitly treats detachment and transport processes on interrill areas. Detachment was described using a semi-empirical relationship between raindrop size and velocity distributions and splash detachment rates. Sediment transport capacity was calculated using the Bagnold equation, which is based on stream power. The model agreed well with the reported experimental data from the study (*13*). Several different relationships have been proposed to represent splash detachment as a function of raindrop properties. Gilley et al. (*11*) evaluated several different functions for splash detachment and recommended kinetic energy times drop circumference as the best, although several others, including kinetic energy alone, were evaluated to predict splash nearly as well. Sharma (*35*) found kinetic energy to relate well to drop splash, but their data indicated a need for the introduction of a critical energy for detachment. Splash studies of Al-Durrah and Bradford (*1*) on nine soils showed a zero intercept between splash and kinetic energy, which contradicts Sharma's data. Sharma (*35*) attributes the different results to the fact that drop heights in his study were shorter; another possible explanation is that not all of the soil splash could be collected in the splash cup for the very low drop heights. In any case, the drop kinetic energies of interest are primarily for drops at or near terminal velocity, and for that case a threshold kinetic energy for splash detachment is not justified.

Rill processes

Rills are defined as concentrated flow channels which are small enough to be tilled over (*9*). Flow in rills acts as a transporting agent to carry sediment from rill and interrill sources downslope. If the shear stress in the rill is high enough, rill flow may also detach a significant amount of soil. A common model for rills describes detachment as a linear function of shear stress, but in fact, experimental results of flume studies show that detachment vs. shear stress relationships are typically non-linear. Figure 6.1 shows the results from a study at the National Soil Erosion Research Laboratory of detachment by surface flow vs. flow shear stress for a Miami silt loam soil. Most erosion models would describe flow detachment as a linear function of flow shear stress (or energy) with a positive intercept on the shear stress axis, which is typically called the critical shear stress of the soil. For instance, the WEPP model uses a function of the form

$$D_c = K_r (\tau - \tau_c) \qquad [2]$$

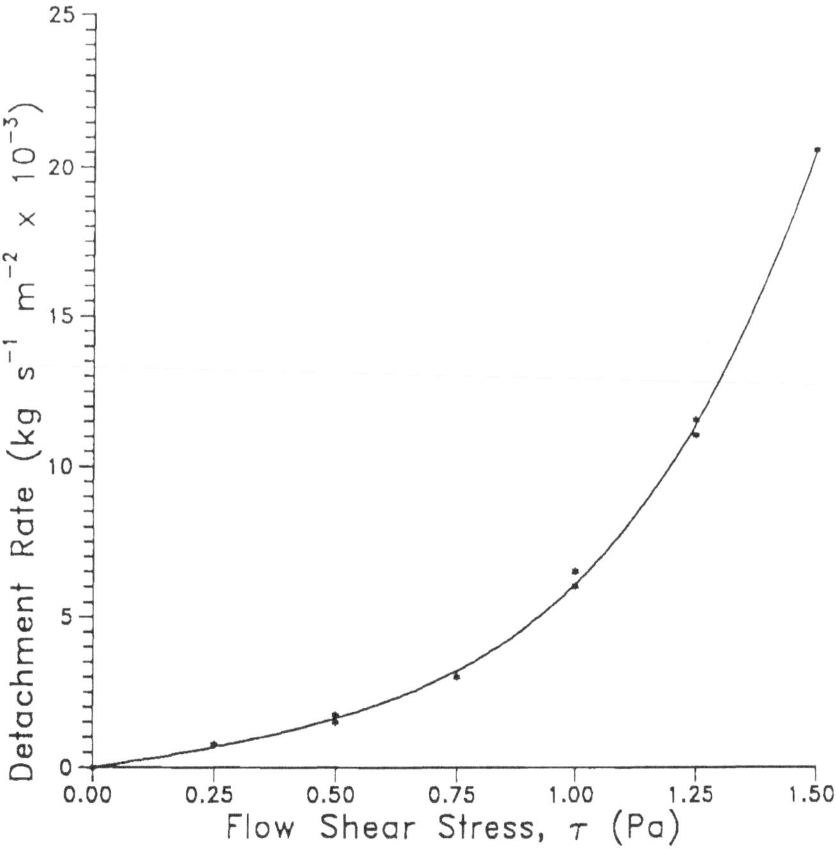

Figure 6.1 Detachment rate by surface flow vs. average hydraulic shear stress as measured in a hydraulic flume on a Miami silt loam soil.

where D_c is the detachment capacity of clear water flow, K_r is the soil's rill erodibility, τ is the shear stress of the flow, and τ_c is the soil's critical hydraulic shear strength.

The concept of the critical shear stress of the soil should not be given more physical significance than is warranted. There has been much discussion and experimentation about the physical meaning of the critical shear stress. Grass (*14*) conceptualizes entrainment as the overlap of two probability distributions, one for the instantaneous and localized flow shear stresses associated with turbulent "bursting" phenomenon and one for the resistance of individual soil particles on the boundary between soil and flow. In Equation 2, critical shear stress of the soil is a mathematical entity which results from the linearization of the model. It is not, and should

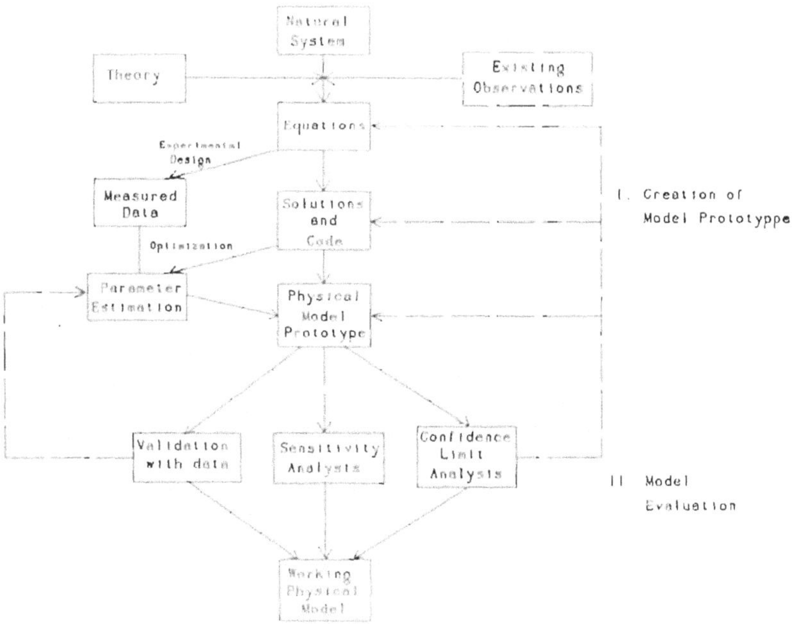

Figure 6.2 An outline for development of a physically-based erosion model.

not be interpreted as being, the threshold level of shear stress below which there is no soil detachment and above which the first soil particles begin to move.

DEVELOPMENT OF A PHYSICALLY-BASED EROSION MODEL

Model development may be divided into two phases. The first is the creation of the physical model prototype and the second is model evaluation. The steps involved in model development are outlined in Figure 6.2. The process begins with conceptualizing the natural system through the use of existing information. An example of the conceptualization process for erosion modeling is the set of equations presented by Meyer and Wischmeier (*24*). They presented a mathematical formulation of the erosion process based on observations by Ellison (*6*) which included descriptions of (a) detachment by rainfall, (b) detachment by flow, (c) transport by rainfall, and (d) transport by flow. The second step in the process of model development is to solve the equations and write the solutions in the form of computer code, assuming that the resultant model is to be computer driven. This step includes development of the overall computer model structure, which involves linking the various components of the technology into a complete working unit. Experimentation for developing a parameter database may begin simultaneously with the development of the computer code, since at this point the fundamental equation structure of the model is set.

After the model code and parameter-experiments are completed, the parameter estimation stage can begin. Parameter estimation involves two distinct steps: (a) parameter identification and (b) development of parameter prediction equations or techniques. Parameter identification is determining the model parameters from an experimental data set (*30*). It involves using the existing computer model and an optimization technique to analyze the experimental data to obtain model parameters for the measured data set. The second step is to develop a method for predicting model parameters for soils or environmental conditions not represented in the measured data set. The completed and tested computer code along with the parameter prediction techniques constitutes the prototype physical model, and represents completion of the first phase of the model development.

The second phase of erosion model development is the model evaluation phase, which includes (a) sensitivity analysis, (b) confidence limit analysis, and (c) validation with data. The results of the model evaluation are used to assess the validity of the model and to make the changes in basic equations, model structure, or parameter estimation procedures necessary to development of the validated working model. It is important that changes dictated by the model evaluation phase are followed through in a complete and logical manner. If changes in model structure are required to produce a valid working model, then the entire model development process must be followed again. The new structure will require a new set of parameters, which means that the measured experimental data will need to be re-analyzed and new parameters identified from the data. Then the parameter estimation procedures will need to be re-evaluated and a new "prototype" model proposed and evaluated. The process is iterative. The model developers must be cognizant of and sensitive to the user's needs in making the decision as to when the process stops and the model is deemed "valid."

There are four major mechanisms for introduction of error in the modeling process. The first is in the formulation of the basic equations. Any mathematical representation of a natural process is approximate, at least when dealing on the scales related to soil erosion, and will cause the introduction of some error in terms of describing the system. These errors can be large, particularly where a minor factor for most cases, and hence neglected in the mathematical descriptions, is a major factor in a specific case. The second source of error is in the solution and coding of the equations. This should be a minor source of error except in certain cases where approximate solution techniques must be used for the sake of computational efficiency. A third source of error is experimental error and variation in experimental data. Experimental data associated with erosion experiments typically have a high degree of variation. A fourth source of error is in the parameter prediction procedure. Any statistical method developed for predicting model parameters for untested situations will have some, and usually a large amount, of error associated with it.

Erosion equations for a steady-state model

The WEPP hillslope profile erosion model is a recent example of a physically-based erosion model. This model has been described in detail elsewhere (*22, 31*) and hence its description here will be brief. The WEPP erosion model uses a steady state sediment continuity equation to describe downslope movement of sediment:

$$dG/dx = D_f + D_i \qquad [3]$$

where x (m) represents distance downslope, G (kg/s/m) is sediment load, D_i (kg/s/m^2) is lateral sediment flow from interrill areas, and Df (kg/s/m^2) is rill erosion or deposition rate. Interrill sediment delivery, D_i, is considered to be independent of x. Rill erosion, D_f, is positive for detachment and negative for deposition.

Interrill erosion in the model is conceptualized as a process of sediment delivery to rills, whereby the interrill sediment is then either carried off the hillslope by the flow in the rill or deposited in the rill. Sediment delivery from the interrill areas is considered to be proportional to the square of rainfall intensity, with the constant of proportionality being the interrill erodibility parameter. The function for interrill sediment delivery also includes terms to account for ground and canopy cover effects, which are discussed below. The interrill function was presented in Equation 1.

Net soil detachment in rills is calculated for the case when hydraulic shear stress exceeds the critical shear stress of the soil and when sediment load is less than sediment transport capacity. For the case of rill detachment

$$D_f = D_c\ [1 - G/T_c] \qquad [4]$$

where Dc is detachment capacity by flow as given in Equation 2 and T_c(kg/s/m) is sediment transport capacity in the rill. Rill detachment is considered to be zero when the hydraulic shear stress is less than critical shear strength of the soil.

Net deposition is computed when sediment load, G, is greater than sediment transport capacity, T_c. For the case of deposition

$$D_f = [V_f / q] [T_c - G] \quad [5]$$

where Vf (m/s) is effective fall velocity for the sediment, and q (m^2/s) is flow discharge per unit width.

Representations of the effects of land use and management practices on erosion control are perhaps the most important part of an erosion prediction tool if the purpose is help plan land and farm management systems to control erosion. Residue management and tillage practices on croplands are the mechanisms through which the farmer usually can most directly impact soil loss and effect erosion control. Table 6.1 shows the effects of soil and residue management on rill and interrill erosion rates as they are represented in the WEPP model. In the WEPP erosion model interrill sediment delivery is adjusted to account for effects of ground cover, dead roots, live roots, and canopy cover. Plant and soil management practices also affect infiltration processes greatly; these effects are discussed in Lane and Nearing (*22*).

The effect of surface cover in rills is probably overall the greatest single management effect on erosion, because it strongly influences both detachment and sediment transport processes. This effect is incorporated into the WEPP model via the hydraulic friction factor terms, which enable partitioning of the flow energy into that acting on the soil from that acting on the surface cover, including residue and rocks. The effect of buried residue is also accounted for in the WEPP model.

Rill erodibility is also affected by disturbance due to tillage. In the WEPP model, baseline rill erodibility for croplands is for the completely disturbed state, which for practical purposes is defined as that found immediately after moldboard tillage. Each tillage implement is defined with a tillage intensity term to reflect the fact that disturbance may be less (or more, if needed) than for the moldboard plow. From the disturbed state the soil consolidates. Computation of consolidation and changes in rill erodibility as a function of time and weathering are made using a fundamentally-based consolidation model of Nearing et al. (*29*).

Interrill erodibility is not adjusted for time in the WEPP model. Various data suggest that interrill erodibility does not greatly change with time (*3, 4*). A further indication of this can be seen by comparing rangeland and cropland interrill erodibilities. Rangelands represent essentially a fully consolidated soil condition and freshly tilled croplands represent a fully un-

Table 6.1 Effects of plant and soil management on erosion.

	Above Ground Residue Cover	Below Ground Biomass	Plant Canopy Cover	Soil Consolidation After Tillage
Interrill Erosion	Ge	Kid,Kil	Ce	**NA**
Rill Erosion	(fc)	Krb	**NA**	K r c

consolidated soil condition. A comparison of average cropland erodibilities for 36 soils (*18*) and 11 rangeland soils (*36*) indicates that interrill erodibility for croplands is about four times that for rangelands. A challenge for the future is to develop a method for predicting which soils will change in interrill erodibility with time, and whether erodibility will increase or decrease for a given soil.

Continuous simulation models

The full benefit of an erosion prediction model is gained through the use of a continuous simulation model. By continuous simulation it is meant that the model "mimics" the processes which are important to erosion prediction as a function of time, and as affected by management decisions and climatic environment. Surface residue, for example, plays an important role in terms of predicting the amount of soil lost during a given rainfall event. An erosion model may use a plant growth and residue decay model to estimate the amount of crop residue present on the soil surface for each day through the year. A certain amount of residue is generated by leaf drop during senescence and by harvesting, and a pass of a given tillage implement will bury a certain percentage of a given type of residue. An erosion model should adjust surface cover as a function of those and other processes which affect residue cover. With a continuous simulation model, the user does not need to specify the amount of residue cover as a function of time. Soil parameter, residue amount, crop growth, soil water content, surface roughness, and essentially all other adjustments to model parameters should be calculated on at least a daily time step.

The output of the continuous simulation model represents time integrated estimates of erosion. In nature, as well as in the model predictions, a large percentage of erosion occurs due to a small percentage of rainfall events. The model simulates some number of years of erosion and sums the total soil loss over those years for each point on the hillslope to obtain average annual values of erosion along the hillslope. The model calculates both detachment and deposition. It predicts where deposition begins and/or ends on a hillslope, which may vary from storm to storm. Certain points on the hillslope may experience detachment during some rainfall events and deposition during other events. The output of the continuous simulation model represents an average over all of the erosion events.

A physically-based erosion model may also be executed in the single-storm mode. In that case, all of the parameters used to drive the hydrology and erosion components of the model must be input by the user, including soil conditions for the day of the rainfall event, crop canopy, surface residue, days since last disturbance, surface random roughness, oriented roughness, etc. In the continuous simulation mode the influence of these user inputs, which represent the initial conditions for the simulation, is small since the model adjusts each of those variables through the continuous simulation. In the single storm mode those inputs have a major influence on the output. The single-storm option of the model requires a great deal more knowledge on the part of the user to interpret and use the output for planning, evaluation, and conservation design purposes. The single-storm model helps in understanding and evaluating the factors which influence erosion on a hillslope; it is of limited value in evaluating conservation systems wherein conditions change as a function of time through the year and from year to year.

Parameter estimation

Soil erodibility for an erosion model is defined relative to the form of the erosion equations used. The approach is to first formulate the erosion equations for the model then to analyze experimental results relative to the particular set of erosion equations used. The equations should as accurately as possible describe the physical processes involved and the effects of various environmental factors on the physical processes, but in the model parameterization process we must recognize that every model is a simplification of reality. The physical significance of the erodibility parameters is limited to the degree to which the erosion equations fully describe the physical processes of erosion. The best alternative for parameterizing a model with experimental data is to use the model to analyze the data. This usually involves using an optimization technique which executes the model and searches for the set of erodibility parameters which provide the best fit between measured and calculated soil loss (*2, 30*).

An example whereby too much physical significance may be placed into the erodibility parameters is with regards to the rill erodibility, K_r. The WEPP model assumes that the primary mode of detachment in rills is scour, i.e., by way of the shear stress acting at the fluid/soil interface along the relatively well defined rill wetted perimeter. We recognize, however, that other mechanisms also act to detach soil in rills, including headcutting and sidewall sloughing. The erodibility parameters in the physically-based models are more fundamental in nature than erodibility parameters for the USLE, for example, but still represent a simplification of the erosion processes.

The discussion above does not imply that more fundamentally-based predictors for erodibility parameters are not needed. Better, more fundamentally based predictors of erodibility will go along with improved erosion equations. Universal, fundamentally derived equations for relating soil properties to soil erodibility are very much needed. Current methodology is to perform as many physical, chemical, mineralogical, mechanical, and erosion tests as possible on as many soils as possible and relate the soil properties to erodibility using statistical regression techniques. The problem with this approach is that it relies on inductive logic, and hence the results are questionable for applications outside the range for which they were derived.

Model analysis

Analysis of an erosion model is a critical step in developing a usable and valid erosion prediction tool, and one often not given the attention and allocation of resources of the total modeling process that it requires. Conventional attitudes are that if the pieces of the entire model are "correct," then the overall model with all components linked will be "correct." Experience shows otherwise. Often it is not possible to test the individual components thoroughly until they are within the framework of the simulation model. Also, interactions between components may cause the model to respond differently than expected a priori. Model analysis is an integral part of the modeling process.

Model analysis consists of three distinct parts: (a) comparisons of model predictions with existing data, (b) sensitivity analysis, and (c) confidence limit analysis. Data that was used to develop the model parameter base should not be used to validate the model. In the case of WEPP, data collected on USLE natural runoff plots are being used to test the model predictions. Historical weather data and crop/management information is being input to the model and comparisons between measured and predicted runoff and soil loss are being made. One comparison to make, obviously, is the overall average of soil loss or runoff volume over the period of data collection. This is a valid comparison, but doesn't give much information on how the model is not working if comparisons are not within the desired range of accuracy, particularly for plots with crop rotations. Storm-by-storm comparisons of measured vs. predicted runoff and erosion provide more useful information. In general, event based comparisons will not give a high correlation for a continuous simulation.

A more important data comparison is that between the frequency distributions of the measured events and predicted events; the critical question is if the model accurately reflects those distributions. Also, from these analyses we may begin to assess the length of simulation necessary for the long-term results to be statistically valid. Figure 6.3 shows a plot of cumulative frequency distributions for a fallow runoff plot at Castana, Iowa, for the years 1960-1969. The predicted values were made using an early version of the WEPP hillslope profile erosion model (April, 1989). For this case the distributions were very similar in form, but the distribution for predicted values was shifted to the right, indicating that the model was overpredicting runoff for this case. This information was used, along with other similar analyses, to evaluate and adjust the model's runoff predictions.

Care must be taken not to make general statements about model validity based on one or two data sets. Erosion is highly variable, and that fact is reflected in measured erosion data. A large number of varied data sets should be evaluated to decide when and how a model must be adjusted to provide more accurate erosion predictions.

Sensitivity analysis is an evaluation of the relative changes in the model's output as a function of relative changes in the values of model input parameters. A detailed evaluation of a model's response can yield a great deal of insight into the nature of the model, and, to the extent that the model accurately represents the physical system which it simulates, sensitivity analysis can provide insight into the factors which influence the response of the physical system. Sensitivity analysis provides a method for examining the response of a model that eliminates the influence of error related to natural variation of the model input parameters. The rationality of the model and the influence of input error can thus be evaluated in detail (*24*).

Three limitations to current methods in sensitivity analysis were discussed by McCuen and Snyder (*24*):

1. The linear form of the sensitivity coefficient does not reflect sensitivity of the variable over the entire range of the parameter because of the non-linear response of the model. However, as pointed out by

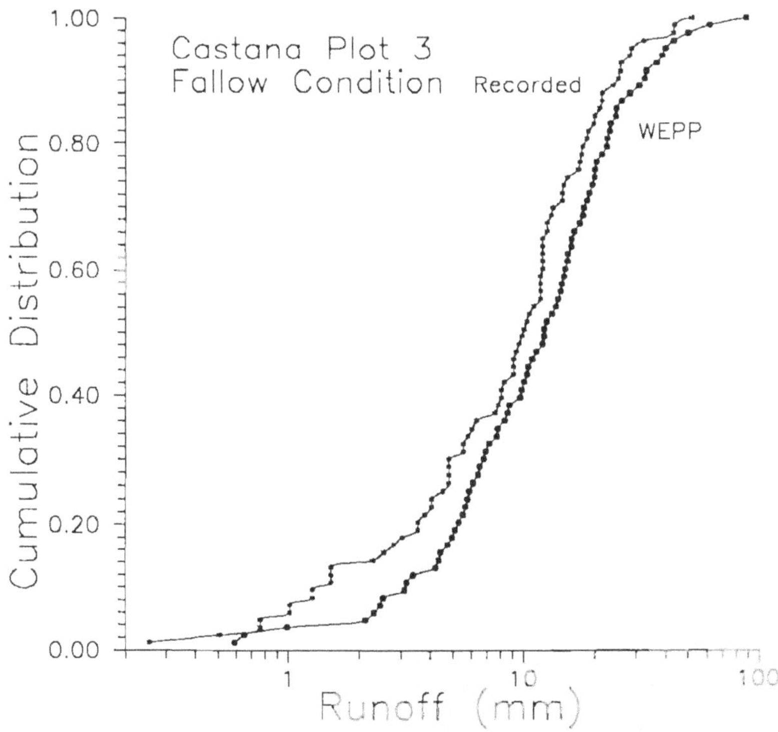

Figure 6.3 Cumulative frequency distributions of measured and predicted runoff volumes for a fallow natural runoff plot located in Castana, Iowa.

McCuen and Snyder (*24*), the sensitivity of the extreme values which represent the physical conditions are often of primary interest.

2. The sensitivity coefficient is a univariate parameter, which implies that there is no interaction between variables. This can be a serious limitation which can lead to misinterpretations of the model. A variable which is insensitive with a given set of companion inputs might be quite sensitive with another set of inputs.
3. The sensitivity coefficient is single valued. A distribution of the output as a function of input parameter distribution might better describe sensitivity. This third point is related to confidence limit analysis discussed below.

In addition to the three limitations listed above the value of sensitivity analysis is limited by the "goodness" of the model. The power of a prediction model is that it integrates a number of interdependent processes to simulate a larger system. The effect is synergistic: the resulting product which a model represents is greater than the sum of the independent components. Still, however, current knowledge dictates the relationships in the model. There is always an inherent bias toward the current scientific knowledge, and specifically toward that knowledge represented in the model.

Nearing et al. (*32*) performed sensitivity analysis on the WEPP hillslope profile erosion model. They evaluated the model's response, relative to soil, slope profile, plant, and hydrologic input parameters. Both the single-storm version of the model and the erosion component alone were used in order to delineate the relative effects of factors on hydrology estimates and erosion estimates. The approach was to use an average linear sensitivity coefficient which represented the overall change in model response relative to the values of input which represent the extremes in the physical conditions represented.

A summary of the results from Nearing et al. (32) of the sensitivity analyses for the erosion model are presented in Table 6.2 and for the single storm model in Table 6.3. Hydrologic factors are key to obtaining good soil loss estimates from the model as shown in both tables. Factors related to rill detachment and transport are also very important. Rill erodibility, critical hydraulic shear of the soil, surface cover in the rills, and rill hydraulic friction factors are major factors in terms of model response. Texture is an important soil property for the model. Much of the sensitivity to texture is introduced through the prediction of rill hydraulic friction factors. Friction factors for both soil roughness and ground cover are very important parameters for erosion calculations, particularly for rills where hydraulic roughness has a major effect on sediment transport calculations. Saturated hydraulic conductivity and interrill erodibility parameters fall into the moderately sensitive range of the parameters. However, as discussed by Nearing et al. (*32*), both of these factors play a greater or lesser role in the predictions depending on the conditions. Interrill erodibility is important on short, flat slopes and in rangeland and no-till conditions. Saturated conductivity is more important for shorter duration, less intense storms and less important for the larger storms. This is because a smaller fraction of the total rainfall becomes runoff for the smaller storms. Interrill cover is important when interrill erosion is important, it responds similarly as the interrill erodibility term. Plant canopy cover is not a dominant factor. Its influence, again, is greater on short flat slopes, but not as great as interrill cover or erodibility. Canopy height is relatively insignificant overall to predicting erosion. Terms related to the suction term of the infiltration equation, those being bulk density and saturation, do not have a major influence on the output. Peak rainfall intensity, time to peak rainfall inten-

Table 6.2 Summary of average sensitivity values (S) for erosion component.

Parameter	Average S value*
Intensity & Runoff	1.1836
Ground Cover (corn)	-0.8134
Runoff	0.8100
Rill Cover (corn)	-0.6504
Ground Cover (wheat)	-0.6080
Rill Erodibility	0.5008
Intensity	0.4697
Critical Shear	-0.4194
Rill Cover (wheat)	-0.4155
Incorporated Residue	-0.3843
Soil Friction Factor†	0.3565
Interrill Erodibility‡	0.2515
Interrill Cover	-0.1101
Canopy Cover	-0.0999
Rill Spacing§	0.0943
Canopy Height	0.0542
Rill Width†	0.0530
Sediment Transportability†	0.0188

*Sensitivity values averaged over sediment loss from slope lengths of 22.13, 50.0, and 200 m at slope gradients of 5 and 9 percent.

†Sensitivity values averaged over sediment loss from slope lengths of 50 m and 5 percent gradient.

‡Sensitivity values averaged over sediment loss from K_i values of $0.5 \times 10^6 - 5.0 \times 10^6$ kg s/m^4

§Sensitivity values averaged over sediment loss from rill space values of 0.5 and 5 m.

sity, rill spacing and width, and sediment transportability (i.e., sediment size distributions) do not play a major role in the predictions of total soil loss in the WEPP model. Obviously, sediment size distributions would play a major role in terms of sediment enrichment.

Confidence limit analysis addresses the issue of how accurate erosion estimates from the model are. Given the expected variation of the data input values, confidence limit analyses evaluate the variation of the model response. Methods such as Monte-Carlo simulation or the point estimate method (*34*) may be used. Figure 6.4 shows the results of an analysis of the coefficient of variation of the WEPP model response as a function of variation in erodibility input parameters, which was calculated from field rainfall simulator experiments on the Miami silt loam soil. Variation was computed using the point-estimate method (*32*). This analysis was done for a fallow soil condition and a relatively highly erodible soil condition. For other conditions where residue is present or where the soil is highly

Table 6.3 Summary of average sensitivity values (S) for detachment per area in single storm component.

Parameter	Average S value
Precipitation	1.12
K_r, Rill Erodibility*	0.81
Rill Cover	-0.794
Sand	-0.455 to -0.630†
Clay	0.245 to 0.630†
Silt	-0.245 to 0.455†
K_s*	-0.43
Duration at constant precipitation	-0.344
K_i, Interrill Erodibility*	0.19
i_p	-0.156
Canopy cover	-0.111
Saturation (init)*	0.1015
Interrill Cover	-0.087
Canopy height	0.0455
Bulk Density*	-0.017
t_p	-0.0130

* End Slope = 0 percent.

†Dependent on fraction of other size classes.

consolidated, as for rangelands and no-till, the response of the coefficient of variation would be much different.

MODELING ON WATERSHED SCALES

The mechanics of detachment, transport, and deposition of soil particles in a watershed is extremely complex to describe in detail. Therefore abstraction is necessary if we are to describe some aspects of erosion and deposition processes on watershed scales. Abstraction consists of replacing the system under consideration with a conceptual model of similar but simpler structure. To develop a conceptual model for describing the erosion and deposition processes on watersheds it is natural to begin with two sets of partial differential equations expressing the conservation of mass and momentum: the free surface flow equation (portraying the flow velocity field) and the sediment continuity equation (portraying the sediment concentration distribution). Conceptually, these equations along with initial and boundary conditions would enable one to describe the erosion-sedimentation system.

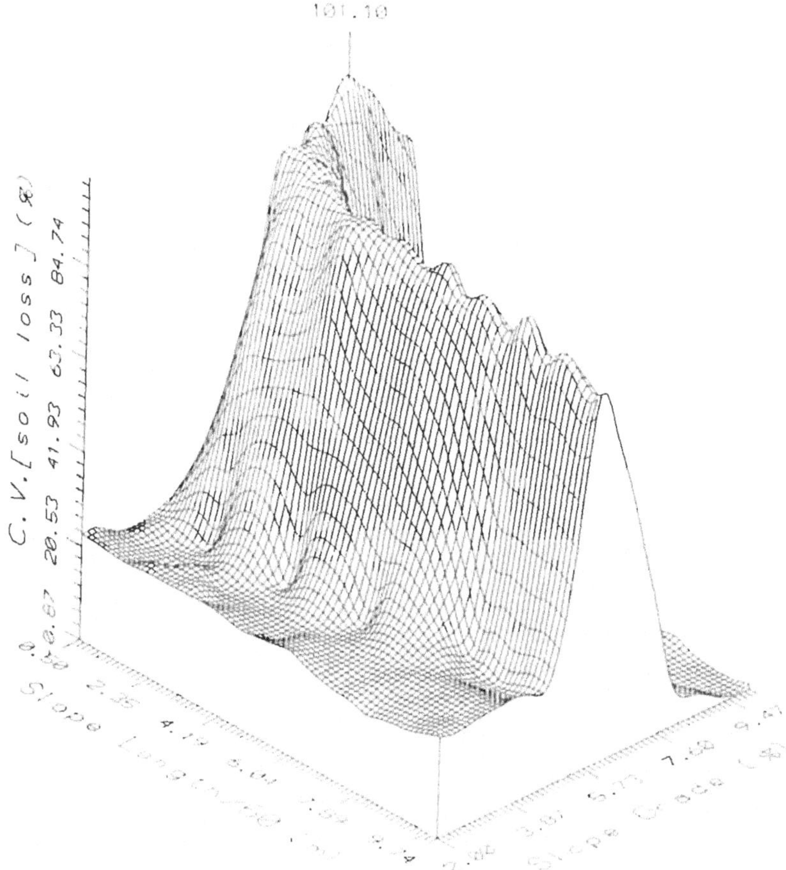

Figure 6.4 Coefficient of variation of WEPP erosion model response as a function of variation in soil erodibility for a Miami silt loam soil plotted over a range of slope lengths and gradients.

According to Woolhiser (*38*) there are four levels of abstractions in modeling of watershed systems: (1) watershed process models, (2) watershed component models, (3) integrated watershed models, and (4) global watershed models.

A watershed process model is a mathematical model describing an individual process in the watershed system, for example, mathematical models of infiltration, unsteady free surface flow, sediment detachment, deposition, etc.

Watershed component models include linked mathematical models of individual processes with a component operator for describing processes occurring in a particular subspace of the watershed system. Examples in this category of mathematical models include models of evapotranspiration, direct surface runoff, and sediment yield. A mathematical model of watershed sediment yield, for example, would include individual process models describing interception, infiltration, soil water content, plant growth, free surface flow, sediment detachment, sediment transport, and deposition. This chapter emphasizes the erosion and sediment yield component of the WEPP model, which is a component model.

An integrated watershed model is an example of a comprehensive watershed model. Often this category of watershed models is developed by a process of synthesis of components and has a well-defined structure that is usually determined by the modeler's concepts of the physical nature of the watershed.

Global watershed models are an alternative to integrated watershed models. Their structure is much simpler. They assume that there is a functional relationship between a set of input and output variables rather than a linkage of individual components.

Erosion models consisting of sets of partial differential equations can be simplified by dropping certain terms from the equations, leading, for example, to steady-state conditions (absence of time variable), plane conditions (absence of one space variable), or uniform conditions (absence of velocity and sediment concentration gradients in the flow direction).

Another simplification consists of reducing the number of independent variables by averaging the equations with respect to the variable to be removed. This procedure is quite common in environmental sciences where the extension and complexity of geometries generally do not allow a detailed and complete description of the processes.

As a further simplification the watershed system might be represented as a "lumped" system where runoff and sediment transport, for example, are assumed to be related to precipitation, without any explicit assumption regarding the internal structure of the runoff-sediment generation system.

The WEPP watershed model is made up of four major components: hillslope, channel, impoundment, and irrigation. The hillslope component is the WEPP hillslope model which calculates erosion and deposition on rill and interrill flow areas. The channel component calculates erosion and deposition within concentrated flow areas which can be represented as permanent channels or ephemeral gullies. The impoundment component calculates deposition of sediment within terrace impoundments and stock tanks. The irrigation component calculates erosion and deposition on border irrigation areas.

Channel flow hydraulics

As in the case of hydraulics of overland flow, the primary purpose of channel hydraulic calculations is to provide flow peak, duration, and shear stress for calculation of detachment, transport, and deposition of sediment within the channel network. The channel hydraulic calculations follow a similar logic to those of the overland flow hydraulics.

A model developed by Lane (*20*) may be used to estimate transmission losses in alluvial streams. This model was developed by regression analysis of observed streamflow data from 14 channel systems in Arizona and Texas. If the volume of inflow to the channel does not satisfy the infiltration capacity of the channel then there will be no flow out of the channel reach. However, if there is inflow at the top of a channel reach, then the model computes the distance that inflow will advance down the reach before stopping.

In the event that upstream inflow is sufficiently large, the flow may traverse the entire channel reach, but its volume and peak rate are reduced by transmission losses. In the WEPP watershed model, the method of estimating the peak discharge at the watershed outlet involves aggregating a watershed made up of a cascade of planes and channels into an equivalent plane and applying regression equations developed from solutions of the kinematic wave equation for a range of plane characteristics and lateral inflow distributions. The watershed is aggregated or reduced to a single plane (or two lateral flow planes and a single channel) by calculating the storage of water on the watershed at kinematic equilibrium and using the storage to calculate an equivalent roughness for routing. The regression equations are derived by obtaining solutions of the kinematic wave equation for a single plane (or two planes and a single channel) for a range of plane and channel characteristics.

Hydraulic shear stress is calculated in two ways. In the case of absence of lateral inflow from adjacent overland flow areas, the channel flow is assumed to be normal flow and the hydraulic shear stress is constant along the channel reach. However, for the case of lateral inflow or in steady flow, the hydraulic radius and friction slope vary with the distance along the channel reach. Because many of the applications of the WEPP model will be for cases where lateral inflow influences the channel discharge, the spatially varied flow equation is used to calculate the friction slope and the hydraulic radius at points along the channel reach.

The steady spatially varied flow equation (*5*) is solved in the WEPP watershed version by regression equations developed by Foster et al. (*8*). A control section is assumed at the channel outlet and can be described by a critical depth, a depth discharge relationship, or normal depth. Given a control section, the regression equations are applied at distance increments upstream from the control section assuming subcritical flow for the entire length of the reach.

Channel erosion and deposition

Channel erosion is based on a steady state sediment continuity equation similar to rill erosion in the overland flow profile component. The sediment load in the channel is a function of the incoming sediment load from adjacent overland flow areas and the ability of the flow to detach soil particles. The ability of the flow to detach depends on the force the flow exerts on the soil and transport capacity of the flow. Net soil detachment will occur when the shear stress of the flow exceeds the critical shear stress of the soil and when the sediment load is less than the transport capacity. Net soil deposition will occur when the sediment load is greater than the transport capacity.

For the channel computations, the channel is divided along its length into segments which are determined by changes in the channel characteristics (slope, surface roughness, etc.) or the contributing hillslope areas. The model computes the initial potential sediment load and the transport capacity for each segment. From this, it can be determined whether detachment, deposition, or both will occur within the segment.

The channel detachment routines assume a rectangular channel cross section for the permanent channels and ephemeral gullies. On ephemeral gullies, detachment initially occurs from the channel bottom until it reaches a nonerodible layer (usually the primary tillage depth). Once the channel reaches this boundary, the channel begins to widen and the erosion rate decreases with time until the flow is too shallow to cause detachment. Channel dimensions are updated after each storm that causes detachment in order to estimate channel hydraulics for subsequent storms.

Impoundment sedimentation

Impoundment terraces are designed to reduce sediment loss from croplands. They function by collecting and detaining runoff from an upslope area for a period of time to allow sediment deposition. The impoundment drains via an underground conduit which connects to the bottom of the impoundment so it drains completely between each storm.

The amount of sediment deposited in the impoundment terrace is a function of the amount of time available for settling. The terrace impoundment routines in WEPP estimate the fraction of each particle size class that leaves the impoundment as a function of the particle size distribution of the incoming sediment, the depth of water in the pond, and the diameter of the outlet pipe.

Reservoir impoundments are designed to collect and store surface runoff for later use. These include stock tanks and farm ponds. As with impoundment terraces, deposition is the main sedimentation process. Runoff from a reservoir impoundment is produced only when it is full and more runoff is introduced. When the pond becomes full, extra runoff is routed over spillways. Estimation of the discharge rate requires either a rating table for the spillway or the spillway dimensions.

Watershed representation

A watershed must be represented by at least one hillslope element. In the WEPP watershed model, hillslope elements can contain up to 10 overland flow sub-elements which may represent changes in cropping patterns (strip cropping), soil variation in the downslope direction, different land use patterns, or changes in grazing intensities. A hillslope element can drain into a channel either at the headwaters or laterally, or into an impoundment. A channel element can receive water and sediment input from hillslope elements, upstream channel elements (up to three channel elements), or an impoundment. An impoundment element can receive input from hillslope or channel elements.

Figure 6.5 illustrates how a typical watershed is represented in the WEPP model and how the hillslope, channel, and impoundment elements are linked together.

Integrated systems for water erosion prediction

Erosion prediction technology must be usable by technicians at the field level. To meet that objective the technology must encompass an integrated system of tools on three levels: database generation, user interface, and simulation models (Figure 6.6). Development on all three levels is a research function. The WEPP landscape profile version erosion model requires four input data files to execute: a soil file, a slope profile file, a crop management file, and a climate file. The user must have file building tools and access to appropriate soil, tillage implement, plant, and climate databases in order to build the four data files. One approach which should be investigated for both database development and user interface development aspects of the prediction technology is expert systems. Engel (7) developed an expert system to interface with an early, single-storm version

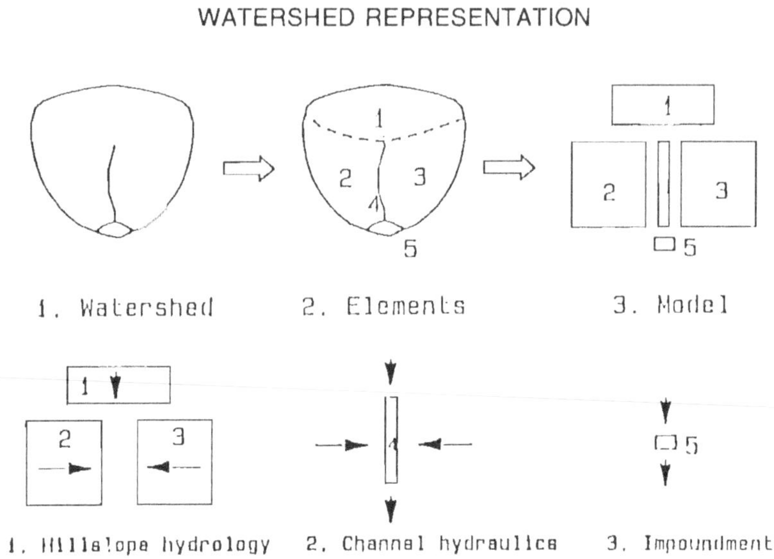

Figure 6.5 Schematic representations for a watershed erosion model.

of the WEPP technology, but such development has not continued. Expert systems are a logical choice to act as the interface between the user, the databases, and the simulation model in order that the user can provide the necessary input and obtain the desired output from the model.

The climate databases and the file building tools required to use the WEPP models within the U.S. are available and will be distributed with the computer model. Likewise, all of the soils information necessary to build the soils data files for within the U.S. will be available when the model is distributed to the user. Plant growth and residue decay parameters for the model are available for only a few crop types. An expert system or some related tool should be developed which is able to communicate with an agronomist who is knowledgeable about a specific crop. The expert system would then translate that knowledge from the agronomist's terminology to the crop parameters required in the crop growth component of the erosion prediction technology. The same approach could be used to build databases for new tillage implements, each of which will have a different effect on soil disturbance, random roughness, and burial of surface residue.

Ultimately, if the computer-based erosion prediction technology is to be usable worldwide, it must be an integrated system of tools. A potential

INTEGRATED SYSTEM FOR WATER EROSION PREDICTION
(I.S.W.E.P.)

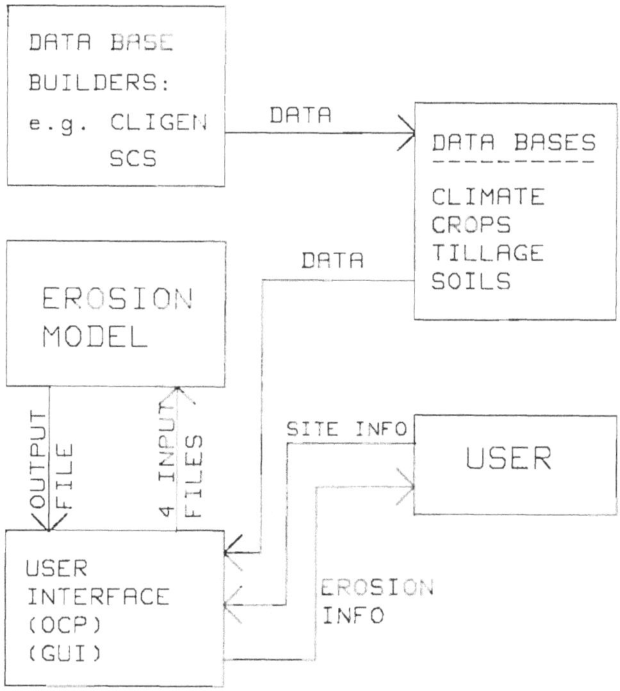

Figure 6.6 Outline for an integrated system of technology for predicting soil erosion by water: OCP and GUI are user interfaces currently under development by the Soil Conservation Service and Agricultural Research Service; CLIGEN is a stochastic weather generator.

user should have the tools to use regional information and expert knowledge, not necessarily in the form of collected data, to build climate, plant, tillage implement, and soils databases. The user interface should be flexible enough for the user to apply it in the new environment with the new databases for the region, or alternatively, technology must be available to readily adapt the interface to the new environment. Obviously, the process of database development and user interfacing must include a major research component, along with aspects of training and technology transfer.

The WEPP prediction technology can be used as an interactive tool for designing conservation systems. The output for the model provides specific information concerning how much soil loss is occurring at each point along the hillslope profile and the time distribution of soil loss. That information allows the user then to "experiment" with alternative management systems, based on the spatial and temporal soil loss distribution estimates, and quickly assess the impact of the proposed systems for the site-specific information. Changes in tillage dates, different tillage implements, new crop rotations, strip cropping, contour farming, buffer strips, terraces, and reduced tillage can all be evaluated for potential in controlling erosion and reducing offsite sediment delivery. Research is needed to provide guidelines and methods for using the technology as an interactive systems design tool for soil conservationists and project planners.

RESEARCH NEEDS FOR EROSION MODELING

Development and transfer of technology is vital to a scientific discipline. Without it there is ultimately no need for research. But technology transfer can also stifle creativity. Research scientists to some degree have their approaches molded by the current trends in scientific thought, which to some extent is good and necessary. But major advances must come from outside the current mold of scientific thought. For that reason it isn't possible to discuss in detail the next generation of soil erosion prediction technology. Some general thoughts for the future of soil erosion modeling are discussed below.

Deposition calculations are very important in terms estimating the sediment delivery rates and enrichment ratios from a slope profile. Good deposition relationships are critical to providing accurate predictions of off-site sediment problems, and much work is needed in the area of predicting deposition on complex slope profiles. This work may be classified into three general areas: 1) If a single effective fall velocity term is to be used to calculate net deposition, improved methods of calculating an effective fall velocity must be developed. 2) Reliable deposition data, particularly for non-uniform slopes in the field, is essentially non-existent. Collection of such data will require innovative techniques and careful experimental procedures. Exact profile descriptions will be essential to interpreting the data. Also, the rate of sediment delivery to the area of net deposition must be accurately measured as a function of time through the experiments. 3) More basic theoretical work needs to be performed to provide better estimates of transport and deposition rates for individual particle size classes. Both CREAMS and WEPP models allow for transport capacity to be shifted between particle size classes. Until more and better data are available, it is difficult to assess the validity of those procedures or to test alternatives.

Physically-based erosion technology represents a major advance in predicting soil movement on complex hillslope profiles. The next generation of technology should be able to represent complex landscape surfaces and the movement of sediment on those surfaces. Digital terrain models can be used to describe landscape surface elevations (*28*). Methods for calculating overland routing on the complex surfaces will need to be developed and then linked to process-based erosion equations (*27*). Data input for soil, topography, crop management, and climate could be accessed through a Geographical Information System.

A continued research effort in terms of technology transfer techniques is essential. Application of knowledge-based engineering to database development and user interfacing is only beginning and should come to fruition with the next generation of erosion prediction technology.

REFERENCES

1. Al-Durrah, M.M., and J.M. Bradford. 1982. *Parameters for describing soil detachment due to single waterdrop impact.* Soil Science Society of America Journal 46: 836-840.
2. Blau, J.B., D.A. Woolhiser, and L.J. Lane. 1988. *Identification of erosion model parameters.* Transactions, American Society of Agricultural Engineers 31: 839-845, 854.
3. Brenneman, L.G. 1988. *The effect of previous crops and time after tillage on rill and interrill soil erodibility.* Ph.D. dissertation. Iowa State University, Ames, Iowa. (Diss. Abstr. 88-26447).
4. Chaves, H.M.L. 1987. *Influence of incorporated crop residue on interrill erosion.* M.S. thesis. Purdue University, West Lafayette, Indiana.
5. Chow, V.T. 1959. *Open-channel hydraulics.* McGraw-Hill Book Co., New York, New York, 680 pp.
6. Ellison, W.D. 1944. *Studies of raindrop erosion.* Agricultural Engineering 25: 131-136, 181-182.
7. Engel, B.A. 1988. *Knowledge engineering in soil erosion.* Ph.D. dissertation. Purdue University, West Lafayette, Indiana (Diss. Abst. 88-74426).
8. Foster, G.R., L.J. Lane, J.D. Nowlin, J.M. Laflen, and R.A. Young. 1980. *A model to estimate sediment yield from field-sized areas: development of model.* In: W.G. Knisel (ed.) *CREAMS: A field-scale model for chemicals, runoff, and erosion from agricultural management systems.* Conservation Research Report No. 26. U.S. Department of Agriculture. 640 pp.
9. Foster, G.R. 1982. *Modeling the erosion process.* In: C.T. Haan, H.P. Johnson, and D.L. Brakensiek (eds.) *Hydrologic modeling of small watersheds.* ASAE Monograph No. 5, American Society of Agricultural Engineers, St. Joseph, Michigan. pp. 295-380.

10. Foster, G.R., L.J. Lane, M.A. Nearing, S.C. Finkner, and D.C. Flanagan. 1989. *Erosion Component.* In: L.J. Lane and M.A. Nearing (eds.) *Water Erosion Prediction Project Landscape Profile Model Documentation.* NSERL Report No. 2. National Soil Erosion Research Laboratory, Agricultural Research Service, U.S. Department of Agriculture, Purdue University, West Lafayette, Indiana.
11. Gilley, J.E., and S.C. Finkner. *Estimating soil detachment caused by raindrop impact.* Transactions, American Society of Agricultural Engineers 28: 140-146.
12. Gilley, J.E., D.A. Woolhiser, and D.B. McWhorter. 1985. *Interrill soil erosion - Part I: Development of model equations.* Transactions, American Society of Agricultural Engineers 28: 147-153.
13. Gilley, J.E., D.A. Woolhiser, and D.B. McWhorter. 1985. *Interrill soil erosion - Part II: Testing and use of model equations.* Transactions, American Society of Agricultural Engineers 28: 154-159.
14. Grass, A.J. 1971. *Structural features of turbulent flow over smooth and rough boundaries.* Journal of Fluid Mech. 50: 233-255.
15. Hairsine, P.B. 1988. *A physically based model of the erosion of cohesive soils.* Ph.D. dissertation. Griffith University, Brisbane, Australia.
16. Kinnell, P.I.A. 1988. *The influence of flow discharge on sediment concentrations in raindrop induced flow transport.* Australian Journal of Soil Research 26: 575-582.
17. Kirkby, M.J. 1980. *Modelling water erosion processes.* In: M.J. Kirkby and R.P.C. Morgan (eds.) *Soil Erosion.* John Wiley and Sons, Chichester, England. p. 183-216.
18. Laflen, J.M., A. Thomas, and R. Welch. 1987. *Cropland experiments for the WEPP project.* ASAE paper no. 87-2544, American Society of Agricultural Engineers, St. Joseph, Michigan.
19. Lane, L.J., H.H. Chang, W.L. Graf, E.H. Grissinger, H.P. Guy, W.R. Osterkamp, G. Parker, S.W. Trimble. 1982. *Relationships between morphology of small streams and sediment yield.* Journal of the Hydraulics Division, Proceedings, American Society of Civil Engineers 108: 1328-1365.
20. Lane, L.J., 1983. *Transmission losses.* In: *SCS National Engineering Handbook,* U.S. Department of Agriculture, Soil Conservation Service, Washington, D.C.
21. Lane, L.J., E.D. Shirley, and V.P. Singh. 1988. *Modelling erosion on hillslopes.* In: M.G. Anderson (ed.) *Modelling geomorphological systems.* John Wiley Publ., New York, New York. pp. 287-308.
22. Lane, L.J., and M.A. Nearing (eds.). 1989. *Water Erosion Prediction Project Landscape Profile Model Documentation.* NSERL Report No. 2. National Soil Erosion Research Laboratory, Agricultural Research Service, U.S. Department of Agriculture, Purdue University, West Lafayette, Indiana.
23. Lopes, V.L., and L.J. Lane. 1988. *Modelling sedimentation processes in small watersheds.* In: *Sediment Budgets, Proceedings, Porto Alegre Symposium, December 1988.* IAHS Publication No. 174. p. 497-508.

24. McCuen, R.H., and W.M. Snyder. 1983. *Hydrologic modeling: statistical methods and applications*. Prentice-Hall, Englewood Cliffs, New Jersey.
25. Meyer, L.D., and W.H. Wischmeier. 1969. *Mathematical simulation of the processes of erosion by water*. Transactions, American Society of Agricultural Engineers 12: 754-758, 762.
26. Meyer, L.D. 1981. *How rainfall intensity affects interrill erosion*. Transactions, American Society of Agricultural Engineers 24: 1472-1475.
27. Moore, I.D., and G.J. Burch. 1986. *Modelling erosion and deposition: topographic effects*. Transactions, American Society of Agricultural Engineers 29: 1624-1630.
28. Moore, I.D., E.M. O'Loughlin, and G.J. Burch. 1987. A *three-dimensional digital topographic model*. Earth Surfaces and Landforms.
29. Nearing, M.A., L.T. West, and L.C. Brown. 1988. *A consolidation model for estimating changes in rill erodibility*. Transactions, American Society of Agricultural Engineers 31: 696-700.
30. Nearing, M.A., D.I. Page, J.R. Simanton, and L.J. Lane. 1989. *Determining erodibility parameters from rangeland field data for a process-based erosion model*. Transactions, American Society of Agricultural Engineers 32: 919-924.
31. Nearing, M.A., G.R. Foster, L.J. Lane, and S.C. Finkner. 1989. *A process-based soil erosion model for USDA-Water Erosion Prediction Project technology*. Transactions, American Society of Agricultural Engineers (in-press).
32. Nearing, M.A., L.D. Ascough, and H.M.L. Chaves. 1989. *WEPP model sensitivity analysis*. In: L.J. Lane and M.A. Nearing (eds.) *Water Erosion Prediction Project Landscape Profile Model Documentation*. NSERL Report No. 2. National Soil Erosion Research Laboratory, Agricultural Research Service, U.S. Department of Agriculture, Purdue University, West Lafayette, Indiana.
33. Renfro, G.W. 1975. *Use of erosion equation and sediment delivery ratios for predicting sediment yields*. In: *Present and Prospective Technology for Predicting Sediment Yields and Sources*. Agricultural Research Service, U.S. Department of Agriculture. ARS-S-40. p. 33-45.
34. Rosenblueth, E. 1975. *Point estimates for probability moments*. Proceedings, National Academy of Science 72: 3812-3814.
35. Sharma, P.P. 1989. *Effect of soil strength on soil detachment by raindrops varying in kinetic energy and momentum*. Ph.D. Thesis, University of Minnesota, St. Paul.
36. Simanton, J.R., L.T. West, M.A. Weltz, and G.D. Wingate. 1987. *Rangeland experiments for water erosion prediction project*. American Society of Agricultural Engineers. Paper no. 87-2545.
37. Williams, J.R. 1975. *Sediment-yield predictions with universal equation using runoff energy factor*. In: *Present and Prospective Technology for Predicting Sediment Yields and Sources*. Agricultural Research Service, U.S. Department of Agriculture. ARS-S-40. p. 244-252.
38. Woolhiser, D.A., and D.L. Brakensiek. 1982. *Hydrologic modeling of small watersheds*. In: C.T. Haan, H.P. Johnson, and D.L. Brakensiek (eds.) *Hydro-*

logic modeling of small watersheds. ASAE Monograph No. 5, American Society of Agricultural Engineers, St. Joseph, Michigan. pp. 1-16.

39. Young, R.A., and J.L. Wiersma. 1973. *The role of rainfall impact in soil detachment and transport*. Water Resources Research 9: 1629-1639.

Chapter 7

RESEARCH PROGRESS on SOIL EROSION PROCESSES and a BASIS for SOIL CONSERVATION PRACTICES

C. W. Rose

Research objectives in all areas of inquiry develop with time. Early objectives commonly deal with mapping the extent to which the variable of interest (here, the rate or amount of soil erosion per unit area) depends upon the range of factors involved. The methodology developed in the United States that led to the universal soil loss equation (USLE) has met this objective successfully (at least for the U.S. Midwest). But this very success may have delayed development of further objectives in the manner that is common in most areas of research.

TYPES AND OBJECTIVES OF SOIL EROSION MODELS

Scientific investigation of agriculturally related questions often begins with a series of experiments in which each of the variables thought to be important is varied or allowed to vary over a significant range. Then, researchers use statistical models to investigate the data obtained, perhaps leading to a concise summary of the major apparent relationships. This sequence of experimental investigation followed by statistical analysis has occurred in the study of soil erosion carried out by the U.S. Soil Conservation Service. The statistical summary of data from field plot experiments in the U.S. Midwest is the USLE (22).

In USLE experiments, the variables of land slope and plot length could be chosen (within limits) and a range of soil types in the geographic region investigated. The experimental program was quite massive because on each soil type the researchers investigated the effect on soil loss of different degrees of cover as well as a suite of land management practices of interest at the time. The significance of environmental and hydrologic characteristics was recognized. Scientists hoped that rainfall characteristics would adequately cover both aspects—a hope not fully fulfilled with the advantages of hindsight.

The USLE summarizes this vast body of regionally derived data, thereby greatly increasing the usefulness of the data base from which it is derived. However, a summary of a data base, whether or not expressed as an equation as in this case, is just that. Thus, researchers increasingly have recognized that the USLE is not universal in its application, partly because it reproduces correlations between rainfall and runoff specific to the data set and partly because of limitations in the range of soil types. Any model based solely on collected data (like the USLE) is a captive of the extent of that data set.

There are more important and general considerations, however. There exists widening recognition that the objective and role of the USLE is not to test a representation of the processes involved in soil erosion. Processes are universal, even though the relative and absolute significance of different processes will vary, as will the particular outcome in any specific set of circumstances. Hence, there is a desire to develop models to represent the processes at work in soil erosion and deposition.

There are quite practical reasons for moving from the purely experimental/statistical approach of the USLE to a process-type approach. First, the USLE deals with "average annual soil loss," a useful concept in the climatic context in which it was developed. For much of the tropical, semitropical, and semiarid world, a far more satisfactory concept is that of a probability distribution of soil loss.

The second reason favoring a move to a process-oriented objective is that in many countries limitations on research resources make it impractical to derive such a probability distribution by direct measurement in all contexts of relevance, despite the historic ability of U.S. scientists to do that for agriculturally important soils in the U.S. Midwest.

BASIC APPROACH TO SEDIMENT EROSION AND TRANSPORT

For this discussion, let us restrict consideration to sediment flow on a sloping, planar land surface (Figure 7.1). Rates of flow per unit strip width

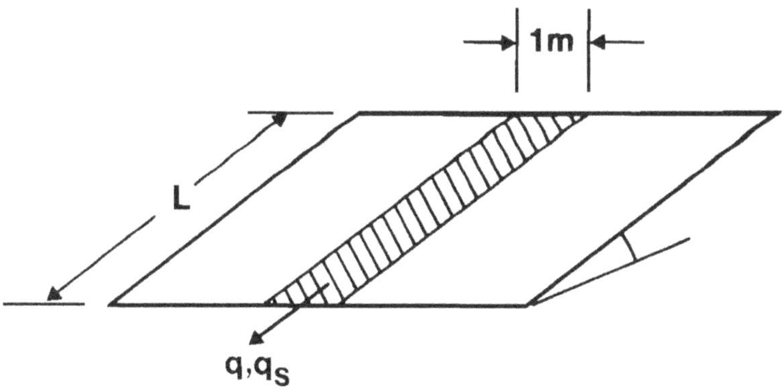

Figure 7.1 Flow of water (q) and sediment (q_s) from a unit strip width on a planar land element of length L.

of plane are called fluxes (Figure 7.1). Sediment flux (q_s) is the mass of sediment flowing per unit time across a unit width perpendicular to the direction of the flux. (This sediment mass is expressed on an oven-dry basis.) Likewise, the volumetric water flux (q) is the rate of volume flow of water per unit strip width (in $m^3\ m^{-1}\ s^{-1}$ or $m^2\ s^{-1}$).

The sediment concentration (c) is expressed as oven-dry mass of sediment per unit volume of suspension. (All symbols are listed in the addendum.) By definition of these terms, it follows that

$$q_s = qc \quad (\text{kg m}^{-1}\ \text{s}^{-1}) \qquad [1]$$

Soil loss from the land area during an erosion event is calculated by summing the time-variable flux q_s at exit from that area (Figure 7.1).

From Equation 1, it follows that a description of soil erosion processes involves description of the hydrology of surface flow (because of term q) and a description of the various erosion processes that add to sediment concentration, c, and deposition, the only process that tends to reduce c. The magnitude of c arises from the balance between these opposing processes. All of these quantities can vary with time and distance down the plane.

The approximate analytic model for overland flow

Rose and associates (*21*) developed the following analytical method for overland flow. Excess rainfall (R) for a land element is defined as follows:

$$R = P - I \quad (\text{m s}^{-1}) \qquad [2]$$

where P is the rainfall rate and I is the infiltration rate into the land surface (all being functions of time, t).

Let Q represent runoff per unit area. Then, from Figure 7.1

$$Q = q / L \quad (\text{m s}^{-1}) \qquad [3]$$

where q is the water flux at x = L, where x is the distance from the top of plane, where it is assumed that q = 0 (Figure 7.1).

If the land element is small, for example, 1 m^2, then the excess rainfall is quickly shed by overland flow from the element, in which case

$$Q = R \quad (\text{m s}^{-1}) \qquad [4]$$

However, if the plane length (L) is substantial and R is time variant (as it normally is), then changes in Q will lag behind those in R because of the time taken for water to gather on the soil surface and flow down the plane. Thus, in general, $R \neq Q$. Using the approximate analytic theory of Rose and associates (*10*), it may be shown that:

$$R \doteq Q + K_p (dQ / dt) \quad (\text{m s}^{-1}) \qquad [5]$$

where the term K_p depends analytically upon the length, slope, and roughness of the plane, on Q, and on how close the overland flow may be to laminar or turbulent.

For an assumed simple time variation in P, the approximate form of the corresponding relationship between R and Q given by Equation 5 is illustrated in Figure 7.2. Note that it follows from Equation 5 that R = Q when Q is a maximum (i.e., dQ/dt = 0).

In general, R cannot be measured. However, Q is measured readily, and Equation 5 allows R to be calculated from Q. With R known, I can be calculated using Equation 2 because P also is measured easily. Hence, infiltration characteristics can be derived allowing I to be estimated from measurements of P (*7*).

The flux q(x) at any x is given from this theory by

$$q(x) = Qx \quad (\text{m}^3 \text{ m}^{-1} \text{ s}^{-1}) \qquad [6]$$

Erosion/deposition process model

A full description of this model is given by Rose and associates (*17, 19, 20*) and Hairsine and associates (*5, 6, 7*). The model relates the sediment flux at any position on a plane and at any time in a runoff event to

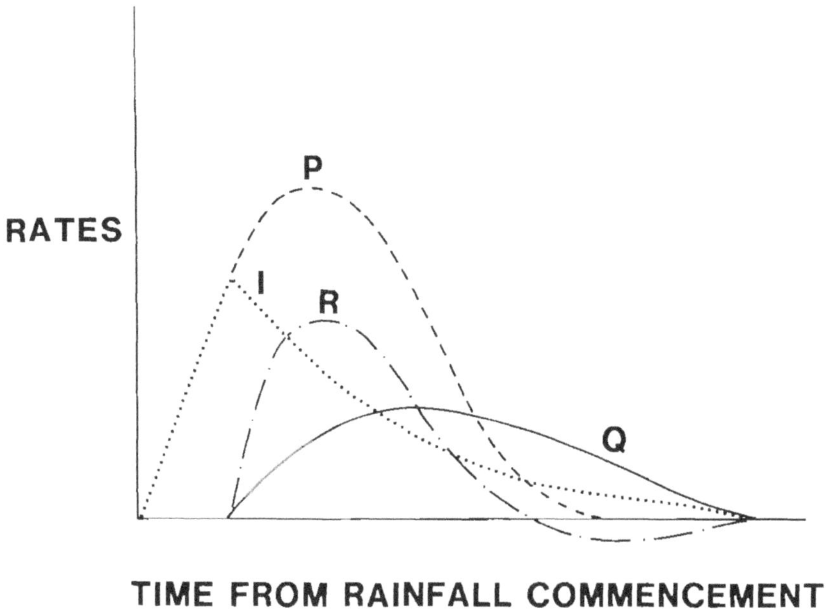

Figure 7.2 Simplified time-variation in rainfall rate (P), infiltration rate (I), rate of runoff per unit area of plane (Q), and the approximate analytic solution for the excess rainfall rate (R).

factors on which this sediment flux depends. The theory also has the capacity, suitably extended, to predict the rate and size distribution characteristics of sediment accumulation elsewhere on the landscape, given information on relevant surface geometry.

Erosion and deposition processes

Excluding landslides or gullies, the following three processes affect sediment concentration:

- *Rainfall detachment*, in which raindrops splash sediment from the soil surface into the water of overland flow.
- *Sediment deposition*, which is the result of sediment settling out under the action of gravity.
- *Entrainment of sediment*, the process whereby overland flow picks up sediment from the soil surface, whether in rills, between rills, or in sheet flow without rills.

Rate of sediment deposition, d. This rate depends upon sediment size distribution and is very rapid for sand and very slow for clay-sized aggregates or particles. Thus, d must be calculated as the sum of d_i, calculated separately for each sediment size class i with settling velocity v_i. It follows that

$$d_i = v_i \; c_i \quad (\text{kg m}^{-2} \text{ s}^{-1}) \qquad [7]$$

where c_i is sediment concentration in size class i.

Rate of rainfall detachment and re-detachment. The impact of raindrops on any unprotected soil surface can be a major cause of erosion due to the high local shear stresses produced. Erosion due to rainfall impact can dominate that due to overland flow in interrill regions of limited extent, or in shallow sheet flow of limited length over gentle slopes (*14*). While terminology in the literature on soil erosion is not standard, the removal of soil for the first time in any given erosion event will hereafter be referred to as "rainfall detachment."

All except very fine sediment detached into a surface water layer will return again to the soil surface because of deposition discussed in the previous section. This deposited sediment can also be removed by raindrop impact, a process called "rainfall re-detachment" by Hairsine and Rose (*5*). Under the same rainfall, re-detachment of previously detached and deposited sediment will occur at a much higher rate than detachment of the original soil if that soil is cohesive.

For simplicity, in what follows it will be assumed that a dynamic equilibrium situation has been achieved, with Hairsine and Rose (*5*) providing more detailed information on equation development and implications.

Some mass per unit area of deposited sediment (M_d) will completely shield the original soil from detachment by rainfall, though this required mass could depend on the depth of the water layer. If the actual mass present per unit area of the deposited layer (m_d) is less than M_d, the ratio will be called the fractional shielding (H) provided to the original soil by the layer, so that

$$H = m_d / M_d \qquad [8]$$

Re-detachment will take place from the deposited layer which shields fraction H of the soil surface. Denote the rate of rainfall re-detachment by e_d, expressed as the mass per unit area and time at which previously deposited sediment is re-detached. Proffitt, Rose, and Hairsine (*15*) have shown the rates of both rainfall detachment and re-detachment to be approximately proportional to rainfall rate (P). Thus

$$e_d = a_d HP \text{ (kg m}^{-2}\text{ s}^{-1}\text{)} \quad [9]$$

where a_d is the detachability of the deposited layer.

With fraction H of the original soil shielded from rainfall impact, detachment of the original soil can take place from the fractional area (1-H) at a rate given by

$$e = a(1-H)P \text{ (kg m}^{-2}\text{ s}^{-1}\text{)} \quad [10]$$

where a is the detachability to rainfall of the original soil matrix.

Rate of entrainment and re-entrainment by overland flow. The equal and oppositely directed shear stress to that which prevents acceleration of water flowing overland in a steady state removes soil in a process which may be called "entrainment" (sometimes called "runoff detachment," terminology not being standard). All but very fine entrained sediment will deposit on the soil surface and be removed once again, a process which will hereafter be referred to as "re-entrainment." The reason for distinguishing between entrainment and re-entrainment is the same as the distinction between rainfall detachment and re-detachment. This is that the work required to re-entrain deposited sediment is much less than that required to entrain cohesive soil.

The source of power available from the flow to either entrain or re-entrain sediment is the rate of working of the shear stress called the "stream power" (Ω) by Bagnold (*1*). Thus

$$\Omega = \tau V \text{ (Wm}^{-2}\text{)} \quad [11]$$

where τ is the shear stress, and V the velocity of flow. For sheet flow where the hydraulic radius R_h is equal to D, the depth of flow,

$$\Omega = \rho g S D V \text{(Wm}^{-2}\text{)} \quad [12]$$

where ρ is the density of water, g the acceleration due to gravity, and S the slope (the sine of the angle of slope).

In general some modest stream power (denoted Ω_0) needs to be exceeded before any sediment is moved by water flowing over it. The difference ($\Omega - \Omega_0$) is called the "excess stream power."

A model of the re-entrainment process and sediment concentration at the "transport limit."

Re-entrainment takes place from sediment deposited during the current erosion event, which shields a fraction H of the soil surface. Thus the excess stream power effective for re-entrainment is

$$HF(\Omega - \Omega_0)(Wm^{-2}) \qquad [13]$$

where F is the fraction of excess stream power assumed available to power entrainment or re-entrainment, found experimentally to be approximately 0.1 (*13*). The re-entrainment process is assumed to be non-selective with respect to size class of the sediment source, which, for re-entrainment, is the deposited layer.

Rose and Hairsine (*19*) and Hairsine and Rose (*6, 7*) have developed a model of the re-entrainment process on this basis making the following further assumptions:

- The effective excess stream power is used simply in lifting sediment against its immersed weight in water, hence implying that the cohesive strength of deposited sediment is negligible.
- Equilibrium or a steady state has been achieved.

The sediment concentration arising from these assumptions appears to be virtually equivalent to the concept of a "transport limit" concentration introduced by Foster (*3*).

As shown in the above references it follows that, for sheet flow, and neglecting Ω_0 in comparison to Ω, the maximum or transport limiting sediment concentration (c_t) will be

$$c_t = \frac{F\rho}{\Sigma v_i / I}\left[\frac{\sigma}{\sigma - \rho}\right] SV(kg\ m^{-3}) \qquad [14]$$

where ρ is the density of water, σ is sediment density, and $\Sigma v_i / I$, called the depositability (*20*), is obtained from the sediment's settling velocity characteristic (*11*). The form of Equation 14, (using particular values of variables) is illustrated by the upper curve in Figure 7.3, labeled $c = c_t$.

The product SV in Equation 14 is the drop in altitude experienced in 1 second by any parcel of fluid, and is thus the rate of decrease in potential energy of unit weight of water (for which mg = 1). Yang (*23*) called the product SV "unit stream power." Summarizing the results of a large number of flume experiments with non-cohesive sediments of limited size range, Yang (*23*) found sediment concentration to be closely proportional to SV, and Equation 14 is in good agreement with his data.

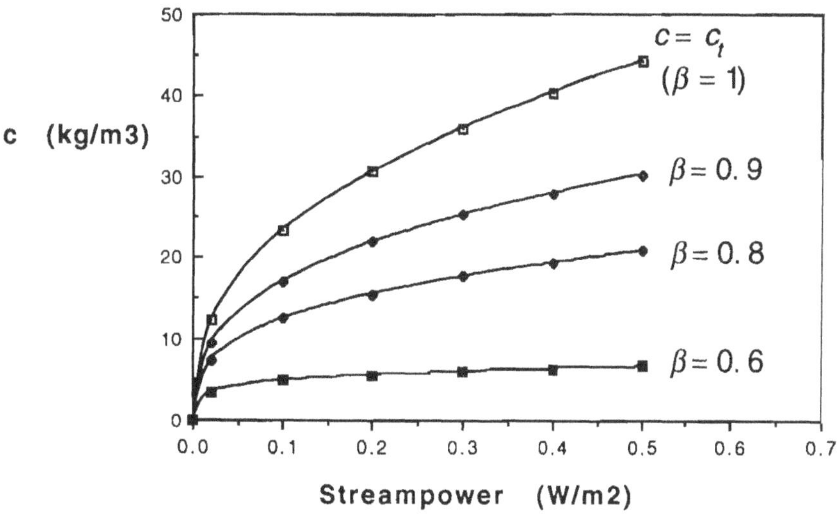

Figure 7.3 Relationship between c and c_t (upper curve) taking F=0.1 and assuming sheet flow as in Equation 14 with S=0.1, σ=2000 kg m^{-3}, and $\Sigma v_i / I$ = 0.1 m s^{-1}. Also shown is the relationship defined below in Equation 16 with β taking the values shown in the figure.

When erosion involves rectangular rills Hairsine and Rose (7) show that, again neglecting Ω_0 in comparison to Ω,

$$c_t = \frac{F\rho}{\Sigma v_i / I}\left[\frac{\sigma}{\sigma - \rho}\right]\frac{\Omega}{gD}\left(\frac{W_b}{W_b + 2D}\right) (\text{kg m}^{-3}) \qquad [15]$$

where W_b is the width of the rill. The reason for the geometrical term involving W_b is as follows. The shear stress acts uniformly around the wetted perimeter (W_b+2D), but re-entrainment will take place from the deposited layer located on the base of the rill of width W_b. Hence W_b/ (W_b+2D) is the proportion of the stream power (Ω) responsible for re-entrainment, which must be the dominant process if c_t is to be achieved.

APPROXIMATE EROSION THEORY AND ITS FIELD APPLICATION

Rose (17) has explained and provided justification for introducing an empirical erodibility parameter β which simplifies analysis of the soil's erodibility. Under some circumstances β is closely related to a more fundamental erodibility parameter defined as the energy per unit mass required to entrain soil (i.e. the specific energy of entrainment).

Erodibility parameter β is defined by

$$\bar{c} = \bar{c}_t^{\beta} \quad [16]$$

where $\bar{c}$ is the average sediment concentration during an erosion event, and $\bar{c}_t$ is the value of c_t calculated using an appropriate average rate of runoff per unit area. Figure 7.3 illustrates the form of relationship given in Equation 16 for selected values of β. It is apparent from Figure 7.3 that the dependence of c on Ω (and thus on slope and slope length) decreases as β decreases. Thus there is no completely general dependence of c (and thus soil loss) on S and L as assumed in the USLE, though the dependence predicted by the theory is in good agreement with the USLE if fairly high values of β are used ($\approx$ 0.9 or greater); such high values of β are consistent with the frequent tillage employed on the USLE experimental plots.

The appropriate average rate of runoff per unit area $\overline{Q}$ to use in the calculation of $\bar{c}_t$ is calculated from the rate of runoff per unit area (Q). Q can be measured as a function of time, for example using tipping bucket technology (*2*). It may be shown that the appropriate average value of Q to use in the calculation of is not the simple time-average, but rather $\overline{Q}$ calculated from Q according to the relation

$$\overline{Q} = \left[\frac{\Sigma Q^{1.4}}{\Sigma Q}\right]^{2.5} \quad [17]$$

The value of $\bar{c}_t$ for use in Equation 16 can be calculated taking the contribution of erosion by rainfall into account which involves adding a rainfall-driven term to Equation 14 or 15. However, because the contribution of rainfall-driven processes to the value of $\bar{c}_t$ is commonly small, and also recognizing the approximate value of the parameter β, it is usually adequate to calculate $\bar{c}_t$ assuming it is due to re-entrainment alone, in which case the value of β acknowledges any erosion by rainfall impact additional to that caused by overland flow.

Thus Equation 14 can be used if the flow is well approximated by sheet flow, Equation 15 is appropriate for rectangular rills, and equations for other rill shapes are given by Hairsine and Rose (*17*).

It follows from Equation 16 that β is conveniently calculated from

$$\beta = \frac{\text{In } \bar{c}}{\text{In } \bar{c}_t} \quad [18]$$

Figure 7.3 shows the relationship between c and Ω for particular values of β, using the same values for other parameters as were used in calculating c_t.

With $\bar{c}$ measured experimentally from runoff plots, and $\bar{c}_t$ calculated using the appropriate equation (Equation 14 or 15) from recorded runoff rate and observations on rilling (if it occurs), β can then be calculated from Equation 18. This methodology is illustrated by application to data from a field experiment at high slope in the tropics.

Unless erosion processes other than those considered are operating, $\bar{c}$ is not expected to exceed $\bar{c}_t$, and if so $\beta > 1$. For example in erosion at high slopes, finding that $\beta > 1$ could be an indication of mass movement.

Application of theory to field experiment

The theory is illustrated using data from an experiment in which soil loss and runoff rate was measured from instrumented plots at the Visayas State College of Agriculture (VISCA) at Baybay, Leyte, the Philippines (latitude 10° 44' N, longitude 124° 48' E at 30 m above sea level, mean annual rainfall 2,000-4,000 mm). Data is from plots of 10 percent slope kept bare of cover by weeding. The soil type was an Oxic Dystropepts, and plot length was 12 m. Most sediment collected at the bottom of the plots deposited in a low-slope concrete trough, and was later measured. Suspended sediment traveled with runoff water, whose flow rate was measured using tipping bucket technology (*2*) and electronic data loggers. Further details of the experimental arrangements are given by M. Escalante and A. Presbtero (personal communication), and methods of data analysis are described by Rose (*17*).

The settling velocity characteristic of the sediment was determined using the modified bottom withdrawal tube method of Lovell and Rose (*11*). From this data the depositability $\Sigma v_i / I$ was calculated using the program of Misra and Rose (*12*) to be 0.060 m s^{-1}.

Table 1 summarizes data for the four storm erosion events in which shallow rectangular rills of width ranging from 0.16 to 0.28 m were formed. On average, some 24 percent of rainfall received was measured as runoff. The effective runoff rate $\overline{Q}$ calculated using Equation 17 was much higher for the event of July 15–16, 1991, compared with other events in Table 1. This high value of $\overline{Q}$ is reflected in this event having the highest calculated mean water depth (D), measured sediment concentration (c), and c_t calculated using Equation 15.

Table 1 also gives the value of β calculated for the four events using Equation 18. The variation in β with time appears to reflect the management history of the plot. Prior to collection of data recorded in Table 1, the

Table 7.1 Data for the field experiment described in the text. SumP is the total rainfall, and Pmax the maximum rate of rainfall for each storm event. SumQ is equivalent depth of runoff. Qbar is the value of Q defined in Equation 17 from which the mean depth of flow (D) is calculated. Sediment concentration at the transport limit (c_t) was calculated using Equation 15, as rills were recorded. Beta (β) was calculated from Equation 18.

Storm No.	Date	SumP(mm)	Pmax(mm/h)	SumQ(mm)	Qbar(mm/h)	D(mm)	c(kg m^{-3})	c_t(kg m^{-3})	Beta
1	08-10/07/91	143.0	76.00	41.6	17.9	3.88	22.6	138.8	0.63
2	15-16/07/91	34.0	137.00	8.6	45.0	6.85	85.2	190.2	0.85
3	19-20/07/91	43.0	76.00	7.2	16.2	3.31	49.5	127.7	0.81
4	02-05/08/91	56.0	91.00	12.8	23.8	3.91	52.7	133.6	0.81

plot was weeded on May 14, 1991, some two months prior to the first recorded event on July 8-10, 1991, and a major rainfall event occurred between that initial weeding and the first recorded event, for which β = 0.63. This was the event with the largest volume of run-off (41.6 mm) and greatest soil loss (9.6 tonne ha^{-1}). Extensive weeding by hoe was carried out only a day prior to the second recorded event on July 15-16, 1991, and this yielded the highest value of β (0.85, Table 1). This event was quickly followed by two further storm events which yielded $\beta = 0.81$, a little lower than the maximum value obtained immediately following the vigorous weeding, which was, in effect, a cultivation.

This small data set illustrates the general experience that soil condition and strength as affected by management history has a strong influence on the erodibility of a given soil, as represented by the value of β. The erodibility parameter described and illustrated in this paper and in Rose (*17*) is related to the erodibility parameter η described in the first edition of this book, and in Rose (*16*). However, it is believed that β, while still empirical, has a better physical basis than η. Seeking relations between β and field soil characteristics such as soil strength measured using portable equipment is an active area of research.

PRINCIPLES OF CONSERVING SOIL FROM WATER EROSION

So far the erosion of soil either by the shear stress imposed on the soil by overland flow or by rainfall impact have been described. The major avenues for conserving soil in an agricultural context lie in providing the bare soil with some form of protection. However there is one commonly used soil conservation method which is applicable whether or not such protective cover is available. This method, used perhaps unconsciously in the small plots typical on tropical hillsides, is to limit the length of slope over which water can flow. This limits the water flow velocity, V in Equation 14, and thus also the sediment concentration and soil loss produced. A conscious use of this method of limiting the maximum flux of sediment achieved is in the construction of graded terraces (*8*) or fanya-ju terraces (*9*) in which downslope overland flow is collected and usually carried via a gently sloping channel across slope for safe disposal, either in a grassed waterway or in some other way. Much of the soil eroded between these structures will be deposited in the low-slope cross channel, which will therefore require sediment removal on occasion to enable the channel to fulfill its function. It may be argued that such structures as graded terraces do not reduce soil loss from land between the structures, but by limiting the maximum sediment flux, they help avoid the development of large gullies which are difficult and expensive to repair or stabilize, and possibly also help avoid the need to re-plant if erosion occurs at a vulnerable time in the cropping cycle.

Equations 14 and 15 predict the sediment concentration at the transport limit which is the "worst event" scenario for sediment concentration. These equations can be used to aid in decision making on terrace spacing; for example by providing a rational basis for varying terrace spacing for lands of different slope. Designing for a "tolerable" soil loss also requires information on surface hydrology.

Effect of surface contact cover and slope on soil loss

In addition to slope length, land slope (S), and the fraction (C_r) of soil surface not protected by cover in direct contact with it are important factors affecting soil loss. Trade-offs between S and C_r exist in practice at the farm level if the objective of limiting soil loss to some tolerable rate is to be achieved. The tolerable rate of soil loss often is called a T value. If a T value represents a soil loss rate that will not lead to soil deterioration and/or production loss in the long term, it is bound to be quite variable, depending, for example, upon soil depth and all the factors affecting the rate of soil formation. Perhaps partly because of the experimental difficulty in determining T values there is still argument about the utility of the concept. However, the concept is used below to illustrate the trade-off between the maximum slope that should be cultivated and the level of cover $(1 - C_r)$ that can be maintained if the soil loss rate is to be restricted to the T value. It should be noted that only protective material, such as stubble mulch, in contact with the soil surface and protecting it from entrainment is considered as contributing to $(1 - C_r)$.

It is important to distinguish between the above-ground or "aerial" cover which typically dominates the view from above, and the ground-hugging cover in intimate contact with the ground surface which is commonly provided by vegetative material but can be provided by stone, gravel, or human-applied protective materials. This distinction is important because aerial cover, despite being so visually obvious, can provide protection only against the impact of raindrops. Since by its definition aerial cover is above the soil surface, it provides no protection at all against erosion due to water flowing over the soil surface, which is typically the major cause of erosion. Trees intercepting rainfall can shed water as large drops which, falling from a height of several meters, can be just as erosive as raindrops. Thus the erosion protection commonly associated with a forest is due, not so much to its aerial foliage, but to the litter layer found on the forest floor, which through decay processes is intimately attached to the soil beneath it. Cover in such close contact with the soil surface can be called "surface contact cover." Let C_r denote the fraction of soil surface not protected by surface contact cover, so that the fraction protected is $(1-C_r)$.

To reduce erosion due to overland flow, surface contact cover is required to impede it and bear a substantial fraction of the shear stress exerted by such flowing water on the surface over which it flows. Thus surface contact cover consists of any material sufficiently close to the soil surface to impede overland flow, and not be carried away by it. Surface contact cover also reduces the likelihood of rill formation. Suppression of rills can very substantially reduce the sediment concentration produced by overland flow.

There have been many studies on the effectiveness of cover in reducing soil loss or sediment concentration using runoff plots. While the distinction between surface contact and aerial cover is not always clear in the reports of such studies, the general form of the relationship showing the effectiveness of surface contact cover in reducing sediment concentration (c) relative to that from a bare plot (c_0) is as illustrated in Figure 7.4. There is some variety in this general form, and Figure 7.4 illustrates a particular data set from Rose and Freebairn (*18*) as well as a series of theoretical curves which illustrate approximately the range of experimental data. The farm-based experiments from which the data of Rose and Freebairn (*18*) come are described in Freebairn and Boughton (*4*). As is common in field-based experimental data there was some scatter about the relationship shown in Figure 7.4. This experimental relationship is not exactly described by any single member of the family of exponential curves in Figure 7.4, but is contained within them.

The form of the exponential relationships in Figure 7.4 can be expressed as

$$c/c_0 = \exp[-k(1-C_r)] \qquad [19]$$

where the rate constant k is derived by fitting the curve to experimental data.

The substantial non-linearity in this relationship and in the experimental data is good news in that only limited fractional cover causes a much greater proportional reduction in sediment concentration and thus in soil loss than if the relationship were linear. A more linear relationship than that shown in Figure 7.4 is possible if sediment concentration is due to rainfall impact alone, though the cause of variation in the type of relationship in Figure 7.4 is worthy of more focused study. The very rapid drop in c/c_0 for small levels of surface contact cover in some cases may be due to the suppression of rilling referred to above.

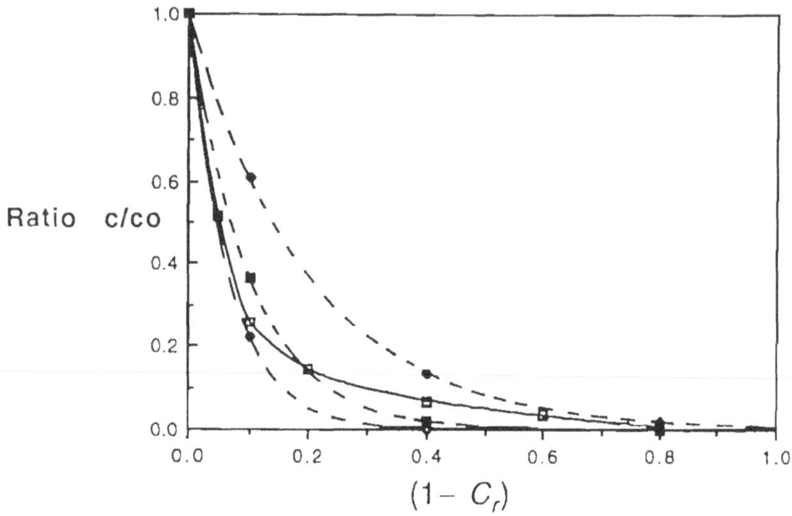

Figure 7.4 The solid curve describes the relationship between the ratio of sediment concentration (c) to that from a bare field (c_0) and the surface contact cover fraction (1-C_r) given in Rose and Freebairn (*18*). The dashed curves are exponential relationships of the type defined in Equation 19 with the constants k equal to 5 (the uppermost curve), 10, and 15 (the lower curve). This set of exponential curves cover most of the experimental data on this relationship given in the literature.

Maintaining an adequately high level of surface contact cover may be all that is necessary to provide an adequate level of protection against soil erosion, since in addition to the protection described earlier, a high level of cover also maintains high rates of infiltration, and therefore less runoff.

It may not be possible at all times to maintain adequate levels of surface contact cover over entire cultivated fields. A supportive technique of soil conservation is then to provide a strip on the contour with a high level of contact cover provided, for example, by a strip of grass, or a densely-planted leguminous hedgerow. A common form of this technique is strip cropping where a strip with crop or crop residue may possibly be assisted in its soil conserving function by reduced tillage and cloddy tilth. Scientific interest has recently been revived in the old conservation practice of using a narrow strip of dense stiff grass (*10*).

From Equation 14 it follows that $c_t \propto SV$, and from Manning's equation $V = S^{0.5}D^{0.667}/n$ (for sheet flow); it then follows that $c_t \propto S^{1.5}$.

Thus from Equation 16 it follows that, if other factors in Equation 14 remain constant, c is approximately proportional to S if $\beta = 1/1.5 = 2/3 = 0.667$. This value of β is within the range given in Table 1 for example, and as β is commonly less than unity the assumption that c is proportional to S may not be greatly in error.

In order to illustrate an approximate but very simple and practical method of presentation of data in a form suitable for farm extension activities, the approximation that $c \propto S$ will be accepted with the justification given in the above paragraph. The data presented is from an experiment on a farm in the Darling Downs of eastern Queensland, Australia, where land slope varied between 5 and 7 percent for different cultivation bays bounded by graded terraces about 50 m apart. The soil was a high clay Pellustert, and the experiments are described by Freebairn and Boughton (*4*). Soil loss in these experimentscould be of the order of 200 tonnes ha^{-1} in a single extreme runoff event on bare soil. The exposure fraction C_r was varied by the use of surface-applied stubble from previous crops.

For this specific site in the Darling Downs, Queensland, D. M. Freebairn (Queensland Department of Primary Industries, personal communication) has obtained the following relationship for average annual relationships between runoff and cover:

$$\int_0^{t_r} (q/L)dt = 59 - 29(1 - C_r) \text{ (mm } y^{-1}) \qquad [20]$$

where t_r is average anual duration of runoff.

Substituting the suite of values shown in Figure 7.4 for the surface cover fraction $(1 - C_r)$ in Equation 20 yields a corresponding suite of values for average annual runoff. Multiplying these by the corresponding value of $\bar{c}$ and using Figure 7.4 gives the average annual soil loss per unit area (Σ q_s/L) expected at this site for any combination of fractional cover $(1 - C_r)$ and land slope S. Such calculations have been restricted to S = 0.1 (or 10 percent) because somewhere beyond this limit landslides or other mechanisms involving gravity that are ignored in this theory may become important. The results of these calculations are given in Figure 7.5.

Figure 7.5 also shows a T value of 1 kg m^{-2} y^{-1} (10 t ha^{-1} y^{-1}), which has been used in some situations. Simply accepting this T value as a desirable upper limit to Σ q_s, then the trade-off can be explored between the maximum land slope that should be cultivated and the fractional cover that can be maintained. Consulting the intersection of the adopted T-value line with the relations in Figure 7.5 we can see that $\Sigma q_s/L < 1$ kg m^{-2} y^{-1} can be achieved for the following illustrative combination of values:

Fractional cover $(1 - C_r)$:	0	0.1	0.2	0.3
Maximum slope S for cultivation (percent):	1	3	5	7

The fractional cover that can be maintained depends upon many crop and management factors. In mechanized agriculture, cover can be maintained at much higher values if suitable stubble-handling machinery is available that minimizes the burial of stubble from the previous crop. Less in-

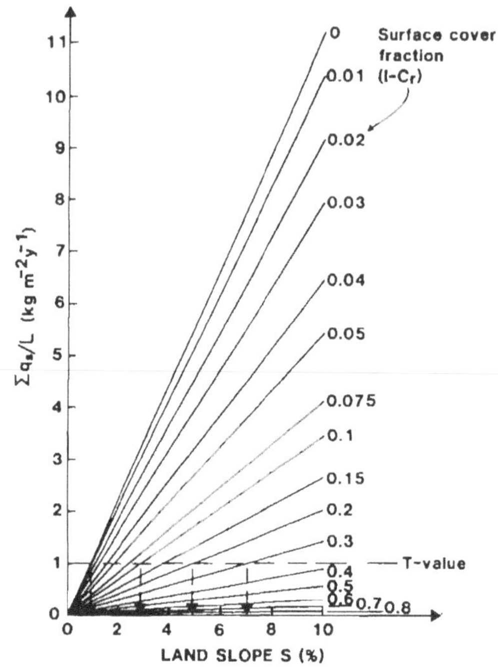

Figure 7.5 For a particular location in the Darling Downs, Queensland, the average annual soil loss ($\sum q_s/L$) varies with land slope S and the fraction ($1\text{-}C_r$) of the soil covered by mulch or other effective contact cover. The T value is a tolerance level corresponding to 1 kg m^{-2} y^{-1} (10 t ha^{-1} y^{-1}). Arrows indicate various trade-offs between maximum cultivated slope and cover if soil loss is not to exceed the T value.

tensively mechanized agriculture would appear to be generally compatible with maintaining a high surface cover by stubble. In suitable climates, intercropping, for example, with a shrub legume, also can be an effective soil-conserving practice.

SOME CONCLUSIONS

Experience in attempting to define erodibility as a soil characteristic has been frustrating. Such frustration has probably arisen for two quite different reasons. First, it is clear that however adequately soil erodibility is defined, it varies for any particular soil in response to management, variation in weather, biological activity, and interactions between these factors. Second, it is not a simple task to establish a well-justified methodology for defining and hence measuring erodibility. The variety of erodibility indices in the literature, such as the K-factor of the USLE or the erodibility parameter β discussed in this chapter are testimony to that fact.

One of the motivations in moving from the type of erosion model that summarizes a large experimental data base toward models which seek to represent the processes at work in erosion and deposition is to provide a more physically-based measure of erodibility. It is believed that such improvement in parameter definition will assist in moving away from the current complete dependence on runoff plots for the measurement of soil erodibility to a more mobile, less-expensive, more efficient methodology. Such a replacement methodology could be approximate and still be useful, since there are also comparable difficulties in predicting the infiltration or runoff-producing characteristics of field soils under rainfall. As Equation 1 reminds us, soil erosion depends on both these characteristics.

REFERENCES

1. Bagnold, R.A. 1977. *Bedload transport by natural rivers.* Water Resources Res. 13:303-311.
2. Bonell, M., and J. Williams. 1986. *The two parameters of the Philip infiltration equation: their properties and spatial and temporal heterogeneity in a red earth of tropical semi-arid Queensland.* Journal of Hydrology 87: 9-31.
3. Foster, G.R. 1982. *Modelling the erosion process.* In: C.T. Hann (ed.) Hydrologic Modelling of Small Watersheds. Am. Soc. Agr. Eng. Monogr. No. 5, 297-379, St. Joseph, Michigan.
4. Freebairn, D.M., and W.C. Boughton. 1981. *Surface runoff experiments on the Eastern Darling Downs.* Australian Journal of Soil Research 19: 133-146.
5. Hairsine, P.B., and C.W. Rose. 1991. *Rainfall detachment and deposition: Sediment transport in the absence of flow-driven processes.* Soil Science Society of America Journal 55: 320-324.
6. Hairsine, P.B., and C.W. Rose. 1992. *Modeling water erosion due to overland flow using physical principles I. Sheet flow.* Water Resources Research 28: 237-243.
7. Hairsine, P.B., and C.W. Rose. 1992. *Modeling water erosion due to overland flow using physical principles I. Rill flow.* Water Resources Research 28: 245-250.
8. Hudson, N. 1973. *Soil Conservation.* B.T. Batsford Ltd. London.
9. Hurni, H. 1986. *Soil Conservation in Ethiopia.* Community Forests and Soil Conservation Development Department, Ministry of Agriculture, Ethiopia.
10. Kemper, D., S. Dabney, L. Kramer, D. Dominick, and T. Kreep. 1992. *Hedging against erosion.* Journal of Soil and Water Conservation 47: 284-288.
11. Lovell, C.J., and C.W. Rose. 1988. *Measurement of soil aggregate settling velocities I. A modified bottom withdrawal tube method.* Australian Journal of Soil Research 26: 55-71.

12. Misra, R.K., and C.W. Rose. 1990. *GUDPRO (Version 1.2) Manual for IBM-PC users.* Division of Australian Environmental Studies Report, Griffith University, Brisbane, Australia, 4111.
13. Proffitt, A.P.B., P.B. Hairsine, and C.W. Rose. 1993. *Modelling soil erosion by overland flow: Application over a range of hydraulic conditions.* Transactions of the American Society of Agricultural Engineers Vol. 36 (In press).
14. Proffitt, A.P.B., and C.W. Rose. 1991. *Soil erosion processes I. The relative importance of rainfall detachment and runoff entrainment.* Australian Journal of Soil Research 29: 671-683.
15. Proffitt, A.P.B., C.W. Rose, and P.B. Hairsine. 1991. *Rainfall detachment and deposition: Experiments with low slopes and significant water depths.* Soil Science Society of America Journal 55: 325-332.
16. Rose, C.W. 1985. *Developments in soil erosion and deposition models.* Advances in Soil Science 2: 1-63. Springer-Verlag, New York.
17. Rose, C.W. 1993. *Erosion and sedimentation.* In: M. Bonnell, M.M. Hufschmidt, and J. S. Gladwell (eds.) *Hydrology and water management in the humid tropics—Hydrological research issues and strategies for water management.* Cambridge University Press, Cambridge, UK. pp. 301-343.
18. Rose, C.W., and D.M. Freebairn. 1985. *A new mathematical model of soil erosion and deposition processes with applications to field data.* In: S.A. El-Swaify, W.C. Moldenhauer, and A. Lo (eds.) *Soil Erosion and Conservation,* Soil and Water Conservation Society, Ankeny, Iowa. pp. 549-557.
19. Rose, C.W., and P.B. Hairsine. 1988. *Process of water erosion.* In W.L. Steffen and O.T. Denmead (eds.) *Flow and Transport in the Natural Environment: Advances and Applications.* Springer Verlag, Berlin. pp. 312-326.
20. Rose, C.W., P.B. Hairsine, A.P.B. Proffitt, and R.K. Misra. 1990. *Interpreting the role of soil strength in erosion processes.* Catena Supplement 17: 153-165.
21. Rose, C.W., J.Y. Parlange, G.C. Sander, S.Y. Campbell, and D.A. Barry. 1983. *Kinematic approximation to runoff on a plane: An approximate analytical solution.* Journal of Hydrology 62: 363-369.
22. Wishmeier, W.H., and D.D. Smith. 1978. *Predicting rainfall erosion losses—a guide to conservation planning.* Agricultural Handbook No. 537. U.S. Department of Agriculture, Washington, D.C.
23. Yang, C.T. 1972. *Unit stream power and sediment transport.* Journal of the Hydraulics Division of the American Society of Civil Engineers 78: (HY10), 1805-1826.

Chapter 8

ERODIBILITY and EROSIVITY

R. Lal and W. Elliot

Soil erodibility and rainfall erosivity are two important physical factors that affect the magnitude of soil erosion. Erodibility, a soil characteristic, is a measure of the soil's susceptibility to detachment and transport by the agents of erosion. Erosivity is an expression of the ability of erosive agents, like water, to cause soil detachment and its transport. Quantification of these two factors is basic to an understanding of soil erosion processes.

SOIL EROSION PROCESSES

The predominant processes that determine erosion are infiltration, run-off, detachment, and transport by raindrops and overland flow (interrill erosion), detachment and transport by concentrated flow (rill erosion), and deposition. Figure 8.1 shows the results of raindrop detachment in an interrill area. Where stones or erosion-resistant clods protect the more erodible soils, pedestals of uneroded soil have formed. Figure 8.2 shows an example of rill erosion where concentrated flow has incised significant rills in a rowcrop.

SOIL ERODIBILITY

The response of a soil to the erosion processes is complex, and is influenced by soil properties such as texture, structural stability, organic matter content, clay mineralogy, and chemical constituents. Some of these properties, like organic matter, can be altered over time by land use, management practices, and farming systems. Erosion of the surface layers can expose less erodible subsoils which may have different properties than the surface. Consequently, the erodibility of a soil can change with time.

Figure 8.1 Interrill erosion is occurring on the ridges, leaving pedestals of uneroded soil beneath small stones and large, unerodible clods.

The soil texture is important in determining erodibility. Sandy soils have lower run-off rates, and are more easily detached, but less easily transported than silt soils. Clay soils are not easily detached, but lower infiltration rates may lead to greater run-off and increased erosion. Silt soils tend to have the greatest erodibilities since particles are easily detached and transported, and consolidation of subsoils, or subsoils with higher clay contents, can lead to greater run-off. The greatest erosion is often associated with high silt loess soils, as found in China's Yellow River watershed.

Erodibility and the universal soil loss equation (USLE)

The USLE is useful for determining the adequacy of conservation measures in farm planning, and for predicting non-point sediment losses in pollution control programs. Despite its simplification of the many variables involved, the USLE continues to be the most widely used method of estimating sediment loss.

The average annual soil loss, as determined by Wischmeier and Smith (*76*) can be estimated from the equation

Figure 8.2 Rill erosion following rows of a crop planted up and down the slope.

$$A = R\ K\ L\ S\ C\ P \quad [1]$$

where A is the average annual soil loss in Mg/ha; R is the rainfall and run-off erosivity index for geographic location; K is the soil erodibility factor; L is the slope length factor; S is the slope steepness factor; C is the cover management factor; and P is the supporting practice factor.

The soil erodibility factor, K, in the USLE is the soil loss from a unit plot per erosive index (EI) unit. A unit plot is defined as a 22-m length of uniform 9 percent slope, continuously clean-tilled up and down the slope, and maintained absolutely free of vegetative cover. There are several different methods for measuring the erodibility factor:

Natural run-off plots. Preferably, researchers should measure soil erodibility under field conditions. Chapter 2 describes the methodology and equipment used for plot establishment and for measurement of run-off and erosion. Figure 8.3 shows a natural run-off plot in the bare-tilled condition.

Practices for managing plots to measure erodibility under natural field conditions include: (a) plowing up and down the slope to normal depth, followed by disking or harrowing until a smooth seedbed is achieved; (b)

Figure 8.3 A natural bare-tilled run-off plot located in eastern Washington. Note the rills incised in the highly erodible silt soil.

plowing operations performed at the normal seeding time for the major crops of the region; (c) all other operations involving surface soil disturbance, for example, cultivation to control weeds and eliminate crust formation, performed on schedule; and (d) if necessary, herbicides used to control weeds.

Plots in tropical climates require more frequent plowing and farm operations for weeding and elimination of surface crust. Plots should be plowed once for each growing season in regions where bimodal rainfall distribution enables two crops per year. Smooth the plot surface soil if rills are progressively deteriorating into gullies. Researchers should run these plots for 10 to 20 years because knowledge of long-term trends or alterations in soil erodibility is important for land use planning and for designing conservation measures. It may be necessary to select a new site after three or four years if accelerated erosion alters the surface soil properties and microrelief.

The topographic factors, L (*76*) and S (*56*), adjust the predicted erosion rates to give greater erosion rates on longer, steeper slopes, when compared to a unit plot. The differences are attributed to increasing rill erosion rates as more run-off accumulates with longer slopes, and greater erosive forces occur with steeper gradients. These factors can be calculated from the equations

$$L = \left[\frac{l}{22}\right]^m \qquad [2]$$

where m is 0.2 for s < 1 percent, 0.3 for 1 percent < s < 3 percent, 0.4 for 3.5 percent < s < 4.5 percent and 0.5 for s > 5 percent; l is the slope length, m; and s is the field slope, percent;

And, for slopes less than 4 m long

$$S = 3.0\,(\sin(\theta))^{0.8} + 0.56 \qquad [3a]$$

For slopes greater than 4 m long, and s < 9 percent

$$S = 10.8\,\sin(\theta) + 0.03 \qquad [3b]$$

For slopes greater than 4 m long, and s ≥ 9 percent

$$S = 16.8\,\sin(\theta) - 0.50 \qquad [3c]$$

where θ is the field slope (= $\tan^{-1}(s/100)$).

Deviation from unit plot dimensions. It is not uncommon to establish run-off plots for measuring soil erodibility on slopes other than nine percent, and with varying plot sizes to suit local relief characteristics and specific budgetary constraints. Under these circumstances, researchers must adjust soil loss to standard conditions of the unit plot. The topographic factors, L and S, are used to correct the factor K by dividing the calculated K value for the plot by the plot's topographic factors. The L and S factors for specific combinations of slope length and gradient may be calculated by Equations 2 and 3. These equations assume uniform slope gradients and do not apply to irregular slopes, such as concave, convex, or complex slope aspects. Irregular slopes are subdivided for computing S factors for the plot.

Use of rainfall simulation on run-off plots. Establishing and maintaining field run-off plots for a minimum period of three to five years are capital-intensive and time-consuming operations. Plots can be set up for only a limited number of soils. Soil erodibility data are, however, needed for a vast number of soils of varying physical, chemical, and mineralogical constituents, and in diverse relief and climatic environments. It is also important to relate field measurements of erodibility using run-off plots to soil properties. Rainfall simulation can facilitate and expedite data procurement in a relatively short time. Chapter 4 describes different types of rainfall simulators for field use. To simulate the effect of additional slope length, clean water is added at the top of the plots, but it should be remembered that clean water is more erosive than sediment-laden water. The K value for a given plot is calculated by dividing the observed erosion rate by the erosivity of the simulated storm and the plot L and S factors.

Researchers also use rainfall simulators under laboratory conditions to evaluate relative erodibility (*18*). Vanelslande et al. (*69*), however, reported significant differences in laboratory-determined values and those measured in the field. Antecedent soil water content, infiltration, and soil matric potential complicate the direct relation between laboratory-measured indices and field behavior. Also important are magnitude and nature of overland flow and the network of rill systems that develop under field conditions. In spite of these limitations, a relationship between laboratory measurements and field responses under natural rainfall conditions can be established (*1*). Laboratory conditions are better suited to studying individual rill and interrill processes than establishing soil erodibility values.

Estimating K from soil properties. The soil-erodibility factor, K, for a series of bench-mark soils was obtained by direct soil loss measurements from fallow plots located in many U.S. states (*76*). An estimate for an unknown K can be calculated from the regression equation

$$K = 2.8 \cdot 10^{-7} M^{1.14} (12 - a) + 4.3 \cdot 10^{-3} (b - 2) + 3.3 \cdot 10^{-3} (c - 3) \quad [4]$$

where M is the particle size parameter = (Percent Silt + Percent Very Fine Sand) · (100 - Percent Clay); a is the percent organic matter; b is the soil-structure code (very fine granular, 1; fine granular, 2; medium or coarse granular, 3; blocky, platy, or massive, 4); c is the profile-permeability class (rapid, 1; moderate to rapid, 2; moderate, 3; slow to moderate, 4; slow, 5; very slow, 6).

On 17 weathered soils located in the southeastern U.S., Barnett and Rogers (*5*), using rainfall simulation, found that observed erosion rates per EI depended on the contents of sand fractions, soil water and field capacity, depth of the A horizon, rock fragments, M (Equation 4), and clay fraction. Bennet (*7*) emphasized the importance of considering erodibility differences due to clay mineralogy as experienced on the more weathered soils of this geographic region.

Measuring other USLE factors. The cover-management factor, C, includes the effects of cover, crop sequence, productivity level, length of growing season, tillage practices, residue management, and the expected time distribution of erosion rainstorms. The value for the cropping and residue variable as a ratio of the soil loss for crops to that for continuous fallow varies from 0.01 for permanent pasture to one for fallow. Rowcrop C factors range from 0.2 to 0.7. Field plots with management practices other than fallow, and rainfall simulation on field plots are used to find C factors.

The supporting practice, P, ranges from one for no practices to 0.5 for contour stripcropping. Terraces reduce the L and S factors as well as reduce the amount of sediment leaving a terrace by up to 95 percent. P factors are difficult to measure in the field as only gross delivery from such fields can be measured. The level of erosion and deposition within the field is difficult to quantify.

Interrill erosion

Interrill erosion is the detachment and transport of soil by raindrop impact and shallow overland flow. Figure 8.1 shows large clods and pedestals of uneroded soil columns on the ridge area, where surrounding, more erodible soil has been eroded by raindrop action. Rainfall intensity, soil properties, and topography determine interrill erosion. It is estimated by the equation (*72*)

$$D_i = K_i\ i^2\ S_f \qquad [5]$$

where D_i = interrill erosion rate, kg/m^2-s; K_i = interrill erodibility, kg-s/m^4; i = rainfall intensity, m/s; S_f= slope factor, dimensionless (= $1.05 - 0.85\ \exp^{(-0.85\ \sin[\theta])}$ [*21*]).

Measuring interrill erodibility. Interrill erodibility can be measured in the field or laboratory with rainfall simulators. Plot size should be limited to under 1 m^2 to ensure that only interrill erosion is occurring. The observed erosion rate is divided by the square of the rainfall intensity and the slope factor to obtain the interrill erodibility. While measuring interrill erodibility, infiltration and crusting properties of soils can also be evaluated for use in process-based erosion prediction equations. Figure 8.4 shows an interrill plot with a rain gauge and a wide mouth bottle collecting run-off and detached sediment for further analysis to find the run-off and interrill erosion rates.

Estimating K_i from soil properties. Like the USLE K value, interrill erodibility can be predicted from measurable soil properties. On a study with 18 soils in Mississippi and Iowa, Meyer and Harmon (*57*) found interrill erodibility was highly correlated to silt fractions, organic carbon, 1500 kPa water content, calcium, sum of the exchangeable bases, cation

Figure 8.4 An interrill erosion plot with an open bottle serving as a raingauge and a wide-mouth bottle collecting sediment-laden run-off

exchange capacity, base saturation, and pH. In an experiment on 36 field sites, the USDA found that interrill erodibility could be estimated by the following equations (*20*):

For Clay > 35 percent

$$\frac{K_i}{10^6} = 2.67 - 0.115\,(\ln(18 - \text{Stab})^2)$$
$$(\text{If } 17 < \text{Stab} < 19, \text{ then } K_i = 2.67) \qquad [6a]$$

For Clay ≤ 35 percent

$$\frac{K_i}{10^6} = -2.92 - 2.71\left(\frac{\text{WDClay}}{\text{Clay}}\right) - 0.5\ \text{Mg} + 0.10\ \text{WDClay}$$
$$+\ 4.19\left(\frac{\text{Clay}}{\text{Fe} + \text{Al}}\right)^{0.16} + 1.24\ \text{Cond} \qquad [6b]$$

where K_i is the interrill erodibility, kg sec m^{-4}; Clay is the clay content, percent; Stab is the stability of aggregates < 0.25 mm, percent; WDClay is the water dispersible clay, percent; Mg is the magnesium content, meq/100 g; Fe is the iron content, meq/100 g; Al is the aluminum content, meq/100 g; and Cond is the electrical conductivity of saturated paste of soil, mmhos.

Rill erosion

Rill erosion is the process of detachment and transport of soil particles by concentrated flow. Rills are eroded channels that can be removed by normal tillage operations. Figure 8.2 shows rills that have followed rows in a crop planted up and down the slope. Rill erosion is measured by rainfall simulation in the field, or in laboratories equipped with large soil boxes (over three m long). Rill detachment rate is a function of the shear of the water flowing in the rill and is calculated from (*51*)

$$D_c = K_r\,(\tau - \tau_c) \qquad [7]$$

where D_c = rill detachment capacity for clean water, kg/m^2•s; K_r = rill erodibility, due to shear s/m; τ_c = critical shear below which no erosion occurs, Pa; τ = hydraulic shear of flowing water, Pa

$$\tau = \gamma\, r\, s \qquad [8]$$

where γ = specific weight of water, N/m^3; r = hydraulic radius of rill, m; and s = hydraulic gradient of rill flow.

Measuring rill erodibility. In order to calculate the rill erodibility, K_r and the critical shear, τ_c, it is necessary to have a range of shear values with concentrated overland flow, measure the erosion rate, and calculate K_r and τ_c by regression analysis. Rills may be preformed and have lengths of 4 to 10 m. If random rills develop on a smooth tilled surface, defining the flow and shape of each rill is more complex. Provision to add clean water at several known flow rates at the top of the plot is necessary. Rill outflow is observed with flumes, or with buckets in which the outflow over a timed period is weighed. Rainfall is applied on the rill until an equilibrium outflow is observed. The shear at equilibrium flow rate from rainfall alone is often below critical shear. Flow should then be added in stepped increments (*21*). For each increment, the flow rate, the velocity, and the concentration of sediment in the outflow are measured. Figure 8.5 shows a rill erosion plot where technicians are measuring the velocity of the flow with dye. The hoses used for introducing additional flow are in the background.

The shape of the rill channel is recorded by measuring the width and depth of the channel, or by observing the shape of the channel immedi-

Figure 8.5 A rill plot with technicians measuring velocity of the flow with dye. Note the hoses for adding additional flow, and the tanks for storing water for the experiment in the background.

ately before or after a given flow measurement. Widths may be measured during the experiment, or recorded on photographs for later analysis. Since the rills are eroding during the experiment, channel shapes can change, so provision must be made to estimate the channel shape for each flow rate. For most cases, it is sufficient to assume that the shape of the rill channel cross section is approximately rectangular. The hydraulic radius of the flow can be calculated from:

$$r = \frac{\text{Cross-Sectional Area}}{\text{Wetted Perimeter}} \qquad [9]$$

Once the radius is known, the shear can be calculated from Equation 8. The rill erosion rate, per rill channel width, is plotted against the calculated shear for each of the flow rates. The slope of the regression line is the rill erodibility, K_r, and the intercept with the horizontal axis is the critical shear, τ_c. Figure 8.6 shows an example from six experimental plots. Added clean water is more erosive than sediment laden water, and under

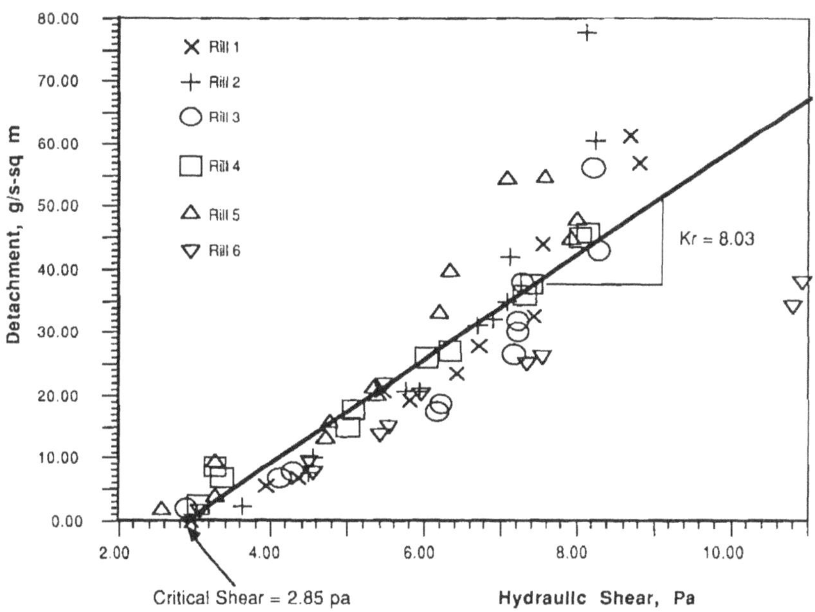

Figure 8.6 Detachment versus hydraulic shear for six rill erosion plots on Portneuf Soil (*21*).

actual eroding conditions, any eroded sediment in the water will result in a reduced erosion rate (*26*). Under highly erosive conditions, the data analysis should consider this occurrence (*20*).

Other methods for measuring K_r and τ_c with jetting devices have also been developed (*37*), but the relationship between these results and rainfall simulation have not been quantified. Critical shear has been estimated in indoor and outdoor flumes, but the saturated soil conditions with no matric potential lead to values different from those experienced under actual field eroding conditions.

Estimating rill erodibility. Field experiments on 36 soils (*20*), found that rill erodibility, K_r, and critical shear, τ_c, could be predicted from the following equations relating soil erodibility to soil properties:

$$\frac{K_r}{10^{-3}} = 196 + 0.015\ M^{0.2}(M^{0.8} - 3500) - \frac{32.7}{CEC^{0.4}} + \frac{0.16}{SAR^{0.75}}\left(\frac{1}{Al} - 50\ SAR^{1.75}\right) + 7.98\,\frac{\left(OM^{1.5} + 3.5\right)}{OM^{0.5}} \quad [10]$$

where K_r is the rill erodibility, s/m; M is defined by Equation 4; CEC is the cation exchange capacity as found by NH_4 titration, meq/100 g; SAR is the sodium absorption ratio $[Na^+] / \sqrt{([Ca^{++}] + [Mg^{++}])/2}$; OM is the organic carbon, percent; $[Na^+]$, $[Ca^{++}]$, and $[Mg^{++}]$ are in meq/l.

For clay content > 30 percent

$$\tau_c = -0.5 - 284 \text{ MC } (\text{MC} - 0.3) \qquad [11a]$$

For clay content < 30 percent

$$\tau_c = -2.85 - \frac{8.87}{(\text{VFS} + 0.1)^{0.2}} - 0.16 \text{ CaCO}_3 + 3.65 \text{ SAR}$$
$$+ 3.79 \text{ SS}^{0.2} + 28.1 \left(\frac{\text{WDClay}}{\text{Clay}}\right)^{0.8} \frac{1}{\text{Sand}^{0.3}} \qquad [11b]$$

where MC is the water content at time of erosive event, g/g; VFS is the very find sand content, percent; $CaCO_3$ is the calcium carbonate content, percent; SAR is the sodium absorption ratio; SS is the specific surface of soil as measured by g EGME retained/g soil; WDClay is the water dispersible clay, percent; Clay is the clay content, percent; and Sand is sand content, percent.

Estimating soil erodibility from laboratory analyses of soil properties. Researchers have used indexes based on routinely measured soil properties to evaluate the relative susceptibility of soils to erosion (Table 8.1). Most indexes are a measure of a soil's detachability or of its resistance to detachment. Indexes are based on properties that govern aggregation and aggregate stability, water transmission and retention properties, raindrop impact, and thermodynamic processes that govern slaking or disruption of aggregates. These indexes are relative measures of detachability and may not reflect the soil's field behavior in response to rainfall and management. The instability index of De Leenheer and De Boodt (*17*), the Henin index (*38*), and the percentage of water-stable aggregates exceeding 0.25 mm are among the indexes related to structural characteristics that have proved applicable in predicting erosion risks for a wide range of soils. These indexes reflect structural properties and a soil's resistance to detachment by rain or overland flow. Also related to the same concept are such indexes as dispersion ratio and colloid ratio. The KE index is a measure of soil strength to resist raindrop impact and is relevant to soil splash.

The choice of an appropriate index depends upon many factors, the most important being the relevance to processes that govern erosion under natural field environments (*12*). In addition, the suitable index should (a)

Table 8.1 Soil erodibility indexes based on parameters that can be measured in the laboratory.

Index	Definition	Limits		Reference
		Erodible	Less Erodible	
A. Structural aggregation and stability				
1. Dispersion ratio	$\frac{\text{amount of (silt + clay) in dispersed state}}{\text{total (silt+clay)}} \times 100$,	>15%	<15%	58
2. Silica: sesquioxide ratio	SiO_2/R_2O_3	9.03	0.52	8
3. Sand ratio or mechanical ratio	Sand/(silt + clay)	-	-	11
4. Surface aggregation ratio	$\frac{\text{Surface area of particles > 0.05 mm}}{\text{aggregated (silt + clay)}}$	-	-	-
5. Instability index (I)	I = mean weight diameter in dry and wet sieving	-	-	17
6. Henin index*	H = (A + L) max/(WS + ES + BS) - 0.9 S.G	-	-	38
7. Percent water stable aggregates	Percent of water stable aggregates >0.5 mm	-	-	12
B. Water transmission properties				
1. Erosion ratio	$\frac{\text{dispersion ratio}}{\text{colloid content/moisture equivalent}}$	>10	<10	58
2. Dispersion-permeability index	E + KD/APp	-	-	6
C. Water retention properties				
1. Erodibility index (E)	$\frac{\text{dispersion coefficient x water holding capacity}}{\text{aggregation}}$	-	-	70
2. Resistance index (I)	$\frac{\text{soil density x range of particle size}}{\text{soil water content}}$	-	-	15
D. Heat of wetting				
1. Temperature profile during infiltration	T (°C) measured at the wet/dry boundary during infiltration	-	-	16
E. Rain drop technique & rainfall simulator				
1. KE index	Kinetic energy required to disrupt an aggregate at pF 4.44	-	-	13
2. Rainulators	Estimating soil erosion using soil trays subjected to standard rainstorms	-	-	33

*Percent aggregation determined by wet sieving following pretreatment with water (WS), ethanol (ES), and Benzene (BS).

be simple and easily adapted for routine measurements, (b) be related to other quantifiable soil properties, and (c) be easily applied for classifying soils into erodibility categories. Considering these factors, there may be no single index that researchers can use for all soils to depict field behavior. A compound factor or combination of several indexes may be required to describe the field behavior appropriately.

Erosivity

The ability of erosion agents to cause soil detachment and transport is erosivity. The erosivity of rainfall is due partly to direct raindrop impact and partly to the runoff that rainfall generates. The ability of rain to cause soil erosion is attributed to its rate and drop size distribution, both of which affect the energy load of a rainstorm. The erosivity of a rainstorm is attributed to its kinetic energy or momentum, parameters easily related to rainfall rate or total amount.

Momentum. Momentum, a product of mass and velocity, is a measure of the impact exerted by rainfall on soil. This impact causes a mechanical stress that leads to breakdown or detachment of soil aggregates. That the erosivity of rainfall relates to momentum has been supported by the work of Rose (*68*) and Williams (*74*). Statistical relationships between momentum and rainfall parameters developed for Australia by Williams (*74*) and Kinnell (*46*) are shown in the following equations:

$$\log\left(\text{momentum (dynes cm}^{-2}\ \text{h}^{-1})\right) = 0.711 \log(i) - 1.461 \qquad [12]$$

$$\text{momentum (dynes cm}^{-2}\ \text{s}^{-1}) = 0.0213\,(i) - 0.62 \qquad [13]$$

where i is rainfall intensity in mm h^{-1}.

Kinetic energy. The kinetic energy of rainfall is a major factor initiating soil detachment. Direct measurements of kinetic energy can be made with pressure transducers or acoustic devices similar to those described by Kowal and Kassam (*49*). Kinetic energy also can be computed by measuring drop size distribution of rains and assuming terminal velocity corresponding to a given drop size. There are various methods of determining drop size distribution, for example, the flour pellet, stain technique, or oil-capturing methods (*41*). In addition, many empirical relations have been developed relating kinetic energy to rainfall intensity or rainfall amount. Kinnell (*48*) described kinetic energy intensity relationships in two ways: (a) the rate of expenditure of the rainfall kinetic energy (E_{RR}), which has the units of energy per unit area per unit time, and (b) the amount of rainfall kinetic energy expended per unit quantity of rain (E_{RA}), which has the units of energy per unit area per unit depth. E_{RA} and E_{RR} are related as follows:

$$E_{RA} = C\, E_{RR}\, i^{-1} \qquad [14]$$

where i is rainfall intensity (depth/time) and C is an empirical constant. Commonly used algebraic equations that relate kinetic energy to rainfall intensity are of the following type:

$$E_{RA} = a + b \log_{10} i \qquad [15]$$

$$E_{RA} = c\,(b - a\, i^{-1}) \qquad [16]$$

$$E_{RA} = bi - a \qquad [17]$$

where a and b are empirical constants. Some commonly used equations are discussed in the section of this chapter dealing with estimation of rainfall erosivity.

Direct measurement of erosivity. Direct measurements of rainfall erosivity involve monitoring energy load and splash simultaneously. The sand splash is measured for sieved, chemically-treated quartz sand of a standard size fraction, maintained at a constant soil water potential and packed to a standard density. The standard technique widely used is the Ellison splash cup method (*41*). The sand splash caused by a rainstorm then is related to simultaneously monitored parameters—kinetic energy, momentum, median drop size, intensity, amount, etc.

Ellison (*22*) defined standard sand as a fine, round-grained material that passes through a 60-mesh sieve but is retained on a 70-mesh sieve. The sand is treated with hydrogen peroxide to remove organic matter, then is washed clean to remove other colloids and binding materials. The detaching or splashing capacity of a rainstorm is then evaluated with the oven-dried sand packed in small aluminum cups, 89 mm (3.5 in.) in diameter and about 50 mm (2 in.) deep (Figure 8.7). A fine-mesh sieve is fixed at the base with overlying cotton or filter paper to retain the sand but to facilitate water movement. The weight of the oven-dried sand packed in the cup is recorded and then saturated by capillary rise. The cup is then exposed to the rain, and the amount of sand splashed is recorded as the loss in oven-dry weight of sand after it has been splashed by falling raindrops. Numerous modifications and improvements have been made since the original design was published more than 40 years ago (*47*).

Measuring size distribution of raindrops. In addition to measuring rainfall intensity, direct measurement of erosivity involves assessment of the drop size distribution. There are several techniques of measuring drop size distribution of rains (*42*). Some of these techniques are briefly outlined below:

a) Stain technique. This technique was developed in the late 19th century and widely used in early 20th century (*42*). The size of raindrop is

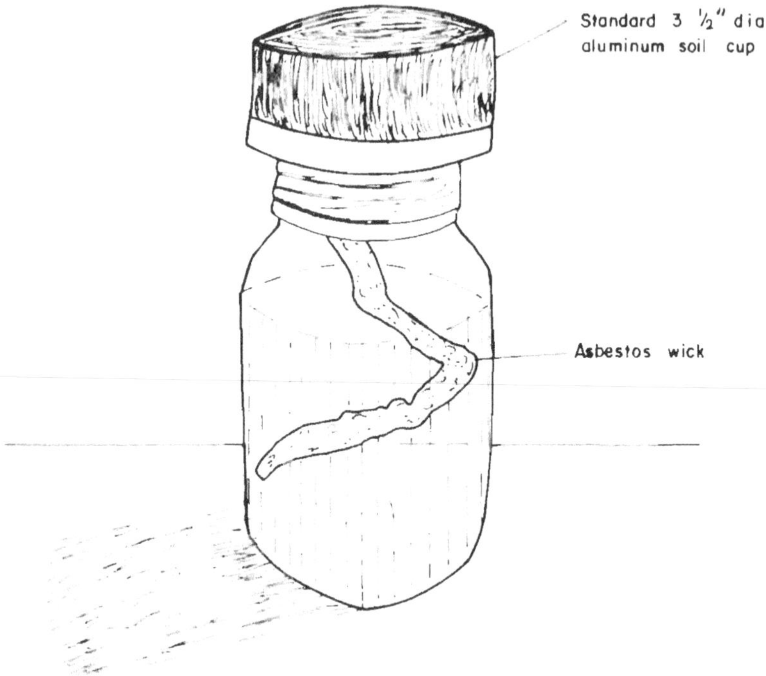

Figure 8.7 Ellison splash cup technique.

estimated from the size of the splash or stain made on an appropriate collecting device, e.g. paper, slate. The collecting device is coated with an appropriate dye that changes color upon getting wet, thereby facilitating accurate measurement of the stain. Procedural details of the technique are described by Neuberger (*63*), Gillespie (*34*), Hall (*36*), and Attle et al. (*4*). Size of the drop is related to size of the stain according to the relation shown in Equation 18.

$$D_r = aS^b \qquad [18]$$

where D_r is the size of raindrop, s is the size of the stain, and a and b are empirical constants.

b) Photographic methods. Photographs of the falling raindrops were also used in early 20th century to measure their size (*54*). In his classic data, Laws (*52*) used this technique to measure velocity of falling raindrops. Atlas and Plank (*3*) developed a technique based on the use of a radar system to measure the drop size distribution within the cloud.

Table 8.2 Frequency distribution of median drop size of rains observed between 1977 and 1979 at UTA, Ibadan, Nigeria

Median Drop Size Range (mm)	Frequency (%)
0.00–0.55	5.0
0.65–1.10	15.0
1.30–1.50	13.2
1.70–1.95	19.2
2.25–2.55	25.1
2.85–3.15	9.0
3.50–4.30	13.7

c) Immersion method. Fuchs and Petajanoff (*32*) developed a method to immerse or capture raindrops in a heavy grade oil. A range of immersion liquids have been used to capture drops with extremes of size ranges (*19, 55, 61*).

d) The flour pellet technique. Bentley (*9*) assessed the size of raindrops from the size of pellets made. The size of raindrop is related to the size of pellets formed according to Equation 19.

$$D_r = \sqrt[3]{\left(\frac{6}{\pi}\right)W} \qquad [19]$$

where D_r is raindrop diameter in mm, and W is average pellet weight in mg. The use of this technique has been described by the classical work of Laws and Parsons (*53*). The technique was also used by Hudson (*39*) and Carter et al. (*14*).

e) The impact assessment. Size of the raindrop is measured from the impact it makes on a tension balance (*62*), on an acoustic device (*45*), or on a pressure transducer (*44*). The data obtained by the use of a pressure transducer (distrometer) from Nigeria is shown in Table 8.2.

Expression of results. The data of drop size distribution are commonly expressed as a single index called D_{50} or the median volume drop diameter. The D_{50} refers to the drop diameter at which 50 percent of the volume of the rain falls in the shape of drops with a smaller diameter and 50 percent as bigger drops. The D_{50} can be computed either on the basis of the volume of rainfall or number of drops. The latter is usually numerically less than the former. The D_{50} is affected by several factors including type, amount, duration, and intensity of rain. The velocity and direction of wind, and air temperature also affect D_{50}. The effect of air temperature on drop size was demonstrated by Zanchi and Torri (*77*) from Italy.

The median volume drop diameter (D_{50}) is related to rainfall intensity (i), terminal velocity, and drop number. Median drop size generally increases with increase in intensity up to a certain limit. This is why the relation between D_{50} and i is expressed as a power equation of the type shown in Equation 20 (*10*).

$$D_{50} = a\ i^{b} \qquad [20]$$

The relationship between D_{50} and rainfall intensity can be easily expressed as a polynomial of the type shown in Equation 21 (*18*).

$$D_{50} = a + bi + ci^{2} + di^{3} + \ldots \qquad [22]$$

The relation between D_{50} and terminal velocity is more complex than that with rainfall intensity. Terminal velocity increases with increase in drop size (*35, 53, 67, 71*). A commonly used relation between drop size and terminal velocity is shown in Equation 22 (*66*).

$$VT = \frac{(\gamma w - \gamma a)\, gd}{1.8} \qquad [22]$$

where γw and γa are water and air density, g is acceleration due to gravity, and d is drop diameter.

There is no direct relationship between number of drops falling per unit area per unit time and the D_{50}. The drop number is, however, related to rainfall intensity. The more the number of drops falling per unit time, the more intense the rain.

Erosivity indexes. Attempts have been made to relate detaching or splashing capacity and kinetic energy of rain to routinely measured rainfall parameters, such as rainfall rate and amount, for use in soil loss prediction. Some of the most commonly used indexes include:

Erosivity index (R) of the USLE. Wischmeier and Smith (*75*) developed a relation between soil loss and a rainfall parameter. The latter is a product (EI_{30}) of the total kinetic energy (E) of the storm times its maximum 30-minute intensity (I_{30}). This parameter is a compound term that reflects the combined potential of raindrop impact and turbulence created in overland flow. The term I_{30} is computed as twice the greatest amount of rain falling in any 30 consecutive minutes. The energy of the storm is calculated by solving the following energy-intensity equation:

$$E = 0.119 + 0.0873 \log_{10} (i) \qquad [23]$$

Table 8.3 Kinetic energy of rainfall expressed in foot-tons per acre per inch of rain.*

Intensity (inch/hr)	0.00	0.01	0.02	0.03	0.04	0.05	0.063	0.07	0.08	0.09
0	-	254	354	412	453	485	512	534	553	570
0.1	585	599	611	623	633	643	653	661	669	677
.2	685	692	698	705	711	717	722	728	733	738
.3	743	748	752	757	761	765	769	773	777	781
.4	784	788	791	795	798	801	804	807	810	814
.5	816	819	822	825	827	830	833	835	838	840
.6	843	845	847	850	852	854	856	858	861	863
.7	865	867	869	871	873	875	877	878	880	882
.8	884	886	887	889	891	893	894	896	898	899
.9	901	902	904	906	907	909	910	912	913	915
	0	0.1	0.2	0.3	0.4	0.5	0.6	0.7	0.8	0.9
1	916	930	942	954	964	974	984	992	1,000	1,064
2	1,016	1,023	1,029	1,036	1,042	1,048	1,053	1,059	1,064	1,069
3	1,074	†								

*Computed by the equation, $E=916+331 \log_{10} I$, where E is kinetic energy in foot-tons per acre per inch of rain and I is rainfall intensity in inches per hour.
†The 1,074 value also applies for all intensities greater than 3 inches/hour.

Table 8.4 Kinetic energy of rainfall expressed in metric ton-meters per hectare per centimeter of rain.*

Intensity (cm/h)	.0	0.1	0.2	0.3	0.4	0.5	0.6	0.7	0.8	0.9
0	0	121	148	163	175	184	191	197	202	206
1	210	214	217	220	223	226	228	231	233	235
2	237	239	241	242	244	246	247	249	250	251
3	253	254	255	256	258	259	260	261	262	263
4	264	265	266	267	268	268	269	270	271	272
5	273	273	274	275	275	276	277	278	278	279
6										
280	280	281	281	282	283	283	284	284	285	
7	286	286	287	287	288	288	289			

*Computed by the equation, $E = 210 + 89\log_{10}I$, where E is kinetic energy in metric ton-meters per hectare per centimeter of rain and I is rainfall intensity in centimeters per hours. The 289 value also applies for all intensities greater than 7.6 cm/h.

Table 8.5 Computation of EI_{30} index for a rainstorm on June 16, 1972. E (foot-ton per acre-inch) = 916 + 331 $\log_{10}$I (where I in inches/hour).

1 Time *Interval*	2 Rainfall Amount *(inch)*	3 Rainfall Intensity *(inch/hr)*	4 Log I	5 331 Log I	6 E *(ft-ton/* *acre-in)*	Total E = 6 × 2 *(ft-ton/acre)*
7.5	0.75	6.0	0.778	257.5	1,173.5	880.1
7.5	0.65	5.2	0.716	236.9	1,152.9	749.9
7.5	0.55	4.4	0.643	212.8	1,128.8	620.8
7.5	0.45	3.6	0.556	184.0	1,100.0	495.0
7.5	0.25	2.0	0.301	99.6	1,015.6	253.9
7.5	0.55	4.4	0.643	212.8	1,128.8	620.8
7.5	0.15	1.2	0.079	26.1	942.1	141.3
7.5	0.30	2.4	0.380	125.8	1,041.8	312.5
7.5	0.20	1.6	0.204	67.5	983.5	196.1
7.5	0.25	2.0	0.301	99.6	1,015.6	253.9
7.5	0.05	0.4	0.398	197.7	784.3	39.2

Total E = 4,563.17 ft-tons/acre; I_{30} (Maximum intensity in 30 minutes) = 4.8 in./hr;

$$\frac{EI_{30}}{100} = \frac{4563.17 \times 4.8}{100} = 219.03$$

where E is kinetic energy in MJ/ha-mm and i is rainfall intensity in mm h^{-1}. The energy equation of the USLE expressed in other units is transformed as follows:

$$E = 210.3 + 89 \log_{10} i \qquad [24]$$

where E is in m-t $(ha\text{-}cm)^{-1}$ and i is in cm h^{-1}, and

$$E = 916 + 331 \log_{10} i \qquad [25]$$

where E is in foot-tons per acre-inch and i is in inches h^{-1}. Tables 8.3 and 8.4 provide tabular solutions to the energy equation for a range of intensities.

The kinetic energy of a rainstorm is calculated from the daily recording rain chart (Figure 8.8) by subdividing the rain into specific intensity ranges. Table 8.5 illustrates an example computation of EI_{30} index. The numerical value of EI_{30} index thus computed varies with the choice of intensity class. The width of intensity class is usually 0.5 inch (10 mm) per hour.

In addition to the energy equation of Wischmeier and Smith (*75*), other algebraic equations have been developed that relate the energy load of a rainstorm to easily monitored storm parameters, for example, rainfall amount and intensity. Kinnell (*46*) related kinetic energy to rainfall rate as follows:

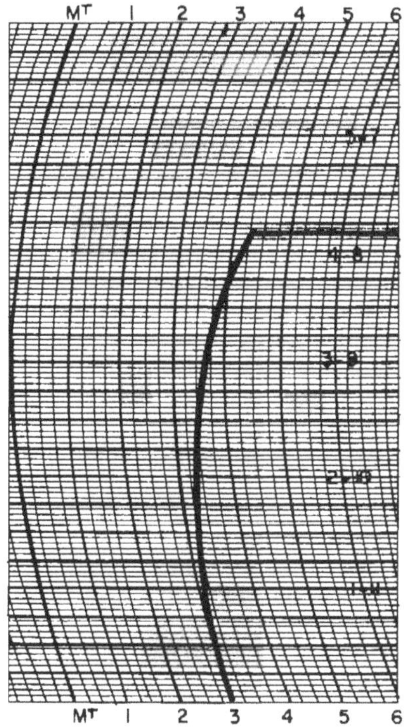

Figure 8.8 Rain gauge chart.

$$\text{KE}\left(\text{ergs cm}^{-2}\,\text{s}^{-1}\right) = 8.37\text{i} - 45.9 \qquad [26]$$

where i is the intensity ranging between zero and 300 mm h^{-1}.

In Zimbabwe, Hudson (*40*) used the following equation to compute kinetic energy using rainfall intensity in mm h^{-1}:

$$\text{KE}\ (\text{J m}^{-2}\ \text{mm}^{-1}) = 29.8 - \frac{127.5}{\text{i}} \qquad [27]$$

In northern Nigeria, Kowal and Kassam (*49*) related the kinetic energy of a rainfall to the rainfall amount per storm (R_a in mm) as follows:

$$\text{KE}\ (\text{ergs cm}^{-1}) = 41.4\ (R_a - 120.0) \times 10^3 \qquad [28]$$

At Ibadan, Nigeria, Lal (*50*) related kinetic energy to rainfall amount P (mm) and to the maximum 30-minute intensity (I_{30} in mm h^{-1}) as follows:

$$\text{KE}\ (\text{J m}^{-2}) = 24.5\ \text{P} + 27.6 \qquad [29]$$

$$\text{KE}\ (\text{J m}^{-2}) = 18.2\ I_{30} + 18.2 \qquad [30]$$

In Zimbabwe, Elwell (*24*) related annual rainfall energy with mean annual rainfall as follows:

$$E\ (\text{J m}^{-2}\ \text{per season}) = 17.368\ (P)\ \text{for regions with morning drizzle} \qquad [31]$$

$$E\ (\text{J m}^{-2}\ \text{per season}) = 18.846\ (P)\ \text{for regions without morning drizzle} \qquad [32]$$

where P is the mean annual/seasonal rainfall in mm.

The EI_{30} factor has been modified to consider the importance of overland flow. The interaction between the depth of overland flow and the drop diameter causes soil detachment. Soil detachment increases with the increase in depth of overland flow up to a threshold almost equal to raindrop diameter. Therefore, the erosivity term should consist of both rainfall and overland flow or runoff components (*59, 65*). Onstad and Foster (*65*) proposed a combined energy term, W, computed as follows:

$$W = 0.5\ EI_{30}\ 15\ Q\ q_p^{1/3} \qquad [33]$$

where Q is storm runoff volume (inches) and q_p is storm peak runoff rate in inches per hour. Further improvements in this equation were made by Foster and associates (*28*).

Varied and often confusing units of energy are used in the EI_{30} index of the USLE. In customary English units, rainfall rate is measured in inches per hour, kinetic energy in foot-tons per acre inch or foot-tons per acre, and the storm erosivity EI_{30} index is expressed in units of foot-ton · inch per acre-hour. The corresponding SI units are mm · h^{-1} for intensity, MJ $(\text{ha} \cdot \text{mm})^{-1}$ for energy, and MJ · mm $(\text{ha} \cdot \text{hr})^{-1}$ for storm erosivity. The conversion factor for changing storm erosivity from English units to SI units is 0.1702.

KE > 1 index. While working on the rainfall erosivity index for rains in southern Africa, Hudson (*41*) observed a threshold value of rainfall intensity below which splash was negligible. He observed that erosion occurred only if rain intensity exceeded about 1 inch per hour. He therefore developed an index that used kinetic energy of rain segments with intensity exceeding 1 inch per hour (KE > 1). Hudson (*41*) obtained better correlation of erosion on soils of Zimbabwe with the KE > 1 index than with EI_{30} index of the USLE. Similar observations have been reported from Sri Lanka by Joshua (*43*).

p^2/P index. Fournier (*31*) developed an erosivity index for river basins on the basis of the relationship between suspended load in rivers and climatic data and relief characteristics. The index, described as climate index C, is defined as follows:

$$C = p^2/P \qquad [34]$$

where p is the rainfall amount in the wettest month and P is the annual rainfall amount. This index subsequently was modified by FAO (*30*) as follows:

$$C_I = \sum_{i=1}^{12} \frac{p_i^2}{P} \quad [35]$$

where p_i is the rainfall in month i and P is the annual rainfall. This index summed for the whole year was found to be linearly correlated with EI_{30} index (R) of the USLE as follows:

$$R = b + a(C_I) \quad [36]$$

where the constants a and b vary widely among different climatic zones. The value of intercept b is -152, -420, -3, and -416 for the United States, the eastern United States, the western United States, and West Africa, respectively. The corresponding values of coefficient a are 4.17, 6.86, 0.66, and 5.44, respectively. This index is an approximation of the EI_{30} index for regions where long-term recording rain gauge records are not available.

Modifications in energy/intensity equation

Mutchler and Murphree (*60*) proposed a modified energy-intensity equation:

$$E = 0.273 + 0.217 \exp(-0.048i) - 0.413 \exp(-0.072i) \quad [37]$$

where E is kinetic energy per 1 mm of rain for a corresponding intensity i (mm/hr) for short time increments. This equation is applicable for intensities exceeding 100 mm/hr whereas the equation proposed by Wischmeier and Smith (*75*) is applicable for intensities < 75 mm/hr.

Universal Index (R). For climatic regions of the east European countries, Onchev (*64*) proposed the Universal Index

$$R = P/S_t \quad [38]$$

where R is Universal Index, P is quantity of rainfall greater than or equal to 9.5 mm with rainfall intensity greater than or equal to 0.180 mm/min and S_t is storm duration with i greater than or equal to 0.180 mm/min. The average annual value of R is computed by summation of indices for all rain events in a year.

Modification in EI_{30}. Two principal modifications have been proposed for using the USLE for a single storm event and for using it on a watershed. Foster et al. (*27, 28*) proposed the index R_m for a single storm event.

$$R_m = 0.5\ R_{st} + 0.35\ V_u\ P_u^{1/3} \quad [39]$$

where R_m is the modified erosivity factor, R_{st} is EI_{30} (N/h) for the storm, V_u is runoff volume (mm), and P_u is the peak runoff rate (mm/hr) from a unit plot of the same soil. The value of the exponent of the function P_u should be increased by 0.1 when rill erosion exceeds the normal level and decreased by 0.1 when rill erosion is less than normal (*29*).

Williams (*73*) proposed modification in R factor of USLE for use on a watershed:

$$R_w = 9.05\ (VQ_p)^{0.56} \quad [40]$$

where V is the volume of runoff (m^3) and Q_p is the peak discharge rate (m^3/s). The modified USLE with the R_w factor is called MUSLE.

In summary

Erosivity is best estimated by direct measurements of a rainstorm's energy load. The data base for these measurements is limited to a few regions only and must be expanded to other agriculturally important areas. Empirical equations that relate rainfall energy with intensity are needed urgently, especially for tropical regions characterized by high-intensity rainstorms. The reliability of the various erosivity indexes discussed depends upon the basic data available. The EI_{30} index can be used reliably for a wide range of climatic regions if the intensity-energy equations for the region are available. Adopting any index to estimate erosivity without evaluating its applicability can lead to grossly erroneous estimates of soil loss.

REFERENCES

1. Aina, P.O., R. Lal, and G.S. Taylor. 1979. *Relative susceptibility of some Nigerian Soils to water erosion.* Nigerian Journal of Soil Science 1: 1-18.
2. Anderson, H.W. 1954. *Suspended sediment discharge as related to streamflow, topography, soil and land use.* Transactions, American Geophysical Union 35: 268-281.
3. Atlas, O., and V.G. Plank. 1953. *Measuring rain with radar.* Journal of Meteorology 10: 291-295.
4. Attle, J.R., D. Oney, and R.A. Swenson. 1980. *Applications of Image Analysis.* American Laboratory.
5. Barnett, A.P., and J.S. Rogers. 1966. *Soil physical properties related to runoff and erosion from artificial rainfall.* Transactions, American Society of Agricultural Engineers 9(1): 123-125, 128.
6. Baver, L.D. 1933. *Some factors affecting erosion.* Agricultural Engineering 14: 51-52.

7. Bennett, H.H. 1939. *Soil Conservation.* McGraw-Hill Book Co., New York, New York.
8. Bennett, H.H. 1926. *Some comparisons of the properties of humid-tropical and humid-temperate American soils, with special reference to indicated relations between chemical composition and physical properties.* Soil Science 21: 349-375.
9. Bentley, W.A. 1904. *Studies of raindrops and raindrop phenomena.* Monthly Weather Review 32: 450-456.
10. Best, A.C. 1950. *The size distribution of raindrops.* Q. J. Roy. Meteorological Society 76: 16.
11. Bouyoucos, G.J. 1935. *The clay ratio as a criterion of susceptibility of soils to erosion.* Journal of American Society of Agronomy 27: 738-741.
12. Bryan, R.B. 1968. *Development, use and efficiency of indices of soil erodibility.* Geoderma 2: 5-26.
13. Bruce-Okine, E., and R. Lal. 1975. *Soil erodibility as determined by raindrop technique.* Soil Science 119: 149-157.
14. Carter, C.E., J.D. Greer, H. Brand, and J. Floyd. 1974. *Raindrop characteristics in south central United States.* Transactions, American Society of Agricultural Engineers 17: 1033-1037.
15. Chorley, R.J. 1959. *The geomorphic significance of some Oxford soils.* American Journal of Science 257: 503-515.
16. Collis-George, N., and R. Lal. 1971. *Infiltration and structural changes as influenced by initial moisture content.* Australian Journal of Soil Research 9: 167-176.
17. De Leenheer, L., and M. De Boodt. 1959. *Determination of aggregate stability by the change in mean weight-diameter.* Mededelingen yan de Landbouwhogeschool, Gent 24: 290-351.
18. De Ploey, J., and D. Gabriels. 1980. *Methods of measuring soil loss.* In: M.J. Kirby and R.P.C. Morgan (eds.) *Soil Erosion.* John Wiley & Sons, Chichester, Engiand.
19. Eigel, J.D., and I.D. Moore. 1983. *A simplified technique for measuring raindrop size and distribution.* Transactions, American Society of Agricultural Engineers 26: 1070-1084.
20. Elliot, W.J., J.M. Laflen, and K.D. Kohl. 1989. *Effect of soil properties on soil erodibility.* Paper No. 892150, American Society of Agricultural Engineers, St. Joseph, Michigan.
21. Elliot, W.J., A.M. Liebenow, J.M. Laflen, and K.D. Kohl. 1989. *A compendium of soil erodibility data from WEPP cropland soil field erodibility experiments 1987 & 88.* NSERL Report No. 3, Agricultural Research Service, U.S. Department of Agriculture, National Soil Erosion Research Lab., West Lafayette, Indiana. 316 pp.
22. Ellison, W.D. 1947. *Soil erosion studies. Part II. Soil detachment hazard by raindrop splash.* Agricultural Engineering 28: 197-201.

23. El-Swaify, S.A., and E.W. Dangler. 1977. *Erodibility of selected tropical soils in relation to structural and hydrologic parameters*. In: *Soil Erosion: Prediction and Control*. Soil and Water Conservation Society, Ankeny, Iowa. pp. 105-114.
24. Elwell, H.A. 1978. *Destructive potential of Zimbabwe/Rhodesia rainfall*. Rhodesian Agriculture Journal 76: 227-232.
25. Food and Agriculture Organization, United Nations. 1977. *Assessing soil degradation*. Soils Bulletin 34, Rome, Italy.
26. Foster, G.R., and L.D. Meyer. 1972. *A closed-form soil erosion equation for upland areas*. In: J.W. Shen (ed.) *Sedimentation: Symposium to Honor Professor H.A. Einstein*. Fort Collins, Colorado. pp. 12.1-12.19.
27. Foster, G.R., L.D. Meyer, and C.A. Onstad. 1977. *An erosion equation derived from basic erosion principles*. Transactions, American Society of Agricultural Engineers 20: 678-682.
28. Foster, G.R., L.D. Meyer, and C.A. Onstad. 1977. *A runoff erosivity factor and variable slope length exponents for soil loss estimates*. Transactions, American Society of Agricultural Engineers 20: 683-687.
29. Foster, G.R. 1982. Modeling the erosion process. In: C.T. Haan, H.P. Johnson, and D.L. Brakensiek (eds.) *Hydrologic Modeling of Small Watersheds*. American Society of Agricultural Engineers, St. Joseph, Michigan. pp. 297-380.
30. Fournier, F. 1956. *The effect of climatic factors on soil erosion: Estimates of solids transported in suspension in runoff*. Assoc. Int. Hydrol. Publ. 38: 6.
31. Fournier, F. 1960. *Climat et erosion: la relation entre l'erosion du sol par l'eau et les precipitations atmospheriques*. Presses Universitaires de France, Paris.
32. Fuchs, N., and I. Petajanoff. 1937. *Microscopic examination of fog, cloud and rain-droplets*. Nature 139: 111-112.
33. Gabriels, D., and M. De Boodt. 1975. *A rainfall simulator for erosion studies in the laboratory*. Pedologie 25: 80-86.
34. Gillespie, T. 1958. *The spreading of low vapor pressure liquids in paper*. Journal of Colloid Science 13: 32-50.
35. Gunn, R., and G.D. Kinzer. 1949. *The terminal velocity of fall waterdrops in stagnant air*. Journal of Meteorology 6: 243-248.
36. Hall, M.J. 1970. *Use of the stain method in determining the drop-size distributions of coarse liquid sprays*. Transactions, American Society of Agricultural Engineers 30: 33-37, 41.
37. Hanson, G.J. 1989. *An in-situ erodibility testing device*. Paper No. 89-2151, American Society of Agricultural Engineers, St. Joseph, Michigan.
38. Henin, S., G. Mournier, and A. Combeau. 1958. *Methode pour l'etude de la stabilite structurale des sols*. Anales Agronomiques 1: 71-90.
39. Hudson, N.W. 1964. *The flour pellet method for measuring the size of raindrops*. Res. Bull. 4, Dept. of Conservation and Extension, Salisbury, Rhodesia.
40. Hudson, N.W. 1965. *The influence of rainfall on the mechanics of soil erosion with particular reference to northern Rhodesia*. M.S. thesis. University of Cape Town, Cape Town, South Africa.

41. Hudson, N.W. 1976. *Soil conservation.* B.T. Batsford, London, England.
42. Hudson, N. 1981. *Instrumentation for Studies of the Erosive Power of Rainfall,* IAHS Bull. 133, IAHS, Washington, D.C. pp. 383-390.
43. Joshua, W.D. 1975. *Erosive power of rainfall in the different climatic zones of Sri Lanka.* Proceedings, Symposium of Erosion and Solid Matter Transport in Inland Waters. Publ. 127. International Association of Hydrological Sciences, Wallingford, England.
44. Joss, V.J., and A. Waldvogel. 1967. *Ein spectrograph fur niederschlagstopher mit automatischer answertung.* Pure and Appl. Geophys. 68: 240-246.
45. Kinnell, P.I.A. 1972. *The acoustic measurement of water drop impacts.* Journal of Applied Meteorology 11: 691-694.
46. Kinnell, P.I.A. 1973. *The problem of assessing the erosive power of rainfall from meteorological observations.* Soil Science Society of America Proceedings 37: 617-621.
47. Kinnell, P.I.A. 1974. *Splash erosion: Some observations on the splash-cup technique.* Soil Science Society of America Proceedings 38: 657-660.
48. Kinnell, P.I.A. 1981. *Rainfall intensity-kinetic energy relationships for soil loss prediction.* Soil Science Society of America Proceedings 45: 153-155.
49. Kowal, J.M., and A.H. Kassam. 1976. *Energy and instantaneous intensity of rainstorms at Samaru, northern Nigeria.* Tropical Agriculture 53: 185-198.
50. Lal, R. 1976. *Soil erosion on Alfisols in Western Nigeria. III. Effects of rainfall characteristics.* Geoderma 16(15): 389-401.
51. Lane, L.J., G.R. Foster, and A.D. Nicks. 1987. *Use of fundamental erosion mechanics in erosion prediction.* Paper No. 87-2540. American Society of Agricultural Engineers, St. Joseph, Michigan.
52. Laws, J.O. 1941. *Measurement of fall velocity of water drops and raindrops.* Transactions, American Geophysical Union 22: 709-721.
53. Laws, J.O., and D.A. Parsons. 1943. *The relationship of raindrop size to intensity.* Transactions, American Geophysical Union 24: 452-460.
54. Mache, H. 1904. *Ueber die Geschuwindigkeit im Grosse der Regentropfer.* Met. Zs. 39: 278.
55. McCool, D.K., M.J. Robinette, J.T. King, M. Molanu, and J.L. Young. 1978. *Raindrop characteristics in the Pacific Northwest.* Transactions, American Geophysical Union 59 (abstract).
56. McCool, D.K., L.C. Brown, G.R. Foster, C.K. Mutchler, and L.D. Meyer. 1987. *Revised slope steepness factor for the Universal Soil Loss Equation.* Transactions, American Society of Agricultural Engineers 30(5): 1387-1396.
57. Meyer, L.D., and W.C. Harmon. 1984. *Susceptibility of agricultural soils to interrill erosion.* Soil Science Society of America Journal (48): 1152-1157.
58. Middleton, H.E. 1930. *Properties of soils which influence soil erosion.* Technical Bulletin 178. U.S. Department of Agriculture, Washington, D.C. 16 pp.
59. Monke, E.J., H.J. Marelli, L.D. Meyer, and J.F. De Jong. 1977. *Runoff, erosion and nutrient movement from interrill areas.* Transactions, American Society of Agricultural Engineers 20: 58-61.

60. Mutchler, C.K., and C.E. Murphree. 1985. *Experimentally derived modification of the USLE.* In: S.A. El-Swaify, W.C. Moldenhauer, and A. Lo (eds.) *Soil Erosion and Conservation.* Soil and Water Conservation Society, Ankeny, Iowa. pp. 523-527.
61. Nawaby, A.S. 1970. *A method of direct measurement of spray droplets in an oil bath.* T. Agric. Eng. Res. 15: 182-184.
62. Neal, J.H., and L.D. Bayer. 1937. *Measuring the impact of raindrops.* Journal of American Society of Agronomy 29: 708-709.
63. Neuberger, H. 1942. *Notes on measurement of raindrop sizes.* Bull. Am. Met. Soc. 23: 274-276.
64. Onchev, N.G. 1985. *Universal index for calculating rainfall erosivity.* In: S.A. El-Swaify, W.C. Moldenhauer, and A. Lo (eds.) *Soil Erosion and Conservation.* Soil and Water Conservation Society, Ankeny, Iowa. pp. 424-431.
65. Onstad, C.A., and G.R. Foster. 1975. *Erosion modeling on a watershed.* Transactions, American Society of Agricultural Engineers 18: 288-292.
66. Park, S.W., J.K. Mitchell, and G.D. Buberger. 1983. *Rainfall characteristics and their relation to splash erosion.* Transactions, American Society of Agricultural Engineers 26: 795-804.
67. Roels, J.M. 1981. Cited by J.M. Eigel and I.D. Moore (1983), p. 1079.
68. Rose, C.W. 1960. *Soil detachment caused by rainfall.* Soil Science 89: 28-35.
69. Vanelslande, A., P. Rousseau, R. Lal, D. Gabriels, and B.S. Ghuman. 1984. *Testing the applicability of soil erodibility nomogram for some tropical soils.* Publication 144. International Association of Hydrological Sciences, Wallingford, England. pp. 468-473.
70. Vosnesensky, A.S., and A.B. Artsruui. 1940. *A laboratory method for determining the anti-erosion resistance of soils.* Soils and Fertilizer 10: 289.
71. Wang, P.K., and H.R. Pruppacher. 1977. *Acceleration to terminal velocity of cloud and raindrops.* Journal of Applied Meteorology 16: 275-280.
72. Watson, D.A., and J.M. Laflen. 1986. *Soil strength, slope, and rainfall intensity effects on interrill erosion.* Transactions, American Society of Agricultural Engineers 29(1): 98-102.
73. Williams, J.R. 1975. *Sediment yield prediction with Universal Equation using runoff energy factor.* In: *Present and Prospective Technology for Predicting Sediment Yields and Sources.* ARS-S-40, U.S. Department of Agriculture, Washington, D.C. pp. 244-252.
74. Williams, M.A. 1969. *Prediction of rainfall splash erosion in the seasonally wet tropics.* Nature 222: 763-765.
75. Wischmeier, W.H., and D.D. Smith. 1958. *Rainfall energy and its relationship to soil loss.* Transactions, American Geophysical Union 39: 285-291.
76. Wischmeier, W.H., and D.D. Smith. 1978. *Predicting rainfall erosion losses—a guide to conservation planning.* Agriculture Handbook 537. U.S. Department of Agriculture, Washington, D.C. 58 pp.
77. Zanchi, C., and D. Torri. 1980. *Evaluation of rainfall energy in central Italy.* In: M. De Boodt and D. Gabriels (eds.) *Soil Erosion Assessment.* John Wiley & Sons, Chichester, England. pp. 133-142.

Chapter 9

ASSESSING VEGETATIVE COVER and MANAGEMENT EFFECTS

M. A. Stocking

Many people believe that vegetative cover is the single most important factor in soil erosion control in the tropics. Soil conservation systems increasingly emphasize the role of organic matter, dead or alive, in arresting erosion. Management for soil conservation is now as much a question of encouraging vegetative growth as it is of constructing physical conservation measures.

As a strategy for soil conservation planning, the promotion of vegetation, or a "biological" approach to soil conservation, has much to offer. Vegetation is the factor most easily manipulated by careful management. Beyond that, better vegetative growth and, hence, better protection of the soil almost always provides direct economic benefits in terms of yield and production. Perhaps the major problem with such a conservation strategy is that it requires continuous, sensitive, and knowledgeable management of both the soil and the crop to be fully effective. But the rewards in terms of reduced soil loss are indisputable,[1] and it is a goal achievable by small or large, rich or poor farmers alike (*39*).

---

[1]It is not my purpose here to review the evidence for the efficacy of vegetative cover and management. Suffice to say, the differences in erosion between good cover/high management plots and clean-tilled, fallow plots can be in orders of magnitude of 100 or more. The reader is referred to Hudson (*19*), Meyer and Mannering (*27*), and many other experiments worldwide.

HOW VEGETATIVE COVER AND MANAGEMENT WORK

Vegetative cover and rainfall

First and foremost, vegetation protects the soil from erosion by intercepting raindrops and absorbing their kinetic energies harmlessly. Some water may be evaporated from the leaves, but most reaches the ground surface either by stemflow or by reforming into droplets that, if the vegetative cover is close-growing, have little chance to pick up speed and gain further kinetic energy.

The German Ewald Wollny pioneered research on rainfall interception (*47*). He demonstrated how planting densities common for many crops allowed nearly 90 percent of the total annual rainfall to reach the ground unimpeded. Even under the highest planting densities, nearly half the rainfall in Wollny's experiments fell directly onto the soil (Table 9.1).

Although this has been a neglected area of research, it is clear from these early experiments and a few later ones, such as that by Sreenivas and associates (*38*), that raindrop interception is the main way that vegetation reduces erosion. Yet, if this were the whole story, a linear relationship would hold between percentage cover and erosion; it does not. Experimental evidence now indicates that the erosion-cover relationship is curvilinear and that erosion is little different whether cover is 100 percent or 60 percent (Figure 9.1). Researchers have found similar curvilinear relationships for runoff (*16, 25*). Such findings are vital to conservation research and planning in helping to design realistic management objectives. But they also raise the question of why a mean seasonal cover of 60 percent is nearly as effective as full, continuous cover.

Vegetative cover and soil

There are many interactive processes between a plant and its soil that affect erosion. Some of these processes include the following:

- The physical binding of soil by plant stems and roots.
- Electrochemical and nutrient bonding between roots and soil.
- Detention of runoff by stalks and organic litter.
- Improved infiltration along root channels.
- Greater incorporation of organic matter into the soil, resulting in better structural and water-holding qualities.
- Increased faunal and biological activity, leading to better soil structure.

Of course, many of these processes themselves contribute to better vegetative cover. This is particularly true in the seasonal tropics where the

Table 9.1 Percentage of total rainfall penetrating a canopy of vegetation at different planting densities on 4-m^2 trial plots (*30, 42, 47*).

Planting Density (plants/m^2)	*Percent Total Annual Rainfall Penetrating Canopy with:* Corn	Soybeans	Oats	Peas
0	100	100	100	100
9	62.9	88.4	-	-
16	60.7	78.2	78.5	-
25	57.0	65.9	78.4	78.9
36	44.5	64.3	78.9	-

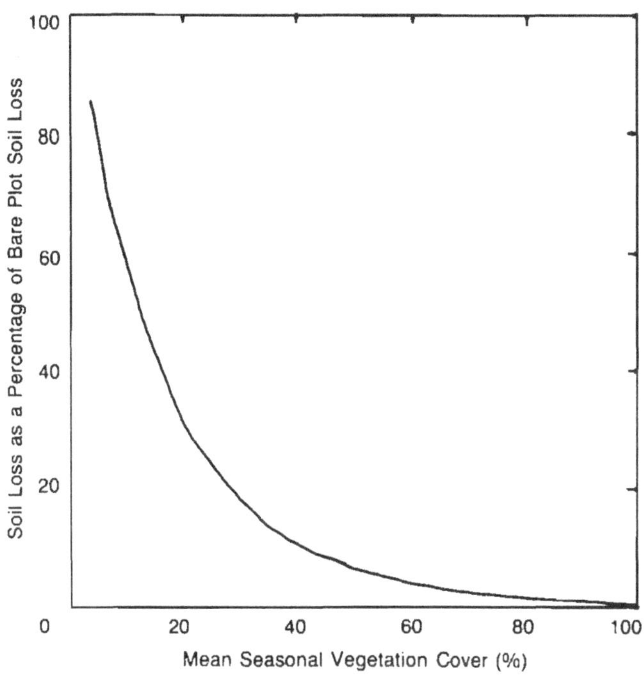

Figure 9.1 Erosion-cover relationship, generalized after Elwell (*13*) and Elwell and Stocking (*15*).

increased infiltration promoted by vegetative cover helps to avoid periodic drought stress and maintains a more vigorous cover. Clearly, provided that vegetation is maintained above a certain level of cover (50–60 percent, but varying according to type of cover and soil), the interactive process between the soil and the plant are sufficient to cope with erosive forces.

Management

All of the above has direct implications for management. But in addition, management works to reduce erosion through fertilization, timely planting, and a whole host of farm practices that encourage vegetative growth. The only exception from normal routine farm practices is weeding, which inadvertently acts to reduce cover. Effects of management on soil characteristics are also important, especially in the maintenance of soil aggregation and the prevention of soil crusting. Aggregates are formed through the bridging effects of organic colloids. It is only through the use of mulches and other supplies of organic matter that aggregation can be kept stable.

Tillage practices are one of the main aspects of management that alter soil porosity and infiltration and incorporate organic residues into the topsoil. In particular, no-till practices and direct drilling techniques effectively reduce erosion, especially when combined with cover crops (*20*).

Ambivalent effects

It would be a mistake to call vegetative cover the panacea for soil erosion control. For example, a mature, natural forest cover is a perfect foil to erosion processes. Rates of erosion under undisturbed forest are usually always well below 500 kg/ha/yr. For example, in Hong Kong, on 29-degree slopes with pine regrowth, soil loss was 450 kg/ha (*24*). The temptation, therefore, is to promote afforestation as a universal conservation measure. However, I know of at least two cases where planted trees accelerated the erosion rate. In Brazil, on the slopes of the Val do Rio Doce, Minas Gerais, the planting of *Eucalyptus* spp. as an erosion control measure stifled ground cover and accelerated sheet erosion. In Mondoro, Zimbabwe, trees planted for gully control appear to have increased the rate of headward recession of gullies so that many of the trees have been undermined by the process they were meant to stop.

What are the causes of the ambivalent effects of vegetation? Three can be identified. First, the height of the vegetative cover above the ground surface is important (*38*). Droplets, often larger in mass than in the original rainfall, may reform on leaves. In falling to the ground, they may accelerate sufficiently to have a sizable kinetic energy. For example, a 2-mm diameter raindrop with a terminal velocity of 6 m/sec has about the same kinetic energy as a 3-mm drop falling at 3.25 m/sec, a speed that is reached well within the first meter of fall. Naturally, it requires more than three 2-mm drops to make a 3-mm one. Nevertheless, the energy impact of droplets falling from vegetation should not be underestimated.

Second, tall-growing vegetation may reduce ground cover completely, either directly by shading or by little-understood chemical effects through the roots of certain woody perennials. Unimpeded surface runoff, coupled with the bombshell effect of large droplets falling from leaves, can cause significant erosion.

Third, ambivalent effects of vegetation also have been demonstrated in laboratory rainfall simulator tests. De Ploey and associates (*9*) found a cover of grass definitely reduced erosion on slopes under five degrees. But above eight degrees, the rate of erosion exceeded the rate on bare soil. The researchers concluded that higher slopes generate turbulent eddies downstream of grass blades that erode more soil. Obviously, there exists a complex interaction among vegetation, slope, soil type, and erosion.

KEY AREAS IN CROP COVER AND MANAGEMENT RESEARCH

Before considering actual research methodologies, it is necessary to pinpoint the nature and type of research that may be needed in crop cover and management over the next few decades. For better or for worse, the greatest effort on a sustained, worldwide basis has gone toward establishing the C and P factor values in the universal soil loss equation. Many now feel that further refinement of such empirical methods through expensive research provides diminishing returns to knowledge and that scientists should pursue new areas of research more vigorously. Following are some areas that need attention.

Soil fertility, productivity, and management

One of the major lessons learned from recent erosion research is the complex nature of the interdependencies among erosion factors. In part, this is, perhaps, a reaction to the more simplistic erosion models, in which all factors are independent (indeed, this is a requirement of empirical-statistical models). Therefore, scientists increasingly have appreciated that the growth of good cover crops not only has an immediate benefit in reducing erosion but a carryover effect into succeeding years as well. For example, the antecedent effect of rotation is such that a maize-following-maize plot would always have higher soil losses than a maize-following-grass/legume plot. The classic analysis of Klemme and Coleman (*21*) on the differential effects of cropping systems on erosion is worth reading in this context. Similarly, it is a sad but true fact that areas which have suffered high erosion tend to continue to have high erosion rates. Some of the blame for this can be ascribed to erosion affecting fertility and fertility affecting crop growth. However, there is an enormous field of research open to investigation of such interactions. The Food and Agriculture Organization recently commissioned a study of erosion and productivity, and specific areas of research relevant to vegetation and management are identified in the report (*41*).

Quality of crop canopy

Vegetation varies significantly in its characteristics and structure, and this has a bearing on its efficacy for soil protection. Aspects might include the value of the vegetation as a mulch, the layout of leaves at different heights so that raindrop interception is maximized, good cover crops for low fertility/high erosion conditions (it is lamentable that nearly all experimental plot research is done under good, high-management conditions on soils that are uneroded), and density of planting as well as experiments with broadcast versus row versus grid planting patterns.

One line of research that needs investigation has to do with the role of stemflow. The structure of certain plants encourages water to quickly channel down the leaves and flow down the stem. Is this dangerous? De Ploey (*8*) demonstrated just how important stemflow is to runoff generation. Studies of the quality of the crop canopy could extend into crop breeding programs where the ability of a plant to produce a quick close-growing canopy is as much a criterion for varietal selection as is yield and resistance to disease. In a similar vein, the development of cover crops that will yield a crop nearly every year under marginal, droughty, and low fertility conditions is important.

Economic aspects of specialty cover crops

One of the major factors responsible for the present high rates of erosion is continuous monocropping with no fallow or grass ley in the rotation. Farmers and landowners argue that it is not economical to have an unproductive year in a rotation and that inorganic fertilizers now make a recuperative year unnecessary. Sufficient evidence exists, however, that shows the deleterious effect of continuous cropping on soil structure and how more and more chemicals have to be applied simply to retain yields at present levels on many soils. More economic analyses are needed that take into account the dynamics of erosion/productivity relationships and incorporate the medium-term beneficial aspects of planted cover crops.

Integrated land use systems

Multiple cropping, agroforestry, alley cropping, interplanting, and stripcropping all have a certain appeal, not the least of which is the improved cover and soil conservation they afford. But the fact remains that, with notable exceptions, there is a dearth of experimental work on these systems, despite sometimes extravagant claims. In particular, there is little on-farm research into integrated land use. It is one matter to demonstrate that a cropping or management system is technically viable on a research station. It is quite another matter to do this under the real management constraints of a small farm. With modern farming techniques, the management of integrated forms of land use is difficult and requires specific skills on the part of the farmer. However, the scope is enormous for innovation and development of farming systems that are acceptable socioeconomically to the farmer while conserving the environment.

Agricultural operations on the farm

A particular case of farming systems research that has many implications for soil erosion is the timing of planting and harvesting. Planting date is critical to yield throughout the tropics. But it is also critical to the degree of protection afforded to the ground over a season by the growing crop. The later planting is delayed, the more opportunity rainfall has to hit bare soil. In the seasonal tropics especially, early season storms can be the most intense and erosive. Similarly, harvesting and disturbance of the soil can cause additional erosion. There are many and varied reasons for late planting. In developing countries, the most widely cited reason is the weakness of draft oxen at the end of the dry season to pull a plow through an often hard and stony soil. Whatever the precise cause, there is reason to examine farming systems with the view to proposing adaptations that would be beneficial to conservation objectives, for example, perhaps a breeding program for improved draft animals or fodder trees.

Tillage and soil structure research

Management also relates to appropriate methods of tillage and maintenance of soil structure. Type, direction, and degree of tillage all have important effects on erosion. There are indications that the same method of tillage year after year is bad for soil structure and that a rotation of implements is desirable, perhaps, in successive years, conventional moldboard plowing followed by zero tillage and direct drilling, light disking, and, finally, chisel plowing. Also, changes in soil properties following agricultural operations have a considerable effect on erodibility.

METHODOLOGIES

The previous discussion highlighted the fact that research on vegetation and management can be extremely diverse in scope and nature, varying from the micromorphological structure of plants, through plot and field experiments, to socioeconomic influences in farming systems. Clearly, it is impossible to detail all methodologies here. What follows is a description of only those approaches most directly relevant to vegetation, management, and erosion.

Vegetative cover

Botanists and ecologists tend to use basal cover—the area of the actual ground surface taken up by stalks of vegetation—as the principal measure of cover. Such measurement methods include line-intercept, wheel-point, and variable-plot techniques (*44*). For purposes of erosion research, however, it is more important to know the proportion of the ground that is covered, even if that cover is some distance off the ground. This gives a measure of the efficiency of the vegetation to intercept raindrops or, alternatively, the proportion of bare ground open to direct raindrop splash.

Techniques. The most simple, most practical, and least costly techniques for direct measurement of vegetative cover are ground-based vertical photography and use of a quadrat sighting frame. Elwell and Gardner (*18*) evaluated the two techniques, concluding that the sighting frame is the better method for routine field measurements. Aerial photography is useful for resource surveys, but it cannot provide the detail necessary for cover determinations; it is also a costly technique for repeated observations throughout a growing season.

Several authors have proposed stereophotography for monitoring vegetative changes (*45*). But this form of ground-based vertical photography from a camera perched on a long pole suffers from excessive radial displacement. Even when used stereoscopically, the percentage of cover is always underestimated because it appears to be spread over a larger area than it actually is. Although Elwell and Gardner (*18*) worked out percentage errors for leaf canopies at different heights, they concluded that even with a camera located more than 5 m above the ground, little more than immediate ground cover can be measured.

In its most simple form, a quadrat sighting frame consists of two horizontal bars set directly above each other. Ten small holes are drilled at regular intervals along each bar so that an observer may peer through a hole in the top bar and see a small area of the ground through the corresponding hole in the lower bar. The observer simply records the presence or absence of a leaf or other item of intercepting vegetation. After a predetermined number of observations, perhaps 100 sights or 10 positions of the frame for a fairly regular cover, or more if cover is discontinuous, the results are expressed as a percentage of hits in the total number of observations. An early user of this method estimated that accuracy with 10 random positions of the frame was ±2 percent of vegetative cover (*4*). However, subsequent experience with the frame indicates that errors are seriously underestimated. Some sources of error include bias in the case of row crops, difficulties in vertical sighting with crops more than 1 m high, problems with partially covered sights, and difficulty of interpretation with heavy shadow. Unpublished tests using three observers on the same field showed that observer error and/or misinterpretation can produce unacceptable errors of ±10 percent. Using similar procedures and statistical analysis, Elwell and Gardner (*18*) found that a total of 1000 sights would be needed to achieve a 2 percent accuracy for uniform cover conditions and 300 sights for a 5 percent accuracy where cover is variable.

An improved instrument has been designed for vegetative cover measurements (Figures 9.2 and 9.3). It retains the ease of operation of the simple sighting frame but removes some of the biases and potential errors and allows the measurement of tall-growing vegetation, such as maize (*14*). Instead of looking vertically down, the new instrument uses adapted gunsights on sliding cursors (to allow adjustment for row crops at different spacings) to look obliquely downward onto a strip of mirror. The observer sees the reflected image of the crop leaves outlined against the sky.

Other techniques for assessing vegetative cover include the use of light meters, point quadrats, shadow measures (*2*), and electronic monitoring devices. One interesting cover study compared botanical methods, including all-contacts-point, first-contact-point, and step-point methods, evaluated against both runoff and sediment production in Colorado; none of the methods gave a good correlation against sediment yield (*3*). Indirect measures include assessing the effectiveness of cover in preventing soil detachment through the use of splash cups. Sreenivas and associates (*38*) proposed the use of a soil cover rating as follows:

$$\text{SCR} = \frac{\text{weight of soil splashed from bare land} - \text{weight of soil splashed under cover}}{\text{weight of soil splashed from bare land}}$$

The technique has not received wide use, however.

Use of cover measurements. A measurement of cover percentage provides a single assessment at one point in the growth period of a crop or vegetation. In other words, if a rainfall event were to occur at that exact time, the cover value is equal to the percentage of rainfall likely to be intercepted before reaching the ground. By itself, such a measurement is of little intrinsic value. Knowledge is required of how the vegetation

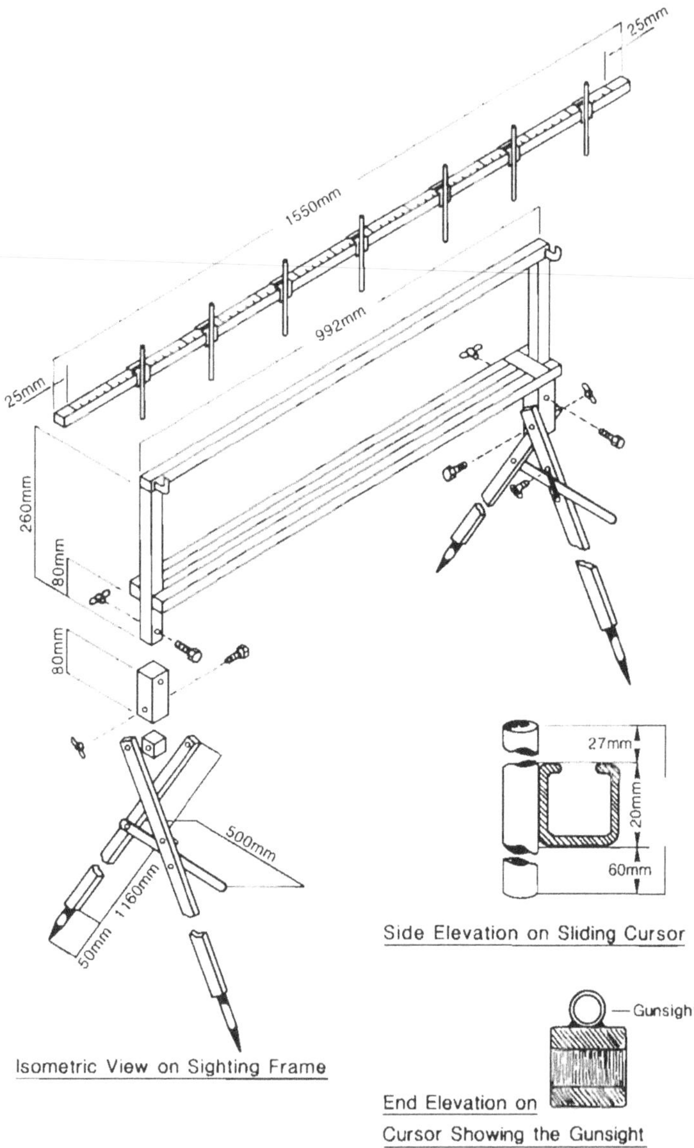

Figure 9.2 Construction details for sighting frame.

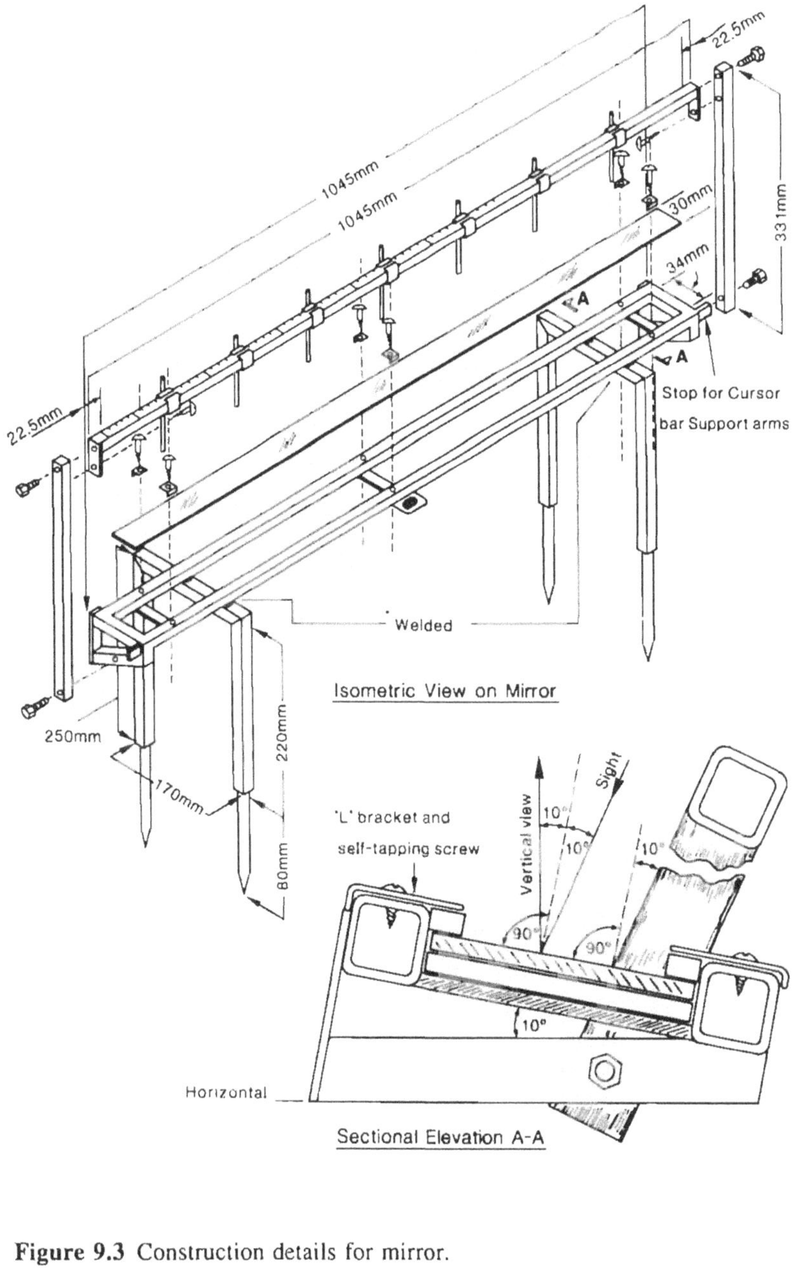

Figure 9.3 Construction details for mirror.

progresses through its growing season and the proportion of the seasonal rainfall that is intercepted. This seasonal interceptive efficiency of vegetation is one of the key parameters used in the soil loss estimation developed in and employed by the Conservation Service in Zimbabwe (*17*). It is similar to the crop protection factor of Sharma and associates (*35*).

To calculate seasonal interception, cover measurements are taken at regular intervals (usually every 10 days) through the growing season. Then a cover curve is constructed (Figure 9.4). Superimposed upon this curve is the mean seasonal distribution of rainfall kinetic energy as measured from a recording rain gauge. For example, if in one 10-day interval the cover averaged 25 percent and the kinetic energy of rainfall was 800 J/m^2, then 200 J/m^2 would be intercepted. Summing these calculations over the whole growth period, it is possible to calculate the percentage of mean seasonal kinetic energy that is intercepted by the vegetation. Typical values of interception range from about 20 percent for a poor crop of maize or tobacco to nearly 100 percent for a dense weed fallow.

Zimbabwe scientists have established a vegetative cover data bank of cover measurements and crop-hazard ratings. For instance, a cotton crop grown under moderate fertility and reasonable management will achieve 38 percent interception of rainfall, whereas a sorghum crop will achieve 52 percent interception. Researchers already know that the erosion-cover relationship is curvilinear (Figure 9.1) and that 38 percent cover is potentially dangerous. If erosion is likely to be severe because of slope or other conditions, then there would be compelling arguments to grow sorghum rather than cotton. The major variables affecting interceptive efficiency are planting date and fertility status of the soil. In Zimbabwe, H. A. Elwell has constructed tables of interception for crops at various expected yield levels and dates of crop emergence (Table 9.2 is an example for one crop).

Other vegetative characteristics and methods

Canopy height, the structure of the plant, and rooting characteristics are measures of vegetation that could influence erosion to varying degrees. Standard length measures and a leaf area index can be used for the former, but root measurements are difficult. Because no single measure can adequately describe all vegetative characteristics that are important in erosion, one method is to classify vegetation into types based upon similarities in soil protective characteristics (Table 9.3).

The problem remains, however, to relate quantitatively types of vegetation to splash detachment and overall erosion. Laboratory and rainfall simulator studies have the most potential. Some possibilities that have been

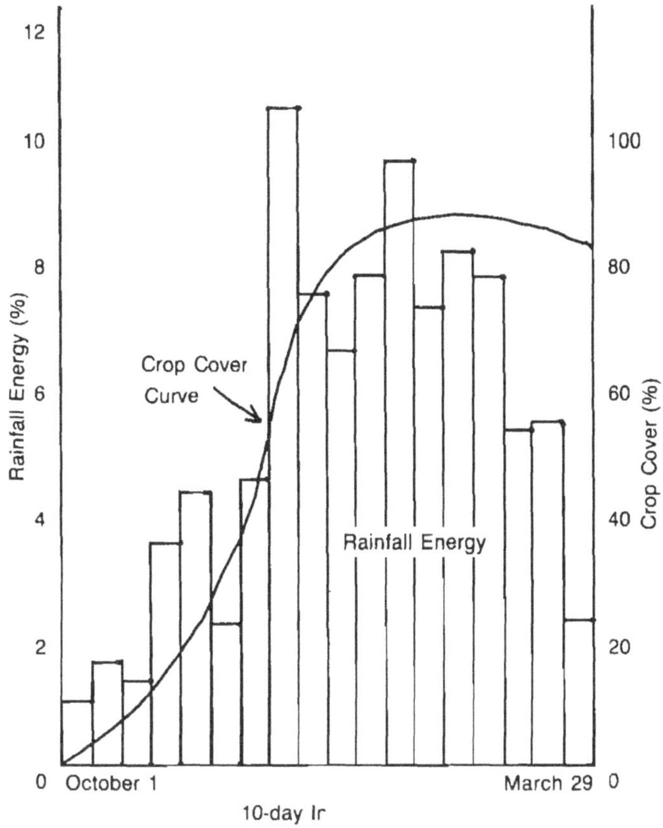

Figure 9.4 Crop cover and rainfall energy distribution (hypothesized) through a growing season, from which seasonal interception of rainfall is calculated.

Table 9.2. Interception values for cotton based on yield and emergence data; data from H.A. Elwell, CONEX, Zimbabwe.

Expected	*Interception Value (%) for Emergence Dates of:*								
Yield (kg/ha)	*1 Sept*	*15 Sept*	*1 Oct*	*15 Oct*	*1 Nov*	*15 Nov*	*1 Dec*	*1 Jan*	*1 Feb*
500	43	41	39	34	29	23	16	7	2
1,000	62	59	55	49	41	32	24	9	2
1,500	72	69	65	57	48	38	28	11	2
2,500	84	79	75	66	56	44	32	13	3
3,500	92	87	82	72	61	48	35	14	3
4,500	95	89	84	74	63	49	36	14	3

discussed at recent meetings, but which remain generally unpublished, include

- Splash detachment in a rainfall simulator using plants grown in trays at various spacings and cover percentages: brussels sprouts, potatoes, and sugar beets have been used at Silsoe, England, for example.
- As above, but with plastic or artificial plants.
- As above, but using an open grid above the soil surface upon which discs or rectangles can be placed to simulate accurately various cover percentages and spacing of the cover. Experiments at the Institute of Agricultural Engineering, Harare, Zimbabwe, used this approach, along with clumps of *Hyparrhenia* grass to represent the stems of row crops (this addition of a basal cover reduces soil loss slightly but consistently).

The classic early experiments of Hudson (*19*) using mosquito gauze to simulate a full cover is another example.

Experiments on the effects of vegetative mulch on erosion are more common, principally because this is an aspect covered specifically by the USLE. Standard soil loss/runoff plot experiments (with or without rainfall simulation) are the usual way of expressing the beneficial aspects of various rates of application and incorporation of organic residues and mulches (*22, 36*).

Plot and field experiments

Apart from experiments aimed at fixing factor values for the USLE (see later section), researchers widely use plot and field experiments to investigate specific crop and management circumstances. The major weakness of such experiments is this very specificity; they are only relevant to the precise conditions of the experiment. Extrapolation must proceed with extreme caution. A few of the many examples that might be quoted include the following:

- Water yield and erosion response to land management on watersheds up to 389 acres (*37*).
- Comparison of the effects of no-till practices and conventional plowing on soil properties and yields of different crops. No-till had negligible soil loss (*23*).
- Effect of narrower row spacing for a sorghum crop. Soil loss was 39 percent less than with conventional spacing (*1*).

All experiments that monitor the relationships among erosion and measures of vegetation and management show that other factors influence the exact nature of the relationship. Some of these factors are the nature of the

Table 9.3 Crop classification based on similarities in soil-protective characteristics (*16*).

Description	*Examples*
A. Row crops	
1. Tall, upright crops generally grown on unridged lands	Annuals: maize, sorghum, sunflowers; perennials: napier fodder, sugarcane
2. Legumes, annuals; short, bunch, and procumbent varieties	Beans: soya, velvet, etc.; groundnuts; cowpeas
3. Tall, upright crops on ridged lands	Tobacco; group 1 crops on ridges
4. Woody, bushy row crops	Cotton
B. Broadcast crops	
1. Tall, upright, for fodder	See A1
2. Short legumes broadcast for fodder and green manure	See A2
3. Medium height plants for fodder, green manure, etc.	Sunhemp, weed fallow
C. Orchards/plantations	
1. Individual trees and bushes planted on regular pattern	Coffee, citrus, deciduous fruit
2. Hedged crops	Tea
3. Thick stands of natural and exotic trees with little to no grass cover	Forestry
D. Grasslands	
1. Stoloniferous grasses planted in rows from runners; permanent pastures	Star, Kikuyu, torpedo
2. Seed established grass; bunch varieties	Love grass, *Sabi panicum*, Rhodes, etc.
3. Species composition related to natural soil and environmental conditions	Natural veld grasses, mixed species, mainly bunch, with some annuals and perennials

rainfall, evapotranspiration, and infiltration rates. In turn, these factors may depend at least partially on the vegetation and management. For the researcher there is often the considerable danger of circularity of argument, as well as lack of adequate control over plot and field experiments.

Vegetation as an indicator of erosion

Within the complexity of natural systems, one variable can effectively integrate several other variables and present an observer with useful information in a simplified manner (*43*). Plant indicators are one of the most powerful integrators, and they have two important areas of application in erosion research: in general survey work for use as dateable markers of old soil levels and through species composition as indicators of the state of the environment or of soil types or conditions that are particularly susceptible to erosion.

Field measurements using vegetation. Dunne and associates (*10*) described several ways of using vegetation as a rapid means of field measurement of erosion. The exposure of tree roots is one common method. But a more accurate assessment used in Tanzania involves use of certain woody species, for example, ant-gall acacia (*Acacia drepanalobium*), that have a physiological mark on their stem (a bulge or branching or color of bark) at the level of the original soil surface (*11*). With suitable species, it is possible to develop a calibration curve between age of tree (by ring count) and circumference at ground level. Therefore, nondestructive sampling can be carried out linking age of tree and depth of erosion since germination of the tree. There are several practical difficulties leading to sources of error with this methodology, such as ambiguity of original surface level or scarcity of suitable shrubs in some places. But it is an unrivalled method for gaining recent historical data on rates of sheet erosion without the need for expensive instrumentation.

Indicator species. As indicators, plants are extremely useful because they are obvious features of the landscape and can react dramatically to small changes in environmental conditions. The reaction of grasslands, in particular, to overgrazing and intensive use is usually by a sequential change in species composition from vigorous, palatable perennial grasses to a reduced density of tufts of less palatable annual species (*32*). In central Zimbabwe, I observed *Panicum Coloratum* with some *Digitaria* spp. on a lightly grazed site and *Brachiaria Humidicola* on an adjacent but intensively used site—an overall serious deterioration of the grassland that can be used as an indirect measure of erosion.

Certain species also tend to be specific to particular soils. Apart from the obvious role of plant indicators for soil surveys, plants can be used to identify certain problem soils from the standpoint of erosion. Throughout central Africa, the mopane tree (*Colophospermum mopane*) is an immediate indicator of sodic soils that are extremely erodible. A less well-known example is the occurrence of mupundu (*Parinari curatellaefolia*), which flourishes on ridge tops in zones with a fluctuating seasonal water table in the root zone. Such conditions form noncalcic hydromorphic soils that have an open-bridge structure, strong when dry but susceptible to sudden collapse when wet. With the use of agricultural machinery, the management of such a soil is especially critical.

Cropping factors in predictive equations

Two USLE factors relate directly to vegetation and management. The cover and management factor, C, is the ratio of soil loss from an area with specified cover and management to that from an identical area in tilled, continuous fallow. The support practice factor, P, is the ratio of soil loss with a support practice, such as contouring, stripcropping, or terracing, to that with straight-row farming up-and-down slope. Field plot measurements form the basis for determining these factor values. Specific procedures for C and P values are laid down in the *Agriculture Handbook 537* (*46*).

Briefly, C values for a particular site are calculated from soil loss ratios representing six cropstage periods (rough fallow, seedbed, establishment, development, maturing crop, and residue/stubble) and three levels of canopy cover at the mature stage. Other variables include crop type; the specific rotation; stage in the rotation; method of plowing—moldboard, conventional, chisel; direction of plowing; use of winter cover crops; dry weight of residues in the spring; effect of chopping stalks for residue; and percentage of soil surface covered by mulch after crop seeding.

Some crops, for example, require further refinements and details to calculate accurate C values. A large number of interrelated variables are covered by this one factor. Consequently, there has to be considerable complexity both in method of determination and in final C values. For example, maize has some 60 C values in the United States to account for the variety of ways it is grown.

The P factor basically describes the effect of those management practices that are conservation-oriented: contouring, ridge planting, stripcropping, and terracing. It is more simple in its application than C, but there is considerable overlap in concept between C and P, with improved tillage, sod-based rotations, fertility treatments, and crop residues relegated to C rather than P.

Considerable research has gone into determining USLE factor values. Noteworthy in this respect are the research programs funded by the French Office for Overseas Scientific Research (ORSTOM) in West Africa and Brazil (*26, 33*). It remains an unfortunate fact that the complexity of tropical farming systems necessitates a vastly greater research effort to establish C and P values than has already been conducted in the United States, an impossible task with present resource constraints. Perhaps for individual high-value crops, the research effort may be worthwhile.

Other models tend not to address the complexities of vegetation and management. One exception is SLEMSA (Soil Loss Estimator for Southern Africa), which uses the concept of rainfall interception described earlier. It also incorporates management effects into soil erodibility on the basis that plowing, residues, and other standard management practices basically affect the susceptibility of the soil to erosion. This is one of the integrative components of SLEMSA that makes it somewhat different than the USLE.

Farming systems approaches

Farming systems research is now seen as a research methodology in its own right, although it has only rarely been applied specifically to soil conservation (*40*). If the objectives of soil conservation are to save soil and maintain productivity, then there are many means to that end. The most obvious and widely used means is to implement physical conservation works and land use planning. But time and again these have failed. What are the alternatives? At least five separate strategies have been identified (*39*), but all of them demand an intimate knowledge of farm practices, social and economic circumstances, labor peaks, and the like before they can be implemented with any confidence that they will be accepted. Farming systems research aims to analyze the farming system background to pinpoint any positive opportunities within the day-to-day life of the farm household in which practices that also meet soil conservation objectives can be incorporated—it is "conservation by stealth!" Notwithstanding the emphasis at some international research centers, farming systems research is not simply a technical analysis of cropping systems.

A key manual on farming systems research methodology is *Planning Technologies Appropriate to Farmers: Concepts and Procedures* (*6*), published by CIMMYT (Centro Internacional de Mejoramiento de Maiz y Trigo). It emphasizes a point that is particularly relevant for soil conservation research and planning: "Few farmers are following in their entirety the recommendations made by researchers and extension workers. Some argue that farmers are at fault, some that extension is ineffective, others that credit is unsuitable, and some that inputs are not available in a timely way. A less frequently heard explanation is that the recommended technologies themselves are simply not appropriate to farmers."

Adoption of a technically proven method of soil conservation by farmers is, in effect, the same procedure as the adoption of any new technology. That adoption hinges upon a number of interrelated factors. In the main, farmers seek technologies that increase their incomes while keeping risks within reasonable bounds. Whether one talks to a rich commercial farmer in a developed country or a shifting cultivator in the tropics, it is of little practical use to appeal to his or her conservation ethics. Farming systems research, in analyzing the interrelationships in decision-making on the farm, hopefully may identify where new technology and what type of technology can be introduced without upsetting the delicate set of conditions that the farmer perceives to be important.

CIMMYT (*6, 29, 34*) has identified a systematic set of procedures that are useful in obtaining information on a farmer's circumstances. In brief, the procedure involves

- Assembling background information, such as agro-climatic, erosion, population data, etc., from published and unpublished secondary sources.
- Making exploratory surveys—informal interviews with farmers, extension workers, merchants, and others with strong local knowledge—as a means of narrowing down the circumstances of farmers. A guiding principle adopted by most farming systems research work is that if a significant number of farmers in a region are using (or not using) a particular practice there is a good reason.
- Making a formal survey—formal interviews by trained interviewers using specific and focussed questions identified in the exploratory survey but needing qualification and verification.

A closely related research approach emphasizing vegetation and management is that adopted by agroforestry research. One of the principal objectives of agroforestry is soil conservation and the sustainability of agricultural/forestry/livestock production in difficult, erosion-vulnerable environments. The International Council for Research in Agroforestry, based in Nairobi, Kenya, has developed "A Diagnostic and Design Methodology for Agroforestry" (*31*) that incorporates elements of farming systems research in the context of rapid rural appraisal to identify the possibilities for the growth of woody plants in conjunction with other land use activities. Indeed, "rapid rural appraisal" (*5*), which has entered into the jargon of tropical development work, is a set of methodologies and techniques designed to maximize the social relevance of project planning activities without resorting to long, time-consuming, expensive survey work—an application especially relevant to soil conservation planning (*7*). Useful listings of plants suitable for soil erosion control and agroforestry purposes have been developed (*12, 28*).

Farming systems research involves a diversity of individual approaches. Of late, there has been a tremendous increase of interest in this line of investigation. If there is one lesson to be drawn from farming systems research that is relevant to all methodologies for the assessment of vegetative cover and management, it is that techniques for research and for planning in soil conservation need to be adapted far more closely to the farmer and his or her local environment. Universal solutions are impossible. The challenge for research is how best to identify human and physical environmental conditions at the local level rapidly and accurately enough to plan conservation effectively.

REFERENCES

1. Adams, J.E., C.W. Richardson, and E. Burnett. 1978. *Influence of row spacing of grain sorghum on ground cover, runoff and erosion.* Soil Science Society of America Journal 42: 959-962.
2. Adams, J.E., and G.R. Arkin. 1977. *A light interception method for measuring row crop ground cover.* Soil Science Society of America Journal 41: 789-792.
3. Branson, F.A., and J.E. Owen. 1976. *Plant cover, runoff and sediment yield relationships on Mancos shale in western Colorado.* Water Resources Research 6: 783-790.
4. Cackett, K.E. 1964. *A simple device for measuring canopy cover.* Rhodesian Journal of Agricultural Research 2(1): 56-57.
5. Chambers, R. 1981. *Rapid rural appraisal: rationale and repertoire.* Public Administration and Development 1: 95-106.
6. Centro Internacional de Mejoramiento de Maiz y Trigo. 1980. *Planning technologies appropriate to farmers: concepts and procedures.* El Batan, Mexico.
7. Collinson, M.P. 1981. *A low cost approach to understanding small farmers.* Agricultural Administration 8(6): 433-456.
8. De Ploey, J. 1982. *A stemflow equation for grasses and similar vegetation.* Catena 9: 139-152.
9. De Ploey, J., J. Savat, and J. Moeyersons. 1976. *The differential impact of some soil loss factors on flow, runoff creep and rainwash.* Earth Surface Processes 1: 151-161.
10. Dunne, T., W.E. Dietrich, and M.J. Brunengo. 1978. *Recent and past erosion rates in semi-arid Kenya.* Zeitschrift for Geomorphologie Supplementband 29: 130-140.
11. Ecosystems, Ltd. 1982. *Southeast Shinyanga Land Use Study. Report to World Bank/Shinyanga RIDEP/Government of Tanzania.* EcoSystems Ltd., Nairobi, Kenya.
12. El-Swaify, S.A., E.W. Dangler, and C.L. Armstrong. 1982. *Soil erosion by water in the tropics.* Research Extension Series 024. College of Tropical Agriculture and Human Resources, University of Hawaii, Manoa.
13. Elwell, H.A. 1980. *Design of safe rotational systems.* Department of Conservation and Extension, Harare, Zimbabwe. 50 pp.
14. Elwell, H.A., and F. Wendelaar. 1977. *To initiate a vegetal cover data bank for soil loss estimation.* Research Bulletin 23. Department of Conservation and Extension, Harare, Zimbabwe.
15. Elwell, H.A., and M.A. Stocking. 1974. *Rainfall parameters and a cover model to predict runoff and soil loss from grazing trials in the Rhodesian sandveld.* Proceedings of the Grassland Society of South Africa 9: 157-164.

16. Elwell, H.A., and M.A. Stocking. 1976. *Vegetal cover to estimate soil erosion hazard in Rhodesia.* Geoderma 15: 61-70.
17. Elwell, H.A., and M.A. Stocking. 1982. *Developing a simple yet practical method of soil loss estimation.* Tropical Agriculture (Trinidad) 59(1): 43-48.
18. Elwell, H.A., and S. Gardner. 1975. *Comparison of two techniques for measuring percent canopy cover of row crops in erosion research programmes.* Research Bulletin 19. Department of Conservation and Extension, Harare, Zimbabwe.
19. Hudson, N.W. 1957. *Erosion control research. Progress report on experiments at Henderson Research Station, 1953-56.* Rhodesian Agricultural Journal 54(4): 297-323.
20. Kemper, B., and R. Derpsch. 1981. *Results of studies made in 1978 and 1979 to control erosion by cover crops and no-tillage techniques in Parana, Brazil.* Soil and Tillage Research 1: 253-267.
21. Klemme, A.W., and O.T. Coleman. 1949. *Evaluating annual changes in soil productivity.* Bulletin 522. Missouri Agricultural Experiment Station, Columbia.
22. Kramer, L.A., and L.D. Meyer. 1969. *Small amounts of surface mulch reduce soil erosion and runoff velocity.* Transactions, American Society of Agricultural Engineers 12(5): 638-641, 645.
23. Lal, R. 1976. *No tillage effects on soil properties under different crops in Western Nigeria.* Soil Science Society of America Journal 48: 762-768.
24. Lam, K.C. 1978. *Soil erosion, suspended sediment and solute production in three Hong Kong catchments.* Journal of Tropical Geography 47: 51-62.
25. Lang, R.D. 1979. *The effect of ground cover on surface runoff from experimental plots.* Journal of the Soil Conservation Service of New South Wales 35(2): 100-114.
26. Leprun, J.C. 1981. *A erosao, a conservacao e o manejo de solo no nordeste Brasileiro: balanco, diagnostico e novas linhas de pesquisas. Recursos de Solos, SUDENE-DRN, 15.* Superintendencia do Desenvolvimento do Nordeste. Recife, Brazil.
27. Meyer, L.D., and J.V. Mannering. 1971. *The influence of vegetation and vegetative mulches on soil erosion.* In: *Biological Effects in the Hydrological Cycle.* Proceedings, Third International Seminar for Hydrology Professors. Purdue University, West Lafayette, Indiana. pp. 355-366.
28. Nair, P.K.R. 1980. *Agroforestry species. A crop sheets manual.* International Council for Research in Agroforestry, Nairobi, Kenya.
29. Norman, L.W. 1980. *The farming systems approach: Relevancy for the small farmer.* Rural Development Paper No. 5. Department of Agricultural Economics, Michigan State University, East Lansing.
30. Peterson, J.B. 1964. *The relation of soil fertility to soil erosion.* Journal of Soil and Water Conservation 19: 15-19.
31. Raintree, J.B. 1982. *A methodology for diagnosis and design of agroforestry land management systems.* International Council for Research in Agroforestry, Nairobi, Kenya.

32. Roberts, B.R., E.R. Anderson, and J.M. Fourie. 1975. *Evaluation of natural pastures; quantitative criteria for assessing condition in the* Themeda *veld of the Orange Free State.* Proceedings of the Grassland Society of South Africa 10: 133-140.
33. Roose, E.J. 1977. *Erosion et ruisellement en Afrique de l'ouest. Vingt annees de mesures en petites parcelles experimentales.* ORSTOM, Paris, France.
34. Shaner, W.W., P.P. Philipp, and W.R. Schmel. 1982. Farming systems research and development. Westview Press, Boulder, Colorado.
35. Sharma, S.S., R.N. Gupta, and K.S. Panwar. 1976. *Concept of crop protection factor evaluation in soil erosion.* Indian Journal of Agricultural Research 10(3): 145-152.
36. Soil and Water Conservation Society. 1979. *Effects of tillage and crop residue removal on erosion, runoff and plant nutrients.* Special Publication No. 25, Ankeny, Iowa.
37. Spomer, R.G., K.E. Saxton, and H.G. Heinemann. 1973. *Water yield and erosion response to land management.* Journal of Soil and Water Conservation 28(4): 168-171.
38. Sreenivas, L., J.R. Johnston, and H.W. Hill. 1947. *Some relationships of vegetation and soil detachment in the erosion process.* Proceedings, Soil Science Society of America 12: 471-474.
39. Stocking, M.A. 1983. *Development projects for the small farmer: Lessons from east and central Africa in adopting soil conservation.* In: *Soil Erosion and Conservation.* Soil and Water Conservation Society, Ankeny, Iowa. pp. 747-758.
40. Stocking, M.A. 1983. *Farming and environmental degradation in Zambia: The human dimension.* Applied Geography 3: 63-77.
41. Stocking, M.A. 1984. *Erosion and soil productivity: A review.* FAO Consultants' Working Paper No. 1. Soil Conservation Programme, Land and Water Development Division, Food and Agriculture Organization, Rome, Italy.
42. Stocking, M.A., and H.A. Elwell. 1976. *Vegetation and erosion: A review.* Scottish Geographical Magazine 92(1): 4-16.
43. Stocking, M.A., and N.O.J. Abel. 1981. *Ecological and environmental indicators for the rapid appraisal of natural resources.* Agricultural Administration 8: 473-484.
44. Walker, B.H. 1978. *An evaluation of eight methods of botanical analysis on grassland.* Journal of Applied Ecology 7: 403-416.
45. Wells, K.F. 1971. *Measuring vegetation changes on fixed quadrats by vertical ground stereophotography.* Journal of Range Management 24(3): 233-235.
46. Wischmeier, W.H., and D.D. Smith. 1978. *Predicting rainfall erosion losses—a guide to conservation planning.* Agriculture Handbook No. 537. U.S. Department of Agriculture, Washington, D.C.
47. Wollny, E. 1890. *Untersuchungen uber das Verhalten der atmospharischen Niederschlage zur Pflanze und zum Boden.* Forsch. Geb. Agriphys. 13: 316-365. (Reported by Baver, L.D. 1938. *Ewald Wollny—a pioneer in soil and water conservation research.* Proceedings, Soil Science Society of America 3: 330-333).

Chapter 10

MONITORING the IMPACT of SOIL EROSION on CROP PRODUCTIVITY

F. J. Pierce and R. Lal

The problem with erosion is not that it occurs. On the contrary, most of the world's soils formed in sediments deposited over geologic time by wind and water or through the action of glaciers and volcanic eruptions. The problem with erosion is found in its consequences. Soil erosion, accelerated relative to natural rates by human activities, reduces crop productivity and causes damage resulting from sediment transport and deposition. Sediment damage to air and water resources is of increasing importance, particularly in developed countries like the United States.

The other chapters in this book deal primarily with erosion processes, particularly in terms of definition and measurement, with an ultimate goal of predicting and controlling erosion. The quantification of the forms, rate, and extent of erosion has been the primary goal of soil erosion-related research worldwide. As a result, much more is known about the erosion process than the consequences of erosion. This is not to say that the emphasis in research on the erosion process was inappropriate. Quantitative knowledge as to the rate and extent of erosion is not only essential, it is still seriously limited relative to the scope of the erosion problem worldwide, in spite of the significant research progress to date. There is a serious paucity of quantitative data on erosion's impact on productivity. This is due, at least in part, to the fact that the early scientific evidence of the detrimental effects of erosion on crop productivity was considered sufficient. Thus the emphasis in research priorities was shifted to quantifying the rate and extent of soil erosion. Global efforts to assess degradation by soil erosion often measure degradation in terms of erosion rate rather than by its impact on productivity or off-site damages. Erosion degradation rates often are estimated from empirical functions relating climate, soils, land cover, and physiography (*24, 25, 26*). Even in the United States, where Stocking (*66*) estimates the majority (85 percent) of erosion-productivity research has been conducted, the level of quantification of erosion's impact on productivity is considered grossly inadequate (*10, 54*). In a comment on the general state of erosion productivity research, Stocking (*66*) states that such "research is often inappropriate—asking the wrong questions (or no questions at all!), using unrepresentative methods, measuring the wrong things, and, not surprisingly, getting the wrong answers." Lal (*43*) was equally pessimistic about our understanding of the problem in his conclusion that in spite of billions of dollars invested in the erosion problem, "we cannot say for sure what effect the loss of a unit of soil depth has on crop yield."

It is clear from the recent literature on the subject that much additional research is needed to quantify the effects of erosion on crop productivity. The data for soils in the tropics, although seriously limited, indicates the rate of decline in productivity to be considerably higher than that for soils in temperate regions (*18, 42*). This suggests that a research emphasis should be directed to soils in the tropics. Monitoring erosion's impact on productivity is very difficult because of the many complexities in the relationship between soils and plants and their management under a variable climate (*10, 32, 43, 54, 66*). The need is further accentuated by the growing concern over groundwater contamination with agrichemicals (*53*), climate change, and global cycles of carbon and nitrogen and their relationship to direct and indirect consequences of soil erosion on productivity.

The objectives of this paper are to establish the need for research on erosion's impact on productivity, and to review existing approaches and outline new approaches for monitoring erosion's impact on crop productivity.

THE NEED FOR EROSION-PRODUCTIVITY RESEARCH

It is widely recognized that erosion of soil by wind and water can reduce crop productivity in significant ways. Although erosion is considered the major agent of soil degradation worldwide (*19, 44, 47*), the large scale effects of erosion on the productivity of soils are not well known. This results from studies limited to relatively few soils (mostly in the United States), of short duration, and conducted under specific weather, cropping, and management conditions and, therefore, site specific (*54*). However, quantifying the impact of erosion on crop productivity is difficult because the response of plants to erosion is complex.

In the first place, erosion can physically damage or destroy plants. The impact of plant damage by erosion on crop productivity is related to the degree of damage, the condition of the plant and its stage of growth (i.e., potential for recovery from damage depends on its condition at the time of damage), the degree to which surrounding plants can compensate for the damaged or destroyed plants, and the environmental (growing) conditions following damage. Direct damage to plants and its impact on productivity are difficult to determine (*17*) and estimates of this effect of wind and water erosion are nearly non-existent.

Secondly, soils vary in their capacity to produce a crop or sequence of crops. Therefore, as erosion alters the physical, chemical, and biological properties that regulate the processes that determine crop productivity for a given soil, the productivity of that soil may change, the magnitude and rate of change depending on the vulnerability of the soil (*56*). The difficulty is that the productive capacity of a given soil varies spatially (*14*) and varies with climate, management, and plant genetics. Even under controlled conditions, it is difficult to identify quantitatively the soil properties that regulate productivity for a given soil volume. In most soil productivity studies, only a percentage of the variability in crop yield is accounted for in measured soil variables. In cases where regression coefficients between soil properties and crop yield are high, the contribution of measured soil property to variation in crop yield often varies with crop and year. Furthermore, these studies often shed little light on the mechanisms involved in the erosion-productivity relationship. Relating soil properties to yield is further confounded by the fact that as management inputs increase or as agriculture becomes more technologically advanced, the contribution of soil to crop yield diminishes (*3*). Managed inputs can often mask erosion damage but to what extent inputs can compensate for erosion damage needs further investigation. Considerable efforts have been directed toward quantifying the relationship between soil properties and crop productivity (*33, 54*). Relating changes in soil properties induced by erosion (real, perceived, or simulated) to crop yield has been a common method for assessing erosion's impact on productivity (*10, 43, 54, 66*). Efforts to quantify erosion's impacts on productivity using this type of approach must clearly recognize these potential limitations. Nonetheless, the majority of erosion productivity research is of this general nature.

Thirdly, erosion can affect crop productivity indirectly. Gully formation, for example, can limit access for agricultural production and thereby prohibit production or inhibit timely field operations. Ephemeral erosion can also change the local hydrology affecting plant-water relations within a field (*21, 43*). Transport of soil-borne pathogens in wind-eroded sediment can affect large areas at some distance from the source through the spread of disease (*32*). These and other indirect effects on productivity are difficult to quantify and are usually ignored.

Quantifying the effects of erosion on crop yields is, therefore, a complex task because it involves the assessment of a series of interactions among soil properties, landscapes, crop characteristics, the prevailing climate, and level of managed inputs. Additionally, the effects of erosion are cumulative and often not observed until long after accelerated soil erosion begins. It is also difficult to separate erosion effects from other degradation processes, for example, compaction or salinization. The spatial relationships and variability of soils within landscapes confound erosion studies. These have not been adequately addressed and may limit the utility of erosion-productivity research (*10*). For example, the phase of eroded soil is identified on the basis of the properties of the soil that remain. The problem is that erosion and soil development both occur at differential rates in the landscape and each can produce similar soil properties. In addition, crop yield, an integrated response to many interacting parameters, is difficult to relate under field conditions to any individual factor. It is, therefore, difficult to establish a one-to-one, cause-effect relationship between rates of soil erosion-induced soil degradation on the one hand and crop yields on the other.

Although complex, it is nonetheless important to assess erosion's impact on crop productivity in order to plan for agricultural development, to assess the adequacy of food resources for the world's population, and to evaluate agricultural policies at the local, regional, and national level (*80*). Knowledge of how erosion affects productivity is key to developing practices and policies for the restoration of eroded soils. Concerns over the quality of surface and ground waters, climate change, and the sustainability of agriculture have increased interest in soil erosion and erosion-induced productivity decline (*1, 12, 20, 23, 43*).

APPROACHES TO THE PROBLEM

There has been a variety of research approaches used to monitor and quantify erosion's effect on productivity. Stocking (*66*) and Lal (*43*) reviewed the subject extensively and provided a comprehensive description of many of the research approaches used throughout the world. Experimental approaches proposed at the Soil Erosion Soil Productivity Research Planning Workshop held in Lafayette, Indiana, (U.S.) in September 1981 were reported and discussed by Williams et al. (*78*), Renard and Follett (*61*), and Meyer et al. (*48*). Most recently, Cassel and Fryrear (*10*) and Hajek et al. (*32*) outlined research approaches proposed at a second national workshop on Research Issues in Soil Erosion/Productivity held in

March 1989 in Minneapolis, Minnesota (U.S.) (*46*). A number of important symposiums, held in the 1980s in the United States, focused specifically on the erosion-productivity problem (*1, 22, 23*). A recent overview of erosion productivity prediction is provided by Pierce (*54*). These publications provide an excellent overview of the problem, describe research approaches used to monitor erosion's impact on productivity, and outline priorities for future research for quantifying erosion's impacts on productivity. The various research approaches used to monitor erosion's impact on productivity are summarized in Table 10.1. General observations surmised by Pierce (*54*) from 50 years of erosion-productivity research are given in Table 10.2. A summary of research issues on erosion-productivity is given in Table 10.3 (*46*).

The research to date has provided sufficient examples demonstrating that erosion reduces crop productivity. It also clearly shows that the nature and extent of productivity changes induced by erosion vary with land use, physiographic region, soil, climate (macro and micro), and management (10). It is clear from Tables 1, 2, and 3 that more studies of this nature are not needed. The emphasis on future research on the erosion-productivity relationship needs to change from a focus on documentation of the occurrence of erosion-induced productivity change to a focus on its prediction in terms of rate and extent. Prediction has been the major focus of erosion process studies. The focus on prediction also reflects the major impetus for erosion-productivity research in the United States, namely Public Law 95-192, the Soil and Water Resources Conservation Act passed by the U.S. Congress in 1977. This law, commonly referred to as the Resources Conservation Act or RCA, requires the secretary of agriculture to assess periodically the status of and trends in use of soil, water, and related resources and their conservation on non-federal lands in the United States. This is to allow for informed, long-range policy decisions regarding the use and protection of these resources (*61*). The development of technology to predict erosion and its impact on productivity was essential to the fulfillment of the law and formed the basis for the RCA appraisals completed in 1982 and 1987 (*73, 74*). To our knowledge, such a mandate does not exist for any other country in the world. Although not explicit, the development of technologies to predict erosion's impact on productivity should also greatly assist in the other major goal of erosion research, that of the restoration and management of eroded lands.

The development of prediction capabilities requires fundamental and applied research on the relationship between soil properties (physical, chemical, and biological) and crop productivity, quantification of how erosion alters those properties, and the integration of this knowledge into predictive models. The ultimate goal is to define the relationship between

Table 10.1 Traditional research approaches used to evaluate erosion's impact on crop productivity (*43*).

Method	Description	Comment
Artificial Soil Removal	Manual removal of soil surface to different depths	Erosion is selective; does not simulate natural condition
Greenhouse	Comparative productivity evaluation under greenhouse conditions for surface vs. subsoil horizons (or mixtures)	Provides information on fertility but cannot simulate soil structure in field; should be validated under field conditions
Long-term Variable Management	Long-term field trials comparing different soil surface management or cropping systems	Difficult to separate management effects from erosion effects
Relating Soil Properties to Crop Yield	Relate erosion-induced alterations in soil properties to crop yield	Alterations in soil properties can be caused by intensive cultivation
Topsoil Depth/Crop Yield	Related crop yield to remaining depth of topsoil	Natural pedogenic factors can produce differential topsoil thickness in a landscape
Reconnaissance Surveys	Relating crop performance and yield to qualitative assessment of **past** soil erosion (e.g., soil erosion class)	Assessments are subjective; degree of past erosion difficult to quantify
Erosion Simulation	Rain and wind simulators used to accelerate rate of soil removal	Does not address long-term soil changes; equipment expensive
Modeling	Prediction of erosion's impact on soil properties and productivity	Existing models poorly validated in field

cumulative soil erosion and crop productivity for individual soils. This relationship is schematically illustrated in Figure 10.1 (*62*) and has been the functional form of the erosion-productivity relationship estimated in regional and national assessments in the United States (*56, 58, 59*). While this exact relationship is likely not known for any soil (*54*), it can be approximated.

While there is agreement on the need for predictive capabilities, there is no consensus on which of the varied approaches used to predict erosion's

Table 10.2 General conclusions drawn from 50 years of erosion and productivity research in the United States (*54*).

1. Yield levels of many of these studies were low relative to present production levels and study durations were for a few years only.
2. Management inputs were sufficient to restore production to levels of undisturbed soils and that the degree to which that was possible was related to the characteristics of the subsoil.
3. Under limited or no fertilizer amendments, yields were often highly related to depth of topsoil.
4. There is a relationship between crop yield and soil depth.
5. The pernancy of losses in crop production from erosion described in these studies is uncertain.
6. The ability to find uneroded sites is uncertain and limits assessment of past erosion.
7. Other effects of erosion have been largely ignored.
8. The effects of erosion on soil productivity are often hard to visualize. They are long-term and, at least temporarily, often masked by technology. The effects can be positive for certain soil conditions.
9. The spatial relationships and variability of soils within the landscape have generally been ignored in soil erosion studies.

impact on productivity is most appropriate. There are two basic approaches to developing predictions: statistical models and physically-based models.

Cassel and Fryrear (*10*) cite three classes of statistical models: 1) regression models in which crop yields are regressed against one or more variables including soil properties, landscape characteristics, and climate variables; 2) multivariate and factor analyses, which use data transformations and mathematical manipulations to analyze the variance structure within a multivariate data set; these often delineate cause-effect relationships not detectable with other statistical techniques and identify soil properties significant in defining crop productivity (*6*), and 3) geostatistical models, which analyze the spatial variance structure of areally distributed data (soil properties and erosion processes) and use the knowledge of spatial variation to predict the areal distribution of properties. The Universal Soil Loss Equation (USLE) (*77*) and the Productivity Index (*39*) are examples of regression type parametric models that were used by Pierce et al. (*55, 56*) to predict the long-term erosion impacts on soil productivity for soils in the Corn Belt region of the U.S.

Table 10.3 Summary of soil erosion and productivity research issues identified at the national workshop on Research Issues in Soil Erosion/Productivity held March 13-15, 1989, in Bloomington, Minnesota (*46*). Research issues are summarized by workgroup subject areas.

I. Landscape Processes Workgroup

- Document within a hillslope system the spatial distribution of soil layers, including restrictive zones, and their physical and chemical properties.
- Develop realistic and comprehensive soil erosion and hydrologic models that use increased detail on properties of soil layers and their spatial distribution.
- Examine the effect of landscape modification by erosion in terms of long-term productivity including: 1) climate, weather, and seasonality factors; 2) infiltration, overland flow, and other surface processes; 3) subsurface processes such as throughflow, deep percolation, anisotropic flow, water table activity, and flow in the shallow saturated zone; and 4) aerodynamics of the surface including the effects of landform materials and surface cover.
- Supplement soil surveys with three-dimensional landscape information including stratigraphy, geomorphology, and hydrology. Also characterize map units by typifying landscapes as well as typifying pedons.
- Improve erosion models so that they simulate redistribution of sediment and chemicals in a landscape at all scales and to predict (a) erosion, transport, and deposition, (b) off-site delivery of sediment and chemicals, and (c) long-term consequences of management alternatives.

II. Erosion and Soil Properties Workgroup

- Techniques to improve soil aggregation and enchance nutrient transformation in eroded soils.
- Predict the effects of erosion and their influence on soil productivity with process-based productivity models.
- Develop a process-based soil development model to predict soil regeneration.
- Assess wind erosion effects on the environment.
- Improve irrigation management to reduce erosion.

III. Productivity Changes Caused by Accelerated Soil Erosion in a Landscape Workgroup.

- Development, refinement and validation of predictive models.
- Development of new research designs for more efficient collection of data and for making precise interpretations.

  Specifically:

  (a) determine interactions among soil properties, weather, and management practices as they affect productivity within specific subareas of a landscape or field;

  (b) determine the impact of technological progress on potential production as well as the response of crop yields on erosional effects;

  (c) coordinate the design and conduct of field experiments with the development of process-based models to expand, more efficiently, our knowledge of proper resource use; and

(d) quantify the economic, environmental and social consequences of accelerated soil erosion.

- Assess productivity variability in map units through:

  a) evaluation of variability within map units using existing or new statistical and geostatistical techniques; and

  b) development of procedures to stratify variability in map units with respect to landscape characteristics and geomorphic components.

- Evaluate alternative technologies for assisting relationships between productivity and erosion across landscapes. Specifically:

  a) development of alternative technologies for acquisition, analyses and display of spatial and temporal data that affect productivity, hydrology, and erosion for all landscape positions on scales as small as a few meters; and

  b) development of accurate digital elevation data layers, which when combined with soil data and updated with remotely sensed land attribute data, can be combined into on-farm GIS (graphic information systems) and models.

IV. Decision Aids to Implement Conservation Workgroup

- Land management helps for conservation. Research to quantify the magnitude of soil conservation that can be achieved with existing erosion control and soil management practices and the potentials for improved technologies.
- Physically-based analytical helps for assessment of in-field erosion and its impact on productivity. Research needs to address spatial and temporal variability, risk assessment, integration of various erosion processes (wind, water, and mechanical), and validation of models.
- Integration of model results and research data for land managers and policymakers. This should include both on and off-site erosion effects as well as economic analyses.
- Building our fundamental research knowledge base. Specifically:

  a) consideration of both systematic and random local nonuniformities in flow characteristics and hydraulic parameters (both wind and water) that affect soil detachment, transport, and deposition.

  b) definition of soil erodibility as dynamic soil properties whose variability can be expressed in such terms as weather, tillage, and cropping management.

  c) determing the importance of soil properties such as soil water content, heat flow, rooting depth, and other properties.

V. Technology Transfer Workgroup

- Identify target audiences and specific needs to resolve conflicts between various users and managers of the soil resource.
- Evaluate the effectiveness of available delivery methods (models, expert systems, GIS, etc.) for technology transfer.
- Scientifically evaluate the impact of technology transfer.
- Coordinate a consistent program delivery. Be certain the information is getting to those who need to use it.

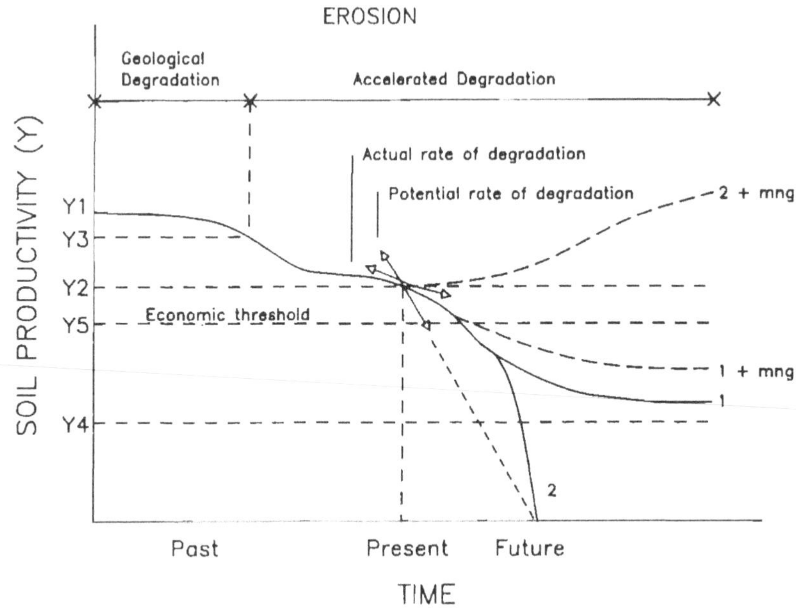

Figure 10.1 Changes in soil productivity as influenced by degradation <–> aggradation processes. Y1, Y2, and Y3 represent soil productivity initially, at present, and at the start of accelerated degradation, respectively. Curve 1 represents the case where cultivation is practiced for several years without fertilization, thus leading to asymptotic low productivity levels Y4. Curve 2 represents accelerated degradation with complete elimination of soil productivity (e.g., complete loss of a shallow soil). Curve 1-mng depicts improved management over curve 1 whereby erosion is stabilized after an initial decrease in soil productivity. Curve 2-mng represents successful management resulting in rebuilding of soil productivity. (*62*).

Multiple regression models are more commonly used to relate measured soil properties to crop yield for specific environment and cultural conditions. For example, Table 10.4 shows that yield of cowpea and maize grown on an Alfisol in Nigeria declined exponentially with cumulative soil erosion and the non-linear regressions coefficients were highly significant (*42*). The extent to which the results of site and management specific studies can be extrapolated to other regions and conditions may be limited. Multivariate and factor analyses are useful in identifying soil properties important to crop productivity but such procedures have found little use in erosion-productivity studies. Geostatistic models are only recently finding use in erosion-productivity studies and offer the important advantage of quantifying the spatial distribution of properties important to ero-

Table 10.4. Relationship between soil erosion, as measured on field runoff plots from 1970 to 1975, and grain yields of maize and cowpea on an Alfisol in southwestern Nigeria (*42*).

Slope %	Regression Equation	Correlation Coefficient
Cowpea		
1	Y = 0.43 exp (-0.036X)*	-0.85^{A}
5	Y = 0.64 exp (-0.06X)	-0.97^{B}
10	Y = 0.49 exp (-0.004X)	-0.91^{A}
15	Y = 0.29 exp (-0.002X)	-0.66
Maize		
1	Y = 6.41 exp (-0.017X)*	-0.99^{C}
5	Y = 6.70 exp (-0.003X)	-0.99^{C}
10	Y = 6.70 exp (-0.003X)	-0.89^{B}
15	Y = 8.36 exp (-0.004X)	-0.86^{A}

*Y is grain yield in ton/ha; X is cumulative soil erosion in ton/ha

A = significant at 5 percent level of probability; B = significant at 1 percent level of probability; C = significant at the 0.1 percent level of probability.

sion and productivity. The failure to account for spatial variability in erosion-productivity studies has raised questions about the validity of past research on erosion-productivity (*10, 14*). Therefore, geostatistical methods (derived from the theory of regionalized variables) should be an important modelling tool in erosion-productivity studies.

Physically based models apply physical laws, principally the conservation of mass and energy (*66*). Erosion prediction is moving away from empirical models like USLE to more physically based erosion prediction models in order to describe more accurately the various erosion processes and thereby improve prediction of soil erosion. In modelling erosion's impact on productivity, where the prediction of crop yield is important, crop simulation models are being developed. Whisler et al. (*75*) provide an excellent review of the need for, building, testing, and application of simulation models, and lists models currently available or under development. The Erosion-Productivity-Impact Calculator (EPIC) is the first simulation model developed for the sole purpose of simulating erosion's impacts on crop productivity. It was used as the basis for a national assessment for the 1987 RCA analysis (*74*). Models were also developed by Stocking and Pain (*67*), Timlin et al. (*71*), and Todorovic et al. (*72*) to describe the erosion-productivity relationship. The Nitrogen-Tillage-Residue Management (NTRM) model of Shaffer et al. (*65*) has been applied to assessing erosion's impact on productivity (*64*). In general, if a crop model effectively describes the important soil-related processes that regulate crop production, then a crop model, along with information about the rates of soil erosion and their effect on soil properties, will allow prediction of erosion's effect on productivity. The major limitation to these models is that they have not been adequately validated using scientifically defensible data (*10*).

According to Nix (*52*), a model or quantitative expression of site quality demands (a) identification of key parameters which specify the multidimensional systems involved, (b) delineation of the spatial and temporal distributions of these key parameters, and (c) a methodology for translating these data into accurate predictions of performance or production. Regardless of the model approach, predictive models must be capable of predicting the spatial and temporal distribution of erosion processes, soil properties, and crop yield (*10, 54*). It has been clearly demonstrated that soil development and erosion processes operate differentially in the landscape and both can produce similar soil properties (*16, 49, 68, 70*). Prediction further requires extensive resource data bases to meet the input needs for these models and an extensive program to monitor changes in the soil resource base (*54*). Soil data bases and monitoring programs also require that the spatial and temporal variation of soil properties be known.

This whole process of quantifying and predicting erosion's impact on crop productivity requires (a) the clear identification of soil properties that regulate crop productivity, (b) a coordinated monitoring program that quantifies the rate and extent of erosion-induced change in soil quality, erosion damage to crops, and indirect effects on crop productivity discussed earlier, (c) a coordinated research program designed to support and/or validate the models, and (d) standardization of field and laboratory methodologies that would allow the establishment of minimum data sets for evaluating erosion effects on soil productivity, regionally or even globally (*54*). A monitoring program can and should be conducted on a broad scale as these data will eventually serve to develop needed regional and national assessments. Erosion measurement and prediction efforts should incorporate such a monitoring program so that prediction models are capable of predicting changes in soil properties, crop damage, and indirect effects in addition to predicting erosion rate.

TOWARD A MINIMUM DATA SET

An erosion-productivity minimum data set (EP-MDS) is important for the development and validation of prediction models, the assemblage of data bases to use in assessment of erosion's impact on productivity over large geographic areas, and in the monitoring of the rate of change in the soil resource base due to soil erosion. We fully recognize that the properties and processes that regulate productivity within a given region and the alteration by erosion of soil properties may be unique to the land use, physiography, soil characteristics, and weather characteristics of that region. Therefore, an EP-MDS satisfactory for all possible soil, plant, and environmental conditions is not likely. However, while certain soil properties may be more prominent in some environments and a preference or a bias toward certain measurements or methodologies may exist, we believe that it is possible to arrive at a regional, if not global, consensus regarding a MDS that includes soil properties important to productivity and/or altered by soil erosion. The exact nature of a MDS for a region would reflect the overall objective for which data were being collected and would be constrained by the financial, social, political, or technical costs involved in obtaining the data.

A working model of the minimum data set concept is the International Benchmark Sites Network for Agrotechnology Transfer (IBSNAT) program funded by the U.S. Agency for International Development (US-AID) and centered at the University of Hawaii in Honolulu, Hawaii. The purpose of the IBSNAT project is to "assemble and distribute a decision software which enables its users to match the biological requirements of crops to the physical characteristics of land so that objectives specified by the user may be obtained" (*36*). A component of the IBSNAT project was the development of a MDS of weather, soil, management, and crop response data for the specified purpose of running and validating crop simulation models developed as another component of the project. The experimental design and data collection procedures for the IBSNAT minimum data set were published in a technical report (*34*). That report provides data recording sheets and detailed instructions for the measurement and recording of data. Details of the field and laboratory procedures for these measurements were also published (*35*). The general content of the IBSNAT minimum data set is listed in Table 10.5.

For purposes of assessing erosion's impact on soil productivity, the IBSNAT MDS contains many items that would be desirable in a MDS for monitoring erosion's impact on productivity. It includes a minimum climate data set and promotes a thorough accounting of the phenology and condition of the growing crop. However, the IBSNAT MDS is incomplete in that it includes limited soils data and excludes most soil properties altered by soil erosion. Therefore, the IBSNAT MDS should be expanded to include soil properties important to erosion-induced productivity changes.

Soil properties identified as having importance to productivity and/or affected by soil erosion have been discussed extensively elsewhere (*32, 43, 45, 66*). Determining a soil property minimum data set is difficult because soil properties that constrain productivity in some soil environments

Table 10.5. The general content of the minimum data set used to run and validate crop simulation models as components of the International Benchmark Sites Network for Agrotechnology Transfer (IBSNAT) Program (*34*).

Institutional Information
Nearby Long-term Climatic Stations
Daily Weather (Required)–temp., precip., solar radiation
Daily Weather (Additional)–dry bulb, wet bulb, and windrun
Experimental Site Data
Experimental Conditions
Experimental Treatments
Experimental Plot Data
Soil Fertility Measurements–pH, NO_3-N, NH_4-N, P, K, Al (%)
Soil Water Contents (preplant and misc. dates)
Cultural Practices–tillage, cultivar, planting, fertilizers, innoculants, amendments, biocides, hormones (type, timing, method, and amount)
Irrigation Data
Crop Damage
Plant Phenology (Growth stages given for maize, grain sorghum, wheat, rice, barley, millet, soybean, dry bean, peanut, ariod, cassava, and potato)
Harvest Data (by plant part)
Plant Nutrient Concentrations–N, P, K, misc. by plant part

may have little or no effect in other environments. There is also a need to quantify criteria for soil degradation by erosion in terms of critical limits beyond which crop growth will be affected (*43*). Additionally, soil data required as input vary with predictive model. The North Central Regional Research Committee NCR-174 in the U.S. has as part of its regional project the effects of erosion on soil productivity to validate the EPIC and NTRM crop simulation models. The MDS developed for this purpose reflects the data input needs for the two models (Table 10.6).

THE EROSION-PRODUCTIVITY MDS

The productive potential of a soil can be evaluated in terms of the environment it provides for root growth and development. The soil rooting environment must be defined in terms of its quantity (volume) and quality. The rooting volume needed for optimum growth depends on the plant species, which vary in their rooting depth and lateral spread of roots (*69*), the genetic potential of the plant, and the environmental conditions imposed by the weather and cultural management practices used in the production system.

Table 10.6. Proposed measurements for the North Central Regional Committee on Soil Productivity and Erosion (NCR-174) regional research project to be used in conjunction with the Erosion-Productivity Impact Calculator (EPIC) and Nitrogen-Tillage-Residue Management (NTRM) simulation models.

Crop Growth
-Planting rate and date
-Dates of 50 percent emergence, silking (corn) or heading (wheat), and maturity (black layer formation for corn and soft dough stage for wheat.
-Plant population at four leaf stage and harvest
-Number of ears and heads per unit at harvest
-Grain and stover or straw yield and moisture content
-Tissue nutrient concentration (N, P, K) at various growth stages
-Notes on pest and disease problems
Soil Measurements
Initial and Final
-Maximum rooting depth
-Horizon thickness
-Bulk density below Ap horizon
-Saturated hydraulic conductivity*
-Particle size analysis
-pH*
-Calcium carbonate equivalent*
-Exchangeable Al*
-Organic C*
-Cation exchange capacity (CEC)*
Annual
-Bulk density after planting (AP)
-Saturated hydraulic conductivity (AP)
-pH, available P and K before planting (BP)
-Soil water contents (AP)*
-Soil surface conditions (crop residue cover, soil roughness, and soil residue reflectence) (AP)
-Total above ground dry matter prior to tillage
-Soil temperatures (AP - 0, and 10 cm, more if possible)
Climate data (daily basis)
-total precipitation (rate if available)
-Max-min air temperature at 2 m
-Wind run at 2 m
-Pan evaporation
Optional Data
-Soil temperature (hourly or daily max-min)
-Soil water content of rooting depth at intervals during the growing season
-Water runoff
-Daily plant water stress

* all horizons or depth increments

Of general interest is the potential depth of rooting in a given soil and the presence of physical, chemical, or biological constraints which regulate the quality of the rooting environment. Chemical constraints to root growth include the presence of inhibitory compounds (e.g. Al saturation, salts), or a nutrient deficiency (either due to low quantity present or high fixation capacity of soil for a particular nutrient such as phosphorus), many of which are pH related. Physical constraints are of two types: those that deal with the suitability of soil to accommodate unobstructed root growth (pore space of sufficient size and continuity for root penetration and expansion and the resistance of the soil matrix to deformation (*31*)), and the capacity of soil for water supply and aeration (also related to quantity and quality of pore space). Biological constraints include the presence of soil-borne pathogens, diseases, and alleleopathic organic compounds. Biological constraints may be more related to past management than to soil development. Munn and Franco (*50*) summarized soil related constraints limiting the use of tropical soils for agriculture (Table 10.7). A number of these constraints can often be alleviated by cultural management practices when they occur at or near the soil surface. Problems in soil productivity arise where inputs are limited or when undesirable soil physical properties occur beyond the depth of management. Erosion can alter the productivity of a soil by altering soil properties to ones that are constraining to root growth or by altering the relative position in the soil profile of soil constraining properties and thereby limiting the effective rooting volume.

We recognize that there is no agreement about which soil properties regulate soil productivity. However, we propose that an EP-MDS should include the IBSNAT MDS and add the soil properties and standardized methodologies and procedures outlined in Table 10.8. These are discussed in the following paragraphs. We recommend that this MDS or some modified version of it be adopted for erosion-productivity research regardless of choice of research objective (Table 10.3) or research approach (Table 10.1). We also recommend that erosion prediction models begin to predict changes in these properties as affected by erosion. The input needs of crop productivity models should be reflected in variations of the EP-MDS.

Soil variability

Each soil property to be included in the EP-MDS is briefly discussed. First, however, we must emphasize the primary objective of experimentation, that is, to control and quantify variation. This is particularly important in field studies since the spatial and temporal variation of many soil properties can be very large and can exceed variation induced by erosion. Recent studies have shown that variation in soil properties is spatially dependent and controlled by landscape position. Variation can be natural due

Table 10.7. Limitations of tropical soils for agriculture (*50*).

Soil Order	Problems and their likelihood of occurrence[a]						
	% Total area	Water	Strength, etc.	Erosion	Salt	Acid	Nutrient
Aridisols	18	***	*	***	***	0	*P,Zn
Vertisols	3	*	*	0	*	0	*P,Zn
Molli- Enti- Inceptisols							
Fine	8	0	0	*	*	0	0
Sandy	18	***	0	**	0	*	*P,K,S,Zn
Ando (ash)	4	*	0	***	0	*	*P
Alfisols	18	**	**	***	0	**	*P,S,Zn,K
Ultisols	12	*	0	***	0	***	***P,S, Mo,K,Zn
Oxisols	20	**	0	*	0	***	*P,Zn, Mo,K,Zn

[a]From 0, little likelihood, to ***, very high likelihood

Table 10.8 Soil properties and standard methodologies for their measurement to be included as part of a minimum data set (EP-MDS) for monitoring erosion's impact on crop productivity.

Soil Parameter	Standard Methodology	Comment
Nutrient Availability	Use analytical soil test for region	
Organic Carbon (C)	Use wet combustion method	(*51*)
Particle Size	Hydrometer method	(*29*)
PAWC	Determined in field best or from water desorption curve	(*5*), (*40*), (*63*), (*11*)
Soil Structure Form	Bulk density from intact soil cores; field measured air permeability or Ksat	(*2*), (*41*), (*13*)
Aggregate Stability	Single sieve wet aggregate stability; dry aggregate stability	(*38*)
Soil Strength	Bulk density or penetration resistance	(*2*), (*4*)
Maximum Rooting Depth	Crop specific—depth of common roots or standard	(*69*)

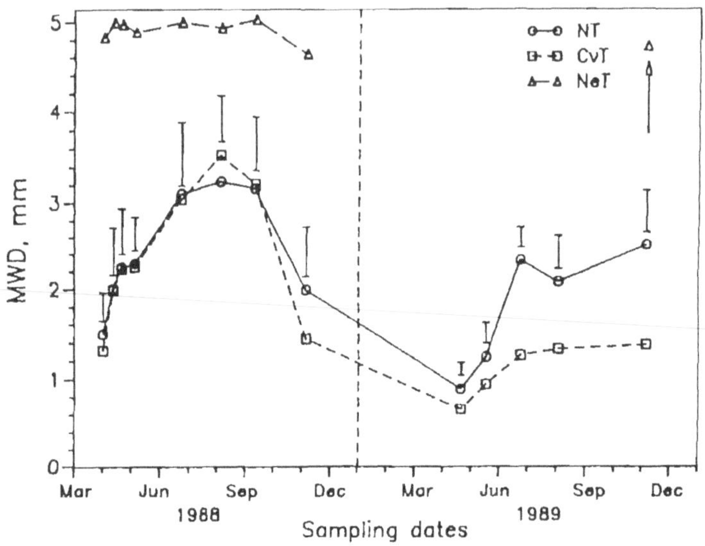

Figure 10.2 Temporal variation of mean weight diameter (MWD) of wet sieved aggregates sampled from the surface 5 cm of a Kalamazoo loam soil (fine-loamy, mixed, mesic Typic Hapludalf) under conventional, no-tillage, and never tilled management in corn (*60*).

to soil forming processes or human-induced by variability in management of soil (*10*). The spatial variation in bulk density in an agricultural field is considerable at close distances due, for example, to alternating presence of wheel tracks (*9*). Large temporal variations may occur in soil properties as illustrated in Figure 10.2 for aggregate stability (*60*). The cyclic pattern of variation was largely attributed to the variation in field water contents at the time of sampling (*60*).

Experimentally, the first choice is to quantify the spatial and temporal variation using appropriate statistical procedures of geostatistics and time series analysis (*10*). However, spatial variation can be controlled and minimized. Be sure that variation due to landscape position is accounted for in comparing erosion effects. Keep sampling locations within an experimental unit in close proximity from one sampling time to another to minimize spatial dependence effects. Minimize temporal variation by sampling in the same period of the temporal cycle. Using the example in Figure 10.2, the appropriate sampling time to minimize the temporal variation in aggregate stability is after the spring thaw and prior to the first soil drying cycle.

Regardless of how variation is approached experimentally, the importance in controlling and quantifying it cannot be overstated.

Nutrient availability

Nutrient availability is an important soil property for productivity and is significantly altered by erosion. This is already included in the IBSNAT MDS. An aspect of nutrient availability that may be important in erosion studies is the measurement of the nutrient supplying potential of a soil, such as N mineralization potential and P buffering capacity. Erosion-induced changes in the nutrient supplying capacity of soils can be significant. Yost et al. (*81*) showed that an important result of surface soil loss in some soils is that residual P that may have accumulated for upwards of 50 years may be lost and its replacement may require large applications over several years to correct the deficiency. Consideration should also be given to monitoring the availability of other nutrients or toxins that might constrain productivity. For example, the IBSNAT MDS includes extractable aluminum (Al). The choice of additional nutrients or toxins should reflect regional soil constraints.

Soil organic carbon (C)

Soil organic C or soil organic matter is greatly affected by soil erosion and identified with productivity. J. F. Power says of organic C that "the quantity of organic C in soil is the one soil property that has an overwhelming effect on soil productivity. . . One might conclude that practices that enhance soil organic C are likely also to enhance productivity" (*32*). Depletion of soil organic matter by erosion is accompanied by the removal of plant nutrients, deterioration of soil structure, and diminished soil workability (*27*). Lower organic matter affects the water holding capacity of soil and alters the efficacy and fate of applied pesticides. Depletion of soil organic matter and erosion are spirally cyclic since a decrease in organic matter increases the susceptibility of a soil to erosion, thereby increasing the rate of depletion of soil organic C. Because of its importance and its susceptibility to change by soil erosion, organic C must be included in the EP-MDS and monitored periodically.

Soil texture

Soil texture in the surface layer of soil is altered because of the selective removal of fine particles during the erosion process, as a result of mixing of subsoil into the surface layer as cumulative erosion reduces the thickness of the surface soil layer, and as a result of deposition of eroded sediments on the soil surface. The implications of a change in the surface soil texture on crop productivity can be significant (*27, 43*). White et al. (*76*) reported a strong relationship between crop yield and thickness and clay content of the A horizon. Frye et al. (*28*) and Stone et al. (*68*) reported that increasing clay content of the Ap horizon correlated well with increasing severity of erosion.

Plant available water-holding capacity

Perhaps the greatest impact erosion has on soil is in the reduction of its available water-holding capacity, PAWC (*78*). Even in humid regions of the world, agronomic crops experience drought sometime during their life cycle nearly every year. Therefore, crop productivity is dependent upon the capacity of a soil to provide water to plants. The importance of plant available water and its measurement are discussed by Ritchie (*63*). Plant available water capacities are required input for nearly all crop simulation models. The PAWC should be determined to the depth of rooting, and temporal changes in PAWC, either natural or induced by management or erosion, should be determined in the surface layers.

Soil structure

We use the term soil structure as broadly defined by Kay (*37*) as having three components. The first is structural form, which refers to the geometry of the soil pore space (porosity, pore size distribution, and pore continuity); the second is aggregate stability, which refers to the size distribution and resistance of aggregates to degradation; and the third is structural resiliency, which refers to the ability of structure to re-form once it has been degraded. All three components of structure are important in quantifying both the processes of erosion and the impacts of erosion on productivity. All measurements of soil structure can be made on the surface layer. Measurements of structural form include bulk density, macroporosity (pore diameter > 60 microns) and saturated hydraulic conductivity. Aggregate stability should be measured using the single sieve method (WAS) of Kemper and Rosenau (*38*). An appropriate measure of resiliency of soil structure is unclear and none is recommended here.

Determination of high strength soil layers that impede root growth are important for the determination of effective soil rooting volume. Either bulk density or penetration resistance measurements (interpreted with respect to water contents) can be used to identify root-impeding layers. To our knowledge, however, only bulk density is currently used in crop simulation models to assess root growth. Soil crusting is an important feature of degraded soil structure and significantly affects water and heat flow in soils. It is not included in the EP-MDS for lack of standardized methodology.

Rooting depth

Soil thickness has been related to crop productivity, particularly in mine reclamation studies (*57*). Soils in which rooting depth is limited by the presence of a physical or chemical constraint are generally less productive. It is well known that as limiting layers are moved closer to the soil surface as erosion removes topsoil, crop productivity generally declines. The depth of rooting is an important parameter to measure in erosion studies. Rooting depth varies by crop species and limits for various species are given by Taylor and Terrel (*69*). Maximum rooting depth should be determined at physiological maturity of the crop species under study.

The EP-MDS proposed here includes climate, crop performance (growth and development), and soil properties. Recent advances in automated weather stations have increased the opportunities for improved climate data acquisition. Crop monitoring programs, such as IPM, already exist in developed countries. Monitoring of soil properties is not routine and poses a significant problem, particularly with regard to erosion-induced changes. Perhaps an even greater problem is the coordination and maintenance of data bases. The IBSNAT program currently receives and maintains all the MDSs collected in their program. There is no repository for EP-MDSs, and that will be required in the long-term. In the short term, however, researchers will have to maintain their own EP-MDSs and share them as needed.

MODEL BUILDING AND VALIDATION

There is somewhat of a paradox in the use of models in studying erosion. The development and use of models has been a primary focus in the study of erosion processes. The Universal Soil Loss Equation (USLE) has been tested and used worldwide. Literally thousands of research papers have been written that either directly or indirectly deal with the development, testing, or use of the USLE. Models dealing with the effects of erosion on crop productivity, however, are few, little used, and are usually severely criticized as being inadequate and poorly validated. Recent reviews, while very critical, identify modeling productivity changes due to erosion as a prime research need (*10, 43, 66*). This paradox is not difficult to understand if one realizes that there is no comparison between the level of research support given erosion process research and that given erosion-productivity research. Dregne's (*18*) recent review of erosion-induced productivity decline in Africa clearly demonstrates that this scenario is true worldwide.

There is no prescription for what comprises the appropriate model. Four major studies have assessed the long-term impacts of erosion on soil productivity in the United States (*56, 58*). All used different models in their assessment. Each predicted that productivity losses, aggregated over all cropland acres, will not be great over the next hundred years under current erosion rates and management practices (*54*).

No one feels comfortable with these low predictions of productivity loss. However, these studies and the Clark et al. (*12*) report on the significance of the off-site effects of erosion caused a shift in policy in the United States toward assessing and controlling the off-site effects of erosion.

Stocking (*66*) suggested that an appropriate or effective model should have (a) readily available inputs, (b) an output that can link directly to economic or conservation planning decisions, (c) physical/mathematical expressions to link the steps connecting erosion to yield losses/fertility decline/productivity, and (d) a suitable structure to guide the planning of field research. A major dilemma in model development is choosing an appropriate level of resolution (*75*). The more mechanistic or high resolution models are those that attempt to represent causality between variables. These often require information not readily available. Empirical or low resolution models are those that describe relationships between variables without referring to any underlying structure between variables. These models do not have good predictive ability. Empirical models like PI (*39, 55*) have proven useful in some situations but are clearly in need of revision in applications in a range of environmental conditions (*43, 66*). Even the higher resolution models like EPIC (*79*) and NTRM (*65*) contain empiricism. In the second RCA appraisal (*74*), the output from EPIC was transformed into a more empirical form such as that illustrated in Figure 10.1 in order to perform the economic analysis (*58*).

Many of the crop simulation models now available (*75*) could be used in predicting the effects of erosion on productivity. These are limited by how effectively the given model describes the important soil related processes that regulate crop production. They also need to be linked to erosion models which predict changes in soil properties. Some models (for example, the IBSNAT family of models) currently do not account for many of the dynamic soil properties induced by tillage and residue management which regulate many important processes that occur at or near the soil surface. There is some advantage to modifying existing crop models which in combination with an EP-MDS could meet the model requirements set forth by Stocking (*66*).

There are a few additional items that models should address. Erosion process models should be modified to predict changes in soil physical properties. This would provide the necessary input for erosion-productivity models and serve as a means of predicting changes in the quality of the soil resource base. Models should also be able to predict the spatial and temporal variation in erosion processes, soil properties, and crop productivity. This will require an understanding of landscape processes and the use of new technologies like geostatistics, geographic information systems (GIS), remote sensing, and satellite based global positioning systems (*10*). Geographically based models of erosion and productivity will also allow the determination of the indirect effects of erosion on productivity, including the effects of deposition and alteration of regional hydrology. Models should also focus on simulating crop phenology so as to predict the timing and nature of crop damage relative to the plant's life cycle. Determination of yield alone often sheds little light on the mechanisms and fails to quantify the full extent of crop damage by erosion.

SUMMARY AND CONCLUSIONS

The need for a major research effort in monitoring erosion's impact on crop productivity is great. Field research that relates soil properties to crop productivity and monitors the alteration of soil properties by erosion must be done on a global scale. The focus of future research should be the development of validated predictive technologies that will allow the assessment of the adequacy and sustainability of soil resources worldwide. A minimum data set (MDS) is described for the evaluation of erosion's impact on productivity. An EP-MDS is necessary for the development and validation of predictive technologies, as input for regional and global assessments of the soil resource, for the determination of the rate of soil degradation, and as a resource to be used in the management and restoration of eroded lands. The coordination and integration of erosion-productivity research suggested here should promote progress in this important research arena and enhance our understanding and protection of our limited soil resource base.

REFERENCES

1. American Society of Agricultural Engineers. 1985. *Erosion and soil productivity*. Publication 8-85, American Society of Agricultural Engineers, St. Joseph, Michigan.
2. Blake, G.R., and K.H. Hartge. 1986. *Bulk Density*. In: A. Klute (ed.) *Methods of Soil Analysis. Part 1*. 2nd ed. Agron. Monogr. 9: 425-444. American Society of Agronomy, Soil Science Society of America, Madison, Wisconsin. p. 363-376.

3. Blakie, P. 1985. *The political economy of soil erosion in developing countries.* Longman Scientific and Technical, London, England.
4. Bradford, J.M. 1986. *Penetrability.* In: A. Klute (ed.) *Methods of Soil Analysis. Part 1.* 2nd ed. Agron. Monogr. 9: 425-444. American Society of Agronomy, Soil Science Society of America, Madison, Wisconsin. p. 463-478.
5. Bruce, R.R., and R.J. Luxmoore. 1986. *Water retention: Field methods.* In: A. Klute (ed.) *Methods of Soil Analysis. Part 1.* 2nd ed. Agron. Monogr. 9: 425-444. American Society of Agronomy, Soil Science Society of America, Madison, Wisconsin. p. 663-686.
6. Bruce, R.R., A.W. White Jr., A.W. Thomas, W.M. Snyder, and G.W. Langdale. 1989. *Characterization of soil-crop yield relations over a range of erosion on a landscape.* Geoderma 43: 99-116.
7. Bruce, R.R., W.M. Snyder, A.W. White Jr., A.W. Thomas, and G.W. Langdale. 1990. *Soil variables and interactions affecting prediction of crop yield pattern.* Soil Science Society of America Journal 54: 494-501.
8. Carter, D.L., R.D. Berg, and B.J. Sanders. *The effect of furrow irrigation erosion on crop productivity.* Soil Science Society of America Journal 49: 207-211.
9. Cassel, D.K. 1983. *Spatial and temporal variability of soil physical properties following tillage of Norfolk loamy sand.* Soil Science Society of America Journal 47: 196-201.
10. Cassel, D.K., and D.W. Fryrear. 1990. *Evaluation of productivity changes due to accelerated erosion.* In: W.E. Larson, G.R. Foster, R.R. Allmaras, and C.M. Smith (adds). *Research Issues in Soil Erosion/Productivity.* University of Minnesota, St. Paul. p. 41-54.
11. Cassel, D.K., and D.R. Nielsen. 1986. *Field capacity and available water capacity.* In: A. Klute (ed.) *Methods of Soil Analysis. Part 1.* 2nd ed. Agron. Monogr. 9: 425-444. American Society of Agronomy, Soil Science Society of America, Madison, Wisconsin. p. 901-926.
12. Clark, E.H., II, J.A. Haverkamp, and W. Chapman. 1985. *Eroding Soils: The Off-farm Impacts.* The Conservation Foundation, Washington, D.C.
13. Corey, A.T. 1986. Air permeability. In: A. Klute (ed.) *Methods of Soil Analysis. Part 1.* 2nd ed. Agron. Monogr. 9: 425-444. American Society of Agronomy, Soil Science Society of America, Madison, Wisconsin. p. 1121-1136.
14. Daniels, R.B., and G.D. Bubenzer. 1990. *Landscape processes and soil erosion and productivity.* In: W.E. Larson, G.R. Foster, R.R. Allmaras, and C.M. Smith (adds). *Research Issues in Soil Erosion/Productivity.* University of Minnesota, St. Paul. p. 1-22.
15. Daniels, R.B., J.W. Gilliam, D.K. Cassel, and L.A. Nelson. 1985. *Soil erosion class and landscape position in the North Carolina Piedmont.* Soil Science Society of America Journal 49: 991-995.
16. Daniels, R.B., J.W. Gilliam, D.K. Cassel, and L.A. Nelson. 1987. *Quantifying the effects of past erosion on present productivity.* Journal of Soil and Water Conservation 42: 183-187.

17. Downes, J.D., D.W. Fryrear, R.L. Wilson, and C.M. Sabota. 1977. *Influence of wind erosion on growing plants.* Transactions, American Society of Agricultural Engineers 20: 885-889.
18. Dregne, H.E. 1990. *Erosion and soil productivity in Africa.* Journal of Soil and Water Conservation 45: 431-436.
19. Dudal, R. 1982. *Land degradation in a world perspective.* Journal of Soil and Water Conservation 37: 345-349.
20. Edwards, C.A., R. Lal, P. Madden, R.H. Miller, and G. House (eds). 1990. *Sustainable Agricultural Systems.* Soil and Water Conservation Society, Ankeny, Iowa.
21. El-Swaify, S.A., E.W. Dangler, and C.L. Armstrong. 1982. *Soil erosion by water in the tropics. Research Extension Series.* HITAHR. College of Tropical Agriculture and Human Resources, University of Hawaii, Honolulu.
22. El-Swaify, S.A., W.C. Moldenhauer, and A. Lo. 1985. *Soil Erosion and Conservation.* Soil and Water Conservation Society, Ankeny, Iowa.
23. Follett, R.F., and B.A. Stewart (eds.). 1985. *Soil Erosion and Crop Productivity.* Soil Science Society of America, Madison, Wisconsin.
24. Food and Agriculture Organization. 1977. *Assessing soil degradation.* Soils Bulletin, 34: 83. Rome, Italy.
25. Food and Agriculture Organization. 1978. *Methodology for assessing soil degradation.* Report on Food and Agriculture Organization Expert Consultation. Rome, Italy.
26. Food and Agriculture Organization. 1979. *A provisional methodology for assessing soil degradation.* Rome, Italy.
27. Frye, W.W. 1987. *The effects of soil erosion on crop productivity.* In: J.M. Harlin and G.M. Berardi (eds.) *Agricultural Soil Loss: Processes, Policies, and Prospects.* Westview Press, Boulder, Colorado. p. 151-172.
28. Frye, W.W., S.A. Ebelhar, L.W. Murdock, and R.L. Blevins. 1982. *Soil erosion effects on properties and productivity of two Kentucky soils.* Soil Science Society of America Journal 46: 1051-1055.
29. Gee, G.W., and J.W. Bauder. 1986. *Particle-size analysis.* In: A. Klute (ed.) *Methods of Soil Analysis. Part 1.* 2nd ed. Agron. Monogr. 9: 425-444. American Society of Agronomy, Soil Science Society of America, Madison, Wisconsin. p. 383-412.
30. Greenland, D.J. 1981. *Soils and crop production in the lowland humid tropics.* In: D.J. Greenland (ed.) *Characterization of soils in relation to their classification and management for crop production: examples from some areas of the humid tropics.* Clarendon Press, Oxford. p. 1-9.
31. Groenevelt, P.H., B.D. Kay, and D.C. Grant. 1984. *Physical assessment of a soil with respect to rooting potential.* Geoderma 34: 101-114.
32. Hajek, B.F., D.L. Karlen, B. Lowery, J.F. Power, T.E. Schumaker, E.L. Skidmore, and R.E. Sojka. *Erosion and Soil Properties.* In: W.E. Larson, G.R. Foster, R.R. Allmaras, and C.M. Smith (adds). *Research Issues in Soil Erosion/Productivity.* University of Minnesota, St. Paul. p. 23-40.

33. Huddleston, J.H. 1984. Development and use of soil productivity ratings in the United States. Geoderma 32: 297-317.
34. International Benchmark Sites Network for Agrotechnology Transfer Project. 1988. *Experimental Design and Data Collection Procedures for IBSNAT.* IBSNAT Technical Report 1. Third edition. Department of Agronomy and Soil Science, College of Tropical Agriculture and Human Resources, University of Hawaii, Honolulu.
35. International Benchmark Sites Network for Agrotechnology Transfer Project. 1986. *Field and laboratory methods for IBSNAT experiments.* IBSNAT Technical Report 2. Department of Agronomy and Soil Science, College of Tropical Agriculture and Human Resources, University of Hawaii, Honolulu.
36. International Benchmark Sites Network for Agrotechnology Transfer Project. 1990. *Proceedings of IBSNAT Symposium: Decision support system for Agrotechnology Transfer, Las Vegas, Nevada. October 16-18, 1989.* Department of Agronomy and Soil Science, College of Tropical Agriculture and Human Resources. University of Hawaii, Honolulu.
37. Kay, B.D. 1989. *Rates of change of soil structure under different cropping systems.* Advances in Soil Science 12: 1-52.
38. Kemper, W.D., and R.C. Rosenau. 1986. *Aggregate stability and size distribution.* In: A. Klute (ed.) *Methods of Soil Analysis. Part 1.* 2nd ed. Agron. Monogr. 9: 425-444. American Society of Agronomy, Soil Science Society of America, Madison, Wisconsin.
39. Kiniry, L.N., C.L. Scrivner, and M.E. Keener. 1983. *A soil productivity index based upon predicted water depletion and root growth.* Research Bulletin 1051, Missouri Agricultural Experiment Station, Columbia.
40. Klute, A. 1986. *Water retention: Laboratory methods.* 635-662. In: A. Klute (ed.) *Methods of Soil Analysis. Part 1.* 2nd ed. Agron. Monogr. 9: 425-444. American Society of Agronomy, Soil Science Society of America, Madison, Wisconsin.
41. Klute, A., and C. Dirksen. 1986. *Hydraulic conductivity and diffusivity: Laboratory measurements.* In: A. Klute (ed.) *Methods of Soil Analysis. Part 1.* 2nd ed. Agron. Monogr. 9. American Society of Agronomy, Soil Science Society of America, Madison, Wisconsin. p. 687-734.
42. Lal, R. 1981. *Soil erosion problems on Alfisols in western Nigeria. VI. Effects of erosion on experimental plots.* Geoderma 25: 215-230.
43. Lal, R. 1987. *Effects of soil erosion on crop production.* CRC Critical Reviews in Plant Science 5(4): 303-367.
44. Lal, R. 1990. *Soil erosion and land degradation: The global risks.* In: R. Lal and B.A. Stewart (eds.) *Soil Degradation.* Advances in Agronomy 11: 129-172.
45. Larson, W.E., T.E. Fenton, E.L. Skidmore, and C.M. Benbrook. 1985. *Effects of soil erosion on soil properties as related to crop productivity and classification.* In: R.F. Follett and B.A. Stewart (eds.) *Soil Erosion and Crop Productivity.* Soil Science Society of America, Madison, Wisconsin. p. 190-213.

46. Larson, W.E., G.R. Foster, R.R. Allmaras, and C.M. Smith (eds.). 1990. *Research Issues in Soil Erosion/Productivity*. University of Minnesota, St. Paul.
47. Larson, W.E., F.J. Pierce, and R.H. Dowdy. 1983. *The threat of soil erosion to long-term production*. Science 219: 458-465.
48. Meyer, L.D., A. Bauer, and R.D. Heil. 1985. *Experimental approaches for quantifying the effects of soil erosion on productivity*. In: R.F. Follett and B.A. Stewart (eds.) *Soil Erosion and Crop Productivity*. Soil Science Society of America, Madison, Wisconsin.
49. Miller, M.P., M.J. Singer, and D.R. Neilsen. 1988. *Spatial variability of wheat yield and soil properties on complex hills*. Soil Science Society of America Journal 51: 1563-1565.
50. Munn, D.N., and A.A. Franco. 1983. *Soil constraints to legume production*. In: P. Graham and S.C. Harris (eds.) *Biological nitrogen fixation technology for tropical agriculture: papers presented at a Workshop held at CIAT*. Centro Internacional de Agricultura Tropical. Cali, Columbia. p. 133-152.
51. Nelson, D.W., and L.E. Sommers. 1982. *Total carbon, organic carbon and organic matter*. In: A.L. Page et al. (eds.) *Methods of soil analysis. Part 2*. Agronomy 9: 539-579.
52. Nix, H.A. 1968. *The assessment of biological productivity*. In: G.A. Steward (ed.) *Land Evaluation*. The Macmillan Company of Australia. p. 77-87.
53. Office of Technology Assessment. 1990. *Beneath the Bottom Line. Agricultural Approaches to Reduce Agrichemical Contamination of Groundwater*. Summary. Office of Technology Assessment (OTA), Congress of the United States. U.S. Government Printing Office, Washington, D.C.
54. Pierce, F.J. 1990. *Erosion productivity impact prediction*. In: R. Lal and F.J. Pierce (eds.) *Soil Management for Sustainability*. Soil and Water Conservation Society, Ankeny, Iowa. p. 6/1-17.
55. Pierce, F.J., W.E. Larson, R.H. Dowdy, and W.A.P. Graham. 1983. *Productivity of soils: Assessing long-term changes due to erosion*. Journal of Soil and Water Conservation 38: 39-44.
56. Pierce, F.J., W.E. Larson, R.H. Dowdy, and W.A.P. Graham. 1984. *Soil productivity in the Corn Belt: An assessment of erosion's long-term effects*. Journal of Soil and Water Conservation 39: 131-136.
57. Power, J.F., F.M. Sandoval, and R.E. Ries. 1981. *Effect of topsoil and subsoil thickness on soil water content and crop production*. Soil Science Society of America Journal 45: 124-129.
58. Putnam, J., J.R. Williams, and D. Sawyer. 1988. *Using the erosion-productivity impact calculator (EPIC) model to estimate the impact of soil erosion for the 1985 RCA appraisal*. Journal of Soil and Water Conservation 43: 321-326.
59. Putnam, J., P. Dyke, and G. Wistrand. 1985. *Erosion-productivity index simulator (EPIS)*. Economic Research Service, U.S. Department of Agriculture, Washington, D.C.

60. Reinert, D.J., F.J. Pierce, and M.C. Fortin. 1990. *Temporal variation in structural stability induced by tillage*. pp. 63–72. In: J.A. Stone, D.A. Angers, and B.D. Kay (eds.) *Soil Structure Research in Eastern Canada.* Herald Press Limited. Windsor, Ontario, Canada. 236 pp.
61. Renard, K.G., and R.F. Follett. 1985. A research strategy for assessing the effect of erosion on future soil productivity in the United States. In: S.A. El-Swaify, W.C. Moldenhauer, and A. Lo (eds.) *Soil Erosion and Conservation.* Soil and Water Conservation Society, Ankeny, Iowa.
62. Riquier, J. 1977. *Philosophy of the world assessment of soil degradation and items for discussion.* In: *Food and Agriculture Organization Assessing Soil Degradation.* Soils Bulletin. Food and Agriculture Organization. Rome, Italy.
63. Ritchie, J.T. 1981. *Soil water availability.* Plant and Soil 58: 327-338.
64. Shaffer, M.J. 1985. *Simulation model for soil erosion-productivity relationships.* Journal of Environmental Quality 14: 144-150.
65. Shaffer, M.J., S.C. Gupta, J.A.E. Molina, D.R. Linden, and W.E. Larson. 1983. *Simulation of nitrogen, tillage and residue management effects on soil fertility.* In: W.K. Lavenroth, G.V. Skogerbo, and M. Flug (eds.) *Analysis of ecological systems: State-of-the-art in ecological modeling.* Elsevier Publishing. Amsterdam, Netherlands. p. 525-544.
66. Stocking, M.A. 1984. *Erosion and soil productivity: A review.* Food and Agriculture Organization Consultants' Working Paper No. 1, Soil Conservation Programme, Land and Water Development Division, Food and Agriculture Organization, Rome, Italy.
67. Stocking, M.A., and A. Pain. 1983. *Soil life and the minimum soil depth for productive yields: developing a new concept.* Development Studies Discussion Paper No. 150, University of East Anglia.
68. Stone, J.R., J.W. Gilliam, D.K. Cassel, R.B. Daniels, L.A. Nelson, and H.J. Kleiss. 1985. *Effect of erosion and landscape position on productivity of Piedmont soils.* Soil Science Society of America Journal 49: 987-991.
69. Taylor, H.M., and E.E. Terrell. 1982. In: M. Rechigl (ed.) *CRC Handbook of Agricultural Productivity, Vol. 1.* CRC Press, Boca Raton, Florida.
70. Thomas, P.J., T.W. Simpson, and J.C. Baker. 1989. *Erosion effects on productivity of cecil soils in the Virginia Piedmont.* Soil Science Society of America Journal 53: 928-933.
71. Timlin, D.J., R.B. Bryant, V.A. Snyder, and R.J. Wagenet. 1986. *Modeling corn grain yield in relation to soil erosion using a water budget approach.* Soil Science Society of America Journal 50: 718-723.
72. Todorovic, P., D.A. Woolhiser, and K.G. Renard. *Mathematical model for evaluation of the effect of soil erosion on soil productivity.* Hydrological Processes 1: 181-198.
73. U.S. Department of Agriculture. 1981. *Soil and Water Resources Conservation Act appraisal. Part I.* Washington, D.C.
74. U.S. Department of Agriculture. 1989. *The Second RCA Appraisal. Soil, Water and Related Resources on Nonfederal Land in the United States, Analysis of Conditions and Trends.* Washington, D.C.

75. Whisler, F.D., B. Acock, D.N. Baker, R.E. Fye, H.F. Hodges, J.R. Lambert, H.E. Lemon, J.M. McKinion, and V.R. Reddy. 1986. *Crop simulation models in agronomic systems.* Advances in Agronomy 40: 141-208.
76. White, A.W., Jr., R.R. Bruce, A.W. Thomas, G.W. Langdale, and H.F. Perkins. 1985. *Characterizing productivity of eroded soils in the southern Piedmont.* In: *Erosion and soil productivity.* Publication 8-85, American Society of Agricultural Engineers, St. Joseph, Michigan. p. 83-95.
77. Wischmeier, W.H., and D.D. Smith. 1978. *Predicting rainfall erosion losses—A guide to conservation planning.* Agriculture Handbook No. 537. U.S. Department of Agriculture, Washington, D.C.
78. Williams, J.R., R.R. Allmaras, K.G. Renard, L. Lyles, W.D. Moldenhauer, G.W. Langdale, L.D. Meyer, and W.J. Rawls. 1981. *Soil erosion effects on soil productivity: A research perspective.* Journal of Soil and Water Conservation 36: 82-90.
79. Williams, J.R., K.G. Renard, and P.T. Dyke. 1983. *EPIC—A new method for assessing erosion's effect on soil productivity.* Journal of Soil and Water Conservation 38: 381-382.
80. Wolman, M.G. 1985. *Soil erosion and crop productivity: a worldwide perspective.* In: R.F. Follett and B.A. Stewart (eds.) *Soil Erosion and Crop Productivity.* Soil Science Society of America, Madison, Wisconsin. p. 9-21.
81. Yost, R.S., S.A. El-Swaify, E.W. Dangler, and A. Lo. 1985. *The influence of simulated soil erosion and restorative fertilization on maize production on an Oxisol.* In: S.A. El-Swaify, W.C. Moldenhauer, and A. Lo. 1985. *Soil Erosion and Conservation.* Soil and Water Conservation Society, Ankeny, Iowa.

Chapter 11

WIND EROSION

E. L. Skidmore

Wind erosion is a serious problem in many parts of the world. Extensive aeolian deposits from past geologic eras also prove it is not a recent phenomenon.

Wind erosion is worse in arid and semiarid areas where the following conditions frequently occur: loose, dry, finely divided soil; a smooth soil surface devoid of vegetative cover; large fields; and strong winds (*44*). Arid and semiarid lands are extensive. Arid lands comprise about one-third of the world's total land area and are the home of one-sixth of the world's population (*37, 50*). Areas most susceptible to wind erosion on agricultural land include much of North Africa and the Near East, parts of southern and eastern Asia, the Siberian Plains, Australia, southern South America, and the semiarid and arid portions of North America (*44*).

Land undergoing desertification becomes vulnerable to wind erosion (*85*). On pastoral rangeland, composition of pastures subject to excessive grazing during dry periods deteriorates, the proportion of edible perennial plants decreases, and the proportion of annuals increases. The thinning and death of vegetation during dry seasons or droughts increase the extent of bare ground, and surface soil conditions deteriorate, increasing the fraction of erodible aggregates on the soil surface. In rainfed farming areas, removal of the original vegetation and fallow expose the soil to accelerated wind and water erosion.

Extensive soil erosion in the U.S. Great Plains during the last half of the 19th century and in the prairie region of western Canada during the 1920s warned of impending disaster. In the 1930s, a prolonged dry spell culminated in dust storms and soil destruction of disastrous proportions in the prairie regions of both western Canada and the Great Plains (*2, 62, 65, 76, 102*).

Wind erosion physically removes from the field the most fertile portion of the soil and, therefore, lowers land productivity (*35, 68*). Some soil from damaged land enters suspension and becomes part of the atmospheric dustload. Hagen and Woodruff (*54*) estimated that eroding land in the Great Plains contributed 244 million and 77 million tons of dust per year to the atmosphere in the 1950s and 1960s, respectively. Jaenicke (*63*) estimated the source strength of mineral dust from the Sahara at 260 million tons a year. Dust obscures visibility and pollutes the air, causes automobile accidents, fouls machinery, and imperils animal and human health. Blowing soil also fills road ditches; reduces seedling survival and growth; lowers the marketability of vegetable crops, such as asparagus, green beans, and lettuce; increases the susceptibility of plants to certain types of stress, including disease; and contributes to transmission of some plant pathogens (*33, 58, 59*).

SOIL ERODIBILITY BY WIND

Scientists recognized early that soil erodibility, the susceptibility or ease of detachment and transport by wind, was a primary variable affecting wind erosion. From wind tunnel tests, Chepil (*17*) determined relative erodibilities of soils reasonably free from organic residues as a function of apparent specific gravity and proportions of dry soil aggregates in various sizes. Clods larger than 0.84 mm in diameter were nonerodible in the range of windspeeds used in the tests. Since then, the nonerodible soil fraction > 0.84 mm, as determined by dry sieving, has been used to indicate erodibility of soil by wind. In an early version of the wind erosion equation (*26*), the nonerodible soil fraction was one of three major factors developed from results obtained principally with a portable wind tunnel (*113, 114, 116*).

A dimensionless soil erodibility index, I, was based on the non-erodible fraction, the percentage of clods > 0.84 mm in diameter (*22, 27*). The quantity of soil eroded in wind tunnel tests is governed by the tunnel's length and other characteristics. Therefore, erodibility was expressed on a dimensionless basis so that for a given soil and surface condition the same relative erodibility value would be obtained regardless of wind-tunnel characteristics (*24*). The soil erodibility index was expressed as follows:

$$I = X_2/X_1 \qquad [1]$$

where X_1 is the quantity eroded from soil containing 60 percent of clods > 0.84 mm and X_2 is the quantity eroded under the same set of conditions from soil containing any other proportion of clods > 0.84 mm. The soil erodibility index, I, gave a relative measure of erodibility, but actual soil loss by wind was not known.

Therefore during the severe wind erosion seasons of 1954-1956, from January through April, Chepil studied 69 fields in western Kansas and eastern Colorado to determine the quantity of soil loss for any field erodibility as determined from various field conditions (*24*). The average depth of soil eroded usually was indicated by the depth to which crowns and roots of plants were exposed.

Seasonal loss was converted to annual soil loss, and relative field erodibility for each field was determined by procedures previously outlined (*23, 26, 27*). The relation between annual soil loss and relative field erodibility was as follows:

$$Y = aX^{b} - 1/cd^{x} \quad [2]$$

where Y is annual soil loss (tons/acre); X is the dimensionless relative field erodibility; and a, b, c, and d are constants equal to 140, 0.287, 0.01525, and 1.065, respectively. Chepil (*24*) recognized that inaccuracies in measuring relatively small annual soil losses from depth of soil removal made converting relative field erodibility to annual soil loss by Equation 2 highly approximate.

When a field is smooth, bare, wide, unsheltered, and noncrusted, its relative erodibility is equal to the erodibility index defined by Equation 1. To obtain potential annual soil loss in tons per acre, I is substituted for x in Equation 2. Equation 2 was multiplied by one-third, then used to generate a table (*109*) for erodibility of soils with different percentages of nonerodible fractions > 0.84 mm (Table 11.1).

A more reliable and technically sound procedure is needed to estimate or predict the erodibility index without making physical measurements. This would save time and expense and provide a means to estimate erodibility more accurately.

In current practice, scientists often estimate soil erodibility by grouping soils, mostly according to predominant soil textural class (Table 11.2).

Because of the utility of predicting soil erodibility from easily obtainable soil properties, Table 11.2, or a similar one, has been used extensively. Problems associated with using Table 11.2 to estimate soil erodibility include the transience of dry soil aggregates > 0.84 mm within a given wind erodibility group (WEG).

Actual erodibility is extremely dynamic and varies seasonally, yearly, and as the result of management operations. In a study of the effects of season on soil erodibility, Chepil (*18*) found erodibility was always higher in the spring than in the previous fall if the soil had received moisture occasionally during the winter. But the increases were not of the same magnitude for all soils. The greatest increase in erodibility from fall to spring occurred in the finest textured soil, the least in the coarsest. Sandy loam was highly erodible in both fall and spring. Clay was least erodible in the fall, but about as highly erodible as sandy loam in the spring. The

Table 11.1 Soil erodibility, I, for soils with different percentages of nonerodible fractions as determined by standard dry sieving (*109*).

	Soil Erodibility by Percentage of Dry Soil Fractions >0.84 mm									
Percent	0	1	2	3	4	5	6	7	8	9
	Mg/ha									
0	-	695	560	493	437	404	381	359	336	314
10	300	294	287	280	271	262	253	244	238	228
20	220	213	206	202	197	193	186	182	177	170
30	166	161	159	155	150	146	141	139	134	130
40	126	121	117	114	112	108	105	101	96	92
50	85	80	75	70	65	61	28	54	52	49
60	47	45	43	40	38	36	36	34	31	29
70	27	25	22	18	16	13	9	7	7	4
80	4	-	-	-	-	-	-	-	-	-

intermediate textured soils had an intermediate erodibility in both spring and fall.

Grouping is discontinuous and results in large, discrete jumps in erodibility with textural change. For example, the soil erodibility indexes for loamy very fine sand (WEG 2) and very fine sandy loam (WEG 3) are 300 and 193 Mg/ha/yr, respectively. All other soils classed in those wind erosion groups have the same erodibility indexes. Do they have the same erodibility? It would be better to predict percentage of aggregates > 0.84 mm and then use Table 11.1. A procedure must be devised for realistically predicting dry soil-aggregate status that accounts for yearly and seasonal fluctuations and dominant soil properties influencing erodibility.

The aggregate status of the soil at any instant in time is the result of many aggregate-forming and degrading processes. These processes comprise a complex interrelationship of physical, chemical, and biological reactions. Aggregation may be the breakdown of clods into more favorable size, or it may be the formation of aggregates from finer materials.

Another factor to be considered in assessing or predicting the aggregate status or erodibility of a soil is the influence of cropping history and tillage. Page and Willard (*82*) found that the degree of aggregation in corn/oats/alfalfa-bromegrass/alfalfa-bromegrass rotation is two to three times greater than that for continuous corn. Cropping systems that included continuous small grain, continuous row crops, and rotations including fallow showed no significant differences in water-stable aggregation (*81*). Soils broken out of native sod lost much of their aggregation in the surface-tilled zone (*81, 92, 100*). Skidmore and associates, in a study of soil physical properties as influenced by management of residues from winter wheat and grain sorghum, found that grain sorghum or wheat management treat-

Table 11.2 Descriptions of wind erodibility groups (*105*).

WEG	Predominant Soil Texture Class of Surface Layer	Dry Soil Aggregates >0.84 mm (%)	Wind Erodibility Index, I (Mg/ha)
1	Very fine sand, fine sand, or coarse sand	1	695
		2	560
		3	493
		5	404
		7	359
2	Loamy very fine sand, loamy fine sand, loamy sand, loamy coarse sand, or sapric soil materials	10	300
3	Very fine sand loam, fine sandy loam, sandy loam, or coarse sandy loam	25	193
4	Clay, silty clay, noncalcareous clay loam, or silty clay loam with more than 35 percent clay content	25	193
4L	Calcareous loam and silt loam or calcareous clay loam and silty clay loam	25	193
5	Noncalcareous loam and silt loam with less than 20 percent clay content or sandy clay loam, sandy clay, and hemic organic soil materials	40	126
6	Noncalcerous loam and silt loam with more than 20 percent clay content or noncalcereous clay loam with less than 35 percent clay content	45	108
7	Silt, noncalcareous silty clay loam with less than 35 percent clay content, and fibric organic soil material	50	85
8	Soils not susceptible to wind	>80	0

ments did not influence most of the soil physical properties measured (*92*). However, the aggregate status differed among crops. Soil aggregates from sorghum plots were smaller, more fragile, less dense, and more wind-erodible than aggregates from wheat plots. Harris and associates (*57*) reported that agronomic systems affect aggregation significantly but that interpreting controlling mechanisms is complicated by the diversity of factors through which the effects are manifest.

Inability to predict both aggregate status and the weather undoubtedly influenced Woodruff and Siddoway's definition of soil erodibility (*109*): "the potential average annual soil loss from a wide, unsheltered, isolated field...for the climate in the vicinity of Garden City, Kansas." In spite of temporal variation of soil aggregate status, Woodruff and Siddoway suggested that soil erodibility can be estimated by standard dry sieving and use of Table 11.1. Use of sieving results assumes that the values determined (percent > 0.84 mm) "characterize" a soil during the critical erosion period for the time domain of the wind erosion equation (*109*).

For determining percentages of dry soil fractions > 0.84 mm, Chepil and Woodruff (*27*) recommended the rotary sieve. A conventional and more readily available flat sieve may be used, but results with it are less accurate than with a rotating sieve.

Researchers should use the following procedure when using a flat sieve:

- Obtain 1 kg samples from the 0- to 2-cm surface layer when soil is reasonably dry. If soil is not near air dryness, dry it in the laboratory before sieving.
- Weigh the sample and sieve it on a 0.84-mm (No. 20), 20.3-cm (8-inch) diameter sieve until the aggregates < 0.84 mm diameter have passed through the sieve. Be careful not to fragment aggregates during sieving. Weigh the amount of sample remaining on the sieve.
- Calculate the mass fraction of the total sample that was retained on the sieve and use Table 11.1 to determine soil erodibility.

Suppose from replicated sievings from a sample site that the total amount of air-dried soil for each sieving was 1035, 945, 850, and 990 grams and the respective amounts retained on the 0.84 mm sieve after sieving were 370, 227, 200, and 250 grams. Therefore, percentages of dry soil fractions > 0.84 mm would be 26.1, 24.0, 20.7, and 25.3, respectively. Corresponding soil erodibility values from Table 11.1 would be 186, 197, 213, and 193 Mg/ha, respectively; the mean would be 197 Mg/ha.

WIND EROSIVITY

Chepil and associates (*25*) proposed a climatic factor to determine average annual soil loss for climatic conditions other than those occurring when the relationship between wind-tunnel erosion and field erosion was obtained. It is an index of wind erosion as influenced by moisture content in surface soil particles and average windspeed. The windspeed term of the climatic factor was based on the rate of soil movement being proportional to average windspeed cubed (*8, 15, 115*). The soil moisture term was developed on the basis that the erodibility of soil varies inversely with the square of the equivalent water content in the near-surface soil, which was assumed to vary as the Thornthwaite index (*20*).

The climatic factor was expressed as follows:

$$C = 386 \frac{\bar{u}^3}{(PE)^2} \qquad [3]$$

where $\bar{u}$ is the mean annual windspeed corrected to 9.1 m and PE is the Thornthwaite (*103*) index. The 386 value indexes the factor to conditions at Garden City, Kansas.

Thornthwaite developed the climatic index to evaluate precipitation effectiveness. An equation was fitted to rather limited data that expressed the P/E ratio to temperature and precipitation as follows:

$$P/E = 0.316 \left(\frac{P}{1.8T + 22} \right)^{10/9} \qquad [4]$$

where P is the mean monthly precipitation in mm, E is the monthly evaporation in mm, and T is temperature in C°. Monthly values were added to obtain an annual value, which was multiplied by 10 to give

$$\text{PE index} = 3.16 \sum_{i=1}^{12} \left(\frac{P_i}{1.8T_i + 22} \right)^{10/9} \qquad [5]$$

Equation 5 was evaluated and used in Equation 3 to determine climatic factors for wind erosion at many locations in the United States (*25, 69, 98*).

As the PE index gets smaller as precipitation declines, as in arid regions, the climatic factor in Equation 3 approaches infinity. In application, an upper limit is set by restricting minimum monthly precipitation to 13 mm (*69*). Monthly climatic factors also were calculated using an annual PE index with monthly mean windspeed (*108*).

The Food and Agriculture Organization (FAO) approached the problem of the climatic factor becoming a large value in arid conditions differently (*45*). Agency researchers modified the Chepil and associates' index (*25*) as follows:

$$C^{1} = 1/100 \sum_{i=1}^{12} \bar{u}^3 \left(\frac{ETP_i - P_i}{ETP_i} \right) d \qquad [6]$$

where $\bar{u}$ is the mean monthly windspeed at a 2-m height, ETP is potential evapotranspiration, P is precipitation, and d is the total number of days in the month. In this case, as precipitation approaches zero, windspeed dominates the climatic factor. Conversely, as precipitation approaches ETP, the climatic factor approaches zero. The influence of soil water in the FAO version is less than the squared influence of soil water demonstrated by Chepil (*20*).

I handled the influence of soil water differently and included a windspeed probability density function as follows (*90*):

$$CE = \rho \int_R^\infty \left[u^2 - R^2\right]^{3/2} f(u)du \qquad [7]$$

where

$$R = u_t^2 + \gamma^1/\rho a^2 \qquad [8]$$

and CE is the wind erosion climatic erosivity, which is directly proportional to mass flow rate of an all erodible material; ρ is air density; u and u_t are windspeed and threshold windspeed, respectively; γ^1 is the cohesive resistance of absorbed water; and a is a combination of constants, $k/\ln(z/z_0)$, for k = 0.41, z = 10 m, z_0 = 0.05 m; thus, a = 0.0774.

The value for γ is approximated as follows:

$$\gamma = 0.5\psi^2 \qquad [9]$$

where ψ is the equivalent soil water content, fraction of water (by mass or volume) in the soil, divided by fraction of water in the same soil at -1,500 J/kg (*20, 90*). It was assumed that equivalent surface water content was approximated by the ratio of precipitation to potential evaporation. The ratio of precipitation to evaporation can be approximated by the Thornthwaite PE index or the inverse of the dryness ratio (*12, 56*). The dryness ratio, D, is defined as follows:

$$D = Rn/(LP) \qquad [10]$$

where Rn is net radiation, L is latent heat of evaporation, and P is precipitation. The dryness ratio at a given site indicates the number of times the net radiative energy could evaporate the precipitation over the same time interval.

The windspeed probability density function, Equation 7, can be expressed as a Weibull distribution:

$$f(u) = (k/c)\,(u/c)^{k-1} \exp\left[-(u/c)^k\right] \qquad [11]$$

where c and k are scale and shape parameters, respectively. Parameter c has units of velocity and k is dimensionless (*3, 66*). Weibull parameters have been determined from windspeed distribution summaries at many locations in the U.S. Great Plains (*53*).

Equation 7, with f(u) defined by Equation 10, can be integrated straightforwardly when k = 2 as follows:

$$CE = 1.33\,\rho c^3 \exp\left[-R/c^2\right] \qquad [12]$$

where R is defined by Equation 8.

The summation procedure for evaluating Equation 7 can be written as follows:

$$CE = r_{u2} \sum_{i+0.5>R}^{n} \left(u_{i+0.5}^{2} - R\right)^{3/2} \left[F(u_{i+1}) - F(u_{i})\right] \qquad [13]$$

where $F(u_i)$ is the cumulative distribution function

$$F(u_i) = 1 - \exp\left[-(u_i/c)^k\right] \qquad [14]$$

When mean windspeed is available but the data from which the mean was calculated are not, the Weibull parameters can be estimated.

Studies have shown that the Weibull scale parameter was about 12 percent larger than mean windspeed and the Weibull shape parameter was a function of the scale parameter (*64, 90*). Thus, if only mean windspeed is known, a reasonable estimate of Weibull distribution can be obtained as follows:

$$c = 1.12\,\bar{u} \qquad [15]$$

and

$$k = 0.52 + 0.23c \qquad [16]$$

Equations 7 and 13 express wind power, W m^{-2}. When multiplied by the time duration in the accounting period represented by f(u), they give erosive wind energy. This is the energy of the wind in excess of that necessary to overcome threshold shear stresses represented by R. Erosive wind energy is a useful parameter to evaluate the climatic factor for the wind erosion equation.

Suppose one wishes to know an appropriate climatic factor for a 30-day period of given conditions: mean windspeed = 5.8 m s^{-1}, average precipitation = 80 mm, net radiation = 490 MJ m^{-2}. Then, from Equations 15 and 16, c and k are estimated to be 6.5 m s^{-1} and 2.0, respectively. The dryness ratio calculated from Equation 10 is 2.5 for heat of vaporization of 2.45 MJ kg^{-1}.

Thus,

$$R = u_t^2 + \gamma/\rho a^2 = 36 + 11 = 47\,m^2\,s^{-2} \qquad [17]$$

Equation 13 could be used to calculate CE. However, because k = 2.0, Equation 12 was used to calculate CE as follows:

$$CE = 1.33\,\rho c^3 \exp\left[-(R/c^2)\right] = 144\ W\,m^{-2} \qquad [18]$$

Therefore, the erosive wind energy for the 30-day period would be as follows:

$$CE \cdot time = 144\ W\,m^{-2} \cdot 8.64 + 10^{4}\ sd^{-1} \cdot 30d = 373\ MJ\,m^{-2} \qquad [19]$$

If the conditions given in this example 30-day period were to prevail for an entire year, then the erosive wind energy would be 4538 MJ m^{-2}. That wind energy, compared to the reference of 8100 MJ m^{-2}, gives a climatic factor of 56. Also, from Figure 11.1, for a dryness ratio of 2.5 and a mean windspeed of 5.8 m s^{-1}, the climatic factor is 56.

RIDGE ROUGHNESS

Chepil and Milne (*32*) investigated the influence of surface roughness on intensity of drifting dune materials and cultivated soils. They found that the initial intensity of drifting was always much less over a ridged surface. Ridging cultivated soil reduced the severity of drifting, but ridging highly erodible dune material was less effective because ridges disappeared rapidly. The rate of flow varied inversely with surface roughness. Armbrust and associates (*7*) studied the effects of ridge-roughness equivalent on the total quantity of eroded material from soils exposed to different friction velocities. From their data, a curve can be constructed showing the relationship between the relative quantity of eroded material and the ridge-roughness equivalent. Presumably, this was the origin of the chart (*109, Figure 11.4*) showing a soil ridge-roughness factor as a function of soil ridge roughness, so that ridging may reduce wind erosion up to 50 percent.

Ridge roughness estimates the fractional reduction of erosion caused by ridges of nonerodible aggregates. It is influenced by ridge spacing and ridge height and is defined relative to a 1:4 ridge-height-to-ridge-spacing ratio.

Tables were prepared of ridge-roughness factors for various combinations of ridge heights and spacings (*88*). Hayes (*60*) suggested evaluating fields as either smooth, semiridged, or ridged and then assigning 1.0, 0.75, and 0.50, respectively, as soil ridge-roughness factors. Williams and associates (*106*) fitted equations to the curve of Woodruff and Siddoway (*109*) to express the ridge-roughness factor as follows:

$$K = 1.0,\ HR^2/IR < 0.57 \qquad [20]$$

$$K = 0.913 - 0.153\ \ln\left(HR^2/IR\right),\ 0.57 < \left(HR^2/IR\right) < 22.3 \qquad [21]$$

$$K = 0.336\ \exp\left(.013\ HR^2/IR\right) \geq 22.3 \qquad [22]$$

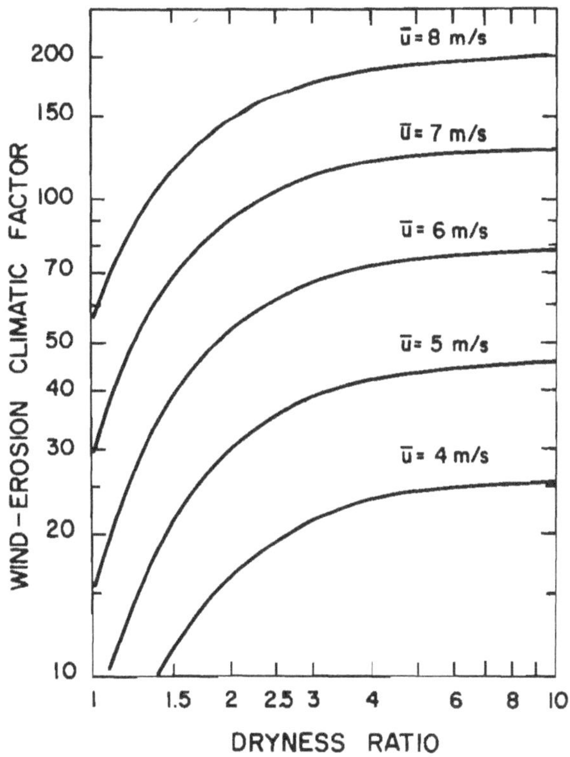

Figure 11.1. Wind erosion climatic factor as influenced by dryness ratio and mean windspeed (*90*).

where HR and IR are ridge height and ridge spacing, respectively, in mm. A field with ridges 100 mm high and spaced 400 mm apart has HR^2/IR = 25. Because 25 > 22.3 and using Equation 22, the ridge-roughness factor K = 0.5.

FIELD LENGTH

Chepil and Milne (*32*) reported that the rate of soil movement started at zero on the windward side of fields or field strips and increased with distance downwind. Later, Chepil (*16*) found that the cumulative rate of soil movement with distance away from the windward edge of eroding fields was the main cause of steadily increasing amounts of erodible particles, increasing abrasion, and gradual reduction in the rate of soil flow with distance downwind "avalanching."

Rate of soil flow increased with distance downwind across an eroding field. If the field was large enough, soil flow reached the maximum that a wind of a given velocity could carry. Beyond that point, the rate of flow remained essentially constant (*21*). That maximum was about the same for soil of any texture—about 50 $gm^{-1}s^{-1}$ for a 17 $m\ s^{-1}$ wind at 10 m. The rate of increase for various soil textures was the same as the order of erodibility for soil texture classes.

The distance required for soil flow to reach the maximum that a wind of a given velocity can carry varies inversely with the erodibility of a field surface. The more erodible the surface, the shorter the distance to reach maximum flow (*23*).

Chepil (*23*) related relative wind erodibility to the distance required for soil flow to reach a maximum. In his earlier work (*16, 21, 32*), he presented data for the rate of soil movement as a function of distance from the windward edge of the field for soils that varied widely in erodibility. He converted the relative surface erodibility, based on four factors—soil cloddiness, crop residue, ridge-roughness equivalent, and soil erodibility—to relative field erodibility, based on additional factors—wind barrier, width of field, and wind direction (*23*). These functional relationships between field erodibility and field width with the many associated factors gave rise to how the field length term was used in the wind erosion equation (*28, 109*).

Originally, field length was considered as the distance across a field in the prevailing wind erosion direction (*109*). However, sometimes almost as much wind occurs from one direction as from another, so there is essentially no prevailing wind erosion direction. In these cases, researchers used the preponderance of wind erosion forces in the prevailing wind erosion direction to assess equivalent field length (*87, 98*). Later, from a more detailed analysis, tables were prepared that give wind erosion direction factors, numbers that when multiplied by field width give median travel distance as a function of preponderance of wind erosion forces in the prevailing direction and deviation of prevailing wind erosion direction from perpendicular to direction of field length (*90*).

In some of the modeling efforts, the procedure for determining L for use in the wind erosion equation was simplified by ignoring wind direction distributions. Cole and associates (*34*) suggested the following:

$$L = \begin{cases} w \sec\theta & L \le \left(l^2 + w^2\right)^{1/2} \\ l \csc\theta & \text{otherwise} \end{cases} \qquad [23]$$

where w and l are the small and large dimensions, respectively, of a rectangular field and θ is the angle between side w and the prevailing wind erosion direction. As θ varies through $\pi/2$ radians, L will range from w to l, with a maximum equal to the main diagonal of the field. The procedure Williams and associates (*106*) used in EPIC, the erosion-productivity impact calculator, was as follows:

$$L = \frac{lw}{l\left|\cos(\pi/2 + \propto - \phi)\right| + w\left|\sin(\pi/2 + \propto - \phi)\right|} \quad [24]$$

where l and w are the large (length) and small (width) dimensions, respectively, of a rectangular field, $\propto$ is the wind direction clockwise from north in radians, and ϕ is the clockwise angle between field length and north in radians. Using Equation 24, L = 236 m for a rectangular field where l = 1000 m, w = 200 m, $\propto = \pi/4$ radians, and $\phi = 0$.

VEGETATIVE FACTOR

Scientists realized early the value of crop residue for controlling wind erosion and reported quantitative relationships (*14*). From wind tunnel tests on plots especially prepared to obtain a range of vegetative cover and soil structure, Englehorn and associates (*39*) found the exponential relationship that best expressed their results. Subsequent studies (*19, 26, 27*) expressed the relationship in the form $x = aI/(RK)^b$, where x is the wind tunnel erodibility; I is the soil erodibility index (percent of clods > 0.84 mm); R is the dry weight of crop residue in pounds/acre; K is the ridge-roughness equivalent; and a and b are constants.

Amounts of wheat straw needed to protect most erodible dune sands and less erodible soils against strong winds were established (*31*). Standing stubble was much more effective than flattened stubble (*29*). Standing sorghum stubble with rows perpendicular to wind direction controlled wind erosion more effectively than rows parallel to wind direction (*39, 97*).

Siddoway and associates (*86*) quantified the specific properties of vegetative covers influencing soil erodibility and developed regression equations relating soil loss by wind to selected amounts, kinds, and orientation of vegetative covers, wind velocity, and soil cloddiness. They found a complex relationship among the relative effectiveness of different kinds and orientation of residue. The relative value of kinds and orientations of residue in controlling erosion must be qualified by soil, wind velocity, and variable characteristics of the residues. Generally, Siddoway and associates concluded that (a) on a weight basis, fine-textured residues were more effective than coarse-textured residues; (b) any orientation of residue, except flattened residue, reduced wind erosion; and (c) fine-leafed crops, such as grasses and cereals, provided a high degree of erosion control per unit weight.

Those studies led to the relationship developed by Woodruff and Siddoway (*109*) showing the influence of an equivalent vegetative cover of small grain and sorghum stubble for various orientations (flat, standing) and heights, then relating soil loss to equivalent vegetative cover.

Efforts to evaluate the protective role of additional crops have continued. In wind tunnel tests, Lyles and Allison (*70, 71*) determined equivalent wind erosion protection provided by selected range grasses and crop residues. They found high simple correlation coefficients from an equation of the form

$$(SG)_e = aX^b \qquad [25]$$

where $(SG)_e$ is the flat small-grain equivalent, X is the quantity of residue or grass to be converted, and a and b are constants. Tables 11.3 and 11.4 show prediction equation coefficients.

It is not practical in testing all combinations of crops and residues to determine their protection value as flat small-grain equivalents. Therefore, a practice is needed to estimate the protection values of crops and residues not tested. Hayes (*59*) suggested that if any residue is not represented, researchers should use a curve for the crop most like the crop in question.

Lyles and Allison (*71*) correlated measurable parameters, which describe crop residues, in several combinations to obtain an equation for predicting the flat small-grain equivalent of flat, random residues as follows:

$$\begin{aligned} (SG)_e &= 0.162\ Rw/d + 8.708\,(Rw/d\gamma)^{1/2} - 271 \\ r^2 &= .92 \end{aligned} \qquad [26]$$

where $(SG)_e$ is the flat small-grain equivalent (kg/ha), Rw is the residue amount to be converted (kg/ha), d is the average stalk diameter (cm), and γ is the average specific weight of the stalk (g/cm^3). Winter wheat, rape, soybeans, cotton, and sunflowers were used in developing Equation 26.

Until recently, all small-grain equivalence data have been limited to dead crop residue or dormant grass. Armbrust and Lyles (*6*) reported flat small-grain equivalents for five growing crops—corn, cotton, grain sorghum, peanuts, and soybeans—as follows:

$$(SG)_e = a_1 Rw^{b_1} \qquad [27]$$

where $(SG)_e$ is the flat small-grain equivalent and Rw is the aboveground dry weight of the crop to be converted, both in kg/ha, and a_1 and b_1 are constant coefficients for each crop. They found that if only rough estimates of $(SG)_e$ are needed, an average coefficient could be used. An average equation determined from pooling all crop data with rows running perpendicular to wind direction gave 8.9 and 0.9, respectively, for a_1 and b_1.

Table 11.3. Coefficients in prediction equation $(SG)_e = aR_w^b$ for conversion of crop residues to an equivalent quantity of flat small-grain residue, both in kg/ha (*71*).

Crop Residue	*Surface Orientation*	*Height (cm)*	*Length (cm)*	*Row Spacing (cm)*	*Row Orientation To Flow*	*Prediction a*	*Equation B*	*Coefficients* r^2
Winter wheat	Standing	25.4	-	25.4	Normal	4.306	0.970	0.997
Rape	Standing	25.4	-	25.4	Normal	0.103	1.400	0.990
Cotton	Standing	34.3	-	76.2	Normal	0.188	1.145	0.998
Sunflowers	Standing	43.2	-	76.2	Normal	0.021	1.342	0.994
Winter wheat	Flat-random	-	25.4	-	-	7.279	0.782	0.993
Soybeans	Flat-random	-	25.4	-	-	0.167	1.173	0.993
Rape	Flat-random	-	25.4	-	-	0.064	1.294	0.997
Cotton	Flat-random	-	25.4	-	-	0.077	1.168	0.998
Sunflowers	Flat-random	-	43.2	-	-	0.011	1.368	0.993
Forage sorghum	Standing	15.9	-	76.2	Normal	0.353	1.124	0.995
Silage corn	Standing	15.9	-	76.2	Normal	0.229	1.135	0.998
Soybeans	1/10 standing	6.4	-	76.2	Normal			
Soybeans	9/10 flat-random	-	25.4	-	-	0.016	1.553	0.991

Table 11.4 Coefficients in prediction equation $(SG)_c$ - aX^b for conversion of range grasses to an equivalent quantity of flat small-grain residue, both in kg/ha (*70*).

Grass Species	*Grazing Management*	*Grass Height (cm)*	*Prediction a*	*Equation b*	*Co-efficients r^2*
Blue grama	Ungrazed	33.0	0.60	1.39	0.98
Buffalograss	Ungrazed	10.2	1.40	1.44	0.97
Big bluestem	Properly grazed	15.2	0.22	1.34	0.99
Blue grama	Properly grazed	5.1	1.60	1.08	0.99
Buffalograss	Properly grazed	5.1	3.08	1.18	0.99
Little bluestem	Properly grazed	10.2	0.19	1.37	0.99
Switchgrass	Properly grazed	15.2	0.47	1.40	0.99
Western wheatgrass	Properly grazed	10.2	1.54	1.17	0.99
Big bluestem	Overgrazed	2.5	4.12	0.92	0.99
Blue grama	Overgrazed	2.5	3.06	1.14	0.99
Buffalograss	Overgrazed	1.5	2.45	1.40	0.99
Little bluegrass	Overgrazed	2.9	0.52	1.26	0.99
Switchgrass	Overgrazed	2.5	1.80	1.12	0.99
Western wheatgrass	Overgrazed	2.5	3.93	1.07	0.99

Suppose one wishes to know the equivalent flat small-grain residue for a field with grain sorghum growing in 400 kg/ha of flat, random winter wheat residue when the dry weight of the growing grain sorghum is 83 kg/ha and the grain sorghum is growing in rows perpendicular to the expected wind. Therefore, $(SG)_c$ for the growing sorghum, from Equation 27, would be as follows:

$$(SG)_e = 8.9(83)^{.9} = 475 \qquad [28]$$

and, from Table 11.3, $(SG)_e$ for the wheat residue would be

$$(SG)_e = 7.3(400)^{.8} = 880 \qquad [29]$$

However, because of nonlinear relationships, the flat small-grain equivalents are not strictly additive. When more than one crop contributes to the residue, it is better to combine the calculation into a single equation as follows:

$$(SG)_c = a_1^{p_1} a_2^{p_2} (Rwt)^{b_1p_1+b_2p_2} \qquad [30]$$

where p_1 and p_2 are fractions of total residue, Rwt, and a_1, a_2 and b_1, b_2 are constant coefficients for respective crops as in Equation 27. For our example as follows

$$(SG)_e = (8.9)^{.172}(7.3)^{.828}(483)^{(.9)(.172)+(.8)(.828)} \\ = 1190 \text{ kg/ha} \qquad [31]$$

Either the equivalent flat small-grain or vegetative factor is needed for the various procedures to estimate wind erosion. The relationship between equivalent flat small-grain and vegetative cover was given graphically by Woodruff and Siddoway (*109*). Williams and associates (*106*) fitted an equation to the graphical relationship as follows:

$$V = 0.2533\,(SG)_e^{\,1.363} \qquad [32]$$

Therefore,

$$V = 0.2533\,(1180)^{1.363} = 3896 \text{ Mg ha}^{-1} \qquad [33]$$

A WIND EROSION MODEL

Researchers and scientists have used a wind erosion equation proposed by Woodruff and Siddoway (109), with various modifications, for the past 20 years. The model was developed as a result of investigations to understand the mechanics of the wind erosion process, to identify major factors influencing wind erosion, and to develop wind erosion control methods. The general functional relationship between the independent variable E, the potential average annual soil loss, and the equivalent variables or major factors is as follows:

$$E = f(I, K, C, L, V) \qquad [34]$$

where I is the soil erodibility index, K is the soil ridge-roughness factor. C is the climatic factor, L is the unsheltered median travel distance of wind across a field, and V is the equivalent vegetative cover. These factors were discussed in more detail earlier.

Solving the functional relationships of the wind erosion equation as presented by Woodruff and Siddoway (*109*) required the use of tables and figures. The awkwardness of the manual solution prompted a computer solution (*43, 99*) and development of a slide rule calculator (*89*).

The model has been adapted for use with personal computers (*55*) and interactive programs (*40*). Cole and associates (*34*) adapted the Woodruff and Siddoway (*109*) model for simulating daily soil loss by wind erosion as a submodel in EPIC (*106*). The latter version was simplified by fitting equations to the figures of Woodruff and Siddoway (*109*).

Solution of the wind erosion equation gives the expected amount of erosion from a given agricultural field. A second application of the equation is to specify the amount of erosion that can be tolerated and then solve the equation to determine the conditions required to limit soil loss to the specified amount, for example, the amount of residue, field width, etc. Conservationists have used the equation widely for both of these applications.

U.S. Soil Conservation Service field workers have used the equation extensively to plan wind erosion control practices (*59*). Hayes (*58*) also used the wind erosion equation to estimate crop tolerance to wind erosion. The equation is a useful guide to wind erosion control principles as well (*13, 80, 111*). Other uses of the equation include (a) determining spacing for barriers in narrow strip-barrier systems (*52*), (b) estimating fugitive dust emissions from agricultural and subdivision lands (*83, 107*), (c) predicting horizontal soil fluxes to compare with vertical aerosol fluxes (*49*), (d) estimating the effects of wind erosion on soil productivity (*67, 106*), (e) delineating those croplands in the Great Plains where various amounts of crop residues may be removed without exposing the soil to excessive wind erosion (*96*), and (f) estimating erosion hazards in a national inventory (*104*).

The following example of how to use the wind erosion equation to predict expected soil loss employs the variables used earlier, that is, I = 197 t ha $^{-1}$yr^{-1}; K = 0.5, C = 56, L = 236 m, and V = 3.9 Mg ha^{-1}. To determine the erosion estimate, however, requires a special combination of the factors. Several approaches are possible to find the solution: graphs, figures, tables, slide rule, or computer. Here, I use the procedure presented by Williams and associates (*106*). This procedure is done stepwise, but it has been simplified computationally by fitting equations to figures of Woodruff and Siddoway (*109*). The first step (E1) is to determine soil erodibility, I. Steps E2 and E3 are determined by multiplying the factors indicated as follows:

$$E2 = IK = 197 \cdot 0.5 = 93 \text{ Mg ha}^{-1}\text{ yr}^{-1} \qquad [35]$$

$$E3 = IKC = 93 \cdot .56 = 52 \text{ Mg ha}^{1}\text{ yr}^{-1} \qquad [36]$$

E4, the inclusion of field length, is

$$E4 = \left(WF^{0.348} + E3^{0.348} - E2^{0.348}\right)^{2.87} = 33 \text{ Mg ha}^{-1}\text{ yr}^{-1} \qquad [37]$$

where

$$WF = E2(1.0 - 0.122(L/Lo)^{-0.383} \exp\left(L/Lo\right) = 64 \qquad [38]$$

and

$$Lo = 1.56 \cdot 10^6 (E2)^{-1.26} \exp(-0.00156\ E2) = 4465\ m \qquad [39]$$

WF is a field length factor; it accounts for the influence of field length on reducing the erosion estimate. Lo is the maximum field length for reducing the wind erosion estimate.

Parameters ψ_1 and ψ_2 are functions of the vegetative cover factor described by the equations

$$\Psi_1 = \exp\left(-0.759V - 4.74 \times 10^{-2} V^2 + 2.95 \times 10^{-4} V^3\right) = 0.026 \qquad [40]$$

$$\Psi_2 = 1 + 8.93 \times 10^{-2} V + 8.51 \times 10^{-3} V^{-2} - 1.5 \times 10^{-5} V^3 = 1.469 \qquad [41]$$

where V is in Mg ha^{-1} and for our example, from Equation 33, has the value of 3.9 Mg ha^{-1}. Therefore,

$$E5 = \Psi_1\ E4^{\Psi_2} = 0.026\ (33)^{1.469} = 4.4\ Mg\ ha^{-1}\ yr^{-1} \qquad [42]$$

The estimate of 4.4 Mg ha^{-1} yr^{-1} given by Equation 42 is the annual rate of expected erosion during the 30-day period represented by the climatic factor C. To determine the expected erosion during the accounting period, it is necessary to multiply the given estimate by the fraction of the average annual total erosive wind energy occurring during the 30-day accounting period.

MANAGEMENT EFFECTS

Rough, cloddy surface

Tillage operations that leave furrows or ridges reduce wind erosion, as discussed earlier. When ridges are nearly gone, vegetative cover is depleted, and the threat of wind erosion continues, a rough, cloddy surface resistant to the force of wind can be created on many cohesive soils with appropriate "emergency tillage." For example, Lyles and Tatarko (*73*) found that chiseling of growing winter wheat on a silty clay soil greatly increased nonerodible surface aggregates without influencing grain yields. Farmers can use listers, chisels, cultivators, one-ways with two or three disks removed at intervals, and pitting machines to bring compact clods to the surface. Emergency tillage is most effective when done at right angles to the prevailing wind direction. Because clods eventually disintegrate, sometimes rapidly, emergency tillage offers only temporary wind erosion control at best (*111, 112*).

Residue

Living vegetation or residue from harvested crops protects the soil against wind erosion. Standing crop residues provide nonerodible elements that absorb much of the shear stress in the boundary layer. When vegetation and crop residues are sufficiently high and dense to prevent intervening soil-surface drag from exceeding threshold drag, soil will not erode. Rows perpendicular to wind direction control wind erosion more effectively than do rows parallel to wind direction (*39, 97*). Flattened stubble, though not as effective as standing stubble, also protects the soil from wind erosion (*29*).

Soon after the disastrous "dirty thirties" in the U.S. Great Plains, researchers demonstrated that stubble-mulching was a feasible method of reducing wind erosion on cultivated land (*38*). Stubble-mulching is a crop residue management system using tillage, generally without soil inversion, usually with blades or V-shaped sweeps (*77, 78*).

Other reduced and modified tillage systems have evolved with efforts to maintain residue on the soil surface. Chemical fallow (*11*) and ecofallow (*41*) systems use herbicides or herbicides and subsurface tillage during fallow periods to conserve a large quantity of residue on the surface.

Directly seeding small grains and other crops into stubble without a fallow period and without tillage is being studied and shows promise. The advantages of this system, compared with the tillage systems designed to preserve residues on the surface, include the following: (a) the standing stubble is needed for erosion control until the seeded crop produces enough cover to control erosion; (b) standing stubble more effectively controls erosion than does an equal quantity of flattened residue; (c) standing stubble, because it is not in direct contact with the soil, is less subject to decomposition than is stubble that has been tilled and mixed with the soil; and (d) without tillage, the soil is not pulverized.

The goal is to leave a desirable quantity of plant residue on the soil surface at all times. Residue is needed for a period of time after the crop is planted to protect the soil from erosion and improve water infiltration. The residue used is generally that remaining from a previous crop. Efforts continue to evaluate the residue needed to control wind erosion (*75, 91, 96*).

Stabilizers

Researchers have evaluated various soil stabilizers to find suitable materials and methods to control wind erosion (*4, 5, 19, 28, 30, 72*). Several tested products successfully controlled wind erosion for a short time, but many were more expensive than equally effective wheat straw anchored with a rolling disk packer (*30*). The following are criteria for surface-soil stabilizers: (a) 100 percent of the soil surface must be covered, (b) the stabilizer must not adversely affect plant growth or emergence, (c) erosion must be prevented initially and reduced for at least two months, (d) the stabilizer should be easy to apply and not require special equipment, and (e) cost must be low enough for profitable use (*5*). Armbrust and Lyles (*5*) found five polymers and one resin-in-water emulsion that met all of these requirements. They added, however, that before soil stabilizers can be used on agricultural land, methods must be developed to apply large volumes rapidly. Also, reliable pre-emergent weed control chemicals for use on coarse-textured soils must be developed, as well as films that are resistant to raindrop impact while allowing water and plant roots to penetrate the soil, without adversely affecting the environment.

Barriers

Use of wind barriers is an effective method of reducing field width to control wind erosion (*9*). Hagen (*51*) and Skidmore and Hagen (*93*) developed a model that when used with local wind data shows wind barrier effectiveness in reducing wind erosion forces. Barriers will reduce wind erosion forces more than they will windspeed. A properly oriented barrier, when winds predominate from a single direction, will reduce wind erosion forces by more than 50 percent from the barrier leeward to 20 times its height; the reduction will be greater for shorter distances from the barrier.

Different combinations of trees, shrubs, tall-growing crops, and grasses can reduce wind erosion. Aside from conventional tree windbreaks (*42, 84, 110*), many other barrier systems are used to control wind erosion. They include annual crops, such as small grains, corn, sorghum, sudangrass, sunflowers (*13, 47, 48, 52, 61*); tall wheatgrass (*1, 10*); sugarcane; and rye strips on sands (John Griffin, SCS agronomist, Gainesville, Florida, personal communication, 1975).

Most barrier systems for controlling wind erosion, however, occupy space that could otherwise be used for crop production. Perennial barriers grow slowly and often are difficult to establish (*36, 110*). Such barriers also compete with crops for water and plant nutrients (*74*). As a result, the net effect of many tree-barrier systems is that their use may not benefit crop production (*46, 79, 94, 95, 101*). Perhaps tree-barrier systems could be designed so that they become a useful crop, furnishing nuts, fruit, or wood.

Stripcropping

The practice of farming land in narrow strips on which the crop alternates with fallow is an effective aid in controlling wind erosion (*21*). Strips are most effective when they are at right angles to the prevailing wind direction, but they also provide some protection from winds that are not perpendicular to the field strip.

Stripcropping reduces erosion damage by reducing the distance the wind travels across exposed soil, localizing drifting that starts at a focal point, and reducing wind velocity across the strip when adjacent fields are covered with tall stubble or crops.

Although each method to control wind erosion has merit and application, establishing and maintaining vegetative cover, when feasible, remains the best defense against wind erosion. However, that becomes a difficult challenge as pressure increases to use crop residues for livestock feed and fuel for cooking.

CONCLUSIONS

Investigations of the factors influencing wind erosion led to the development of a wind erosion equation. The two-fold purpose of the wind erosion equation is to predict average annual soil loss from a field for specified conditions and to guide the design of wind erosion control practices.

Principles suggested by the wind erosion equation for controlling wind erosion include stabilizing erodible surface soil with various materials; producing a rough, cloddy surface; reducing field width or the distance wind travels in crossing a field unprotected with barriers and strips of crops; and establishing and maintaining sufficient vegetative cover. This last item is sometimes referred to as the "cardinal rule" for controlling wind erosion.

Although the wind erosion equation is extremely useful and widely applicable, users are cautioned that the value obtained for E is an estimate of average annual potential soil loss. The actual soil loss may differ from the potential because of (a) variation from the average of wind and precipitation, (b) inaccuracies in converting from relative field erodibility to annual soil losses, (c) relationships among variables not well defined for all combinations of field and climatic conditions, (d) seasonal variation of field erodibility, and (e) uncertainties inherent in the empiricism used in developing the equation.

Research in progress to improve the accuracy and applicability of the wind erosion equation includes

- Determining the percentage of eroding soil that can be suspended during erosion under a wide range of field conditions and the residence time and fate of the various sizes of particles suspended by wind erosion.

- Refining the soil moisture term of the climatic factor, C, in the wind erosion equation. The current procedures assume that effective moisture of the surface soil particles varies with the PE index or dryness ratio, but surface moisture content is transient. Drying rate and dryness of particles, as a function of soil hydraulic properties and climatic variables, need to be examined and then related to the wind erosion process.
- Converting wind erosion prediction from a deterministic to a stochastic model by incorporating probability functions for some of the dynamic variables.
- Developing more applicable flux equations that can be integrated over time and space to predict soil erosion during single windstorms. Soil flux from fields that contain nonerodible elements decreases with time, which suggests that a time function is needed in the prediction equation.

REFERENCES

1. Aase, J.K., F.H. Siddoway, and A.L. Black. 1976. *Perennial grass barriers for wind erosion control, snow management, and crop production.* In: *Shelterbelts on the Great Plains, Proceedings of the Symposium.* Publication No. 78. Great Plains Agricultural Council, Lincoln, Nebraska. pp. 69-78.
2. Anderson, C.H. 1975. *A history of soil erosion by wind in the Palliser Triangle of western Canada.* Historical Series No. 8. Research Branch, Canada Department of Agriculture, Ottawa, Ontario. 25 pp.
3. Apt, K.E. 1976. *Applicability of the Weibull distribution to atmospheric activity data.* Atmospheric Environment 10: 777-782.
4. Armbrust, D.V., and J.D. Dickerson. 1971. *Temporary wind erosion control: Cost and effectiveness of 34 commercial materials.* Journal of Soil and Water Conservation 26: 154-157.
5. Armbrust, D.V., and L. Lyles. 1975. *Soil stabilizers to control wind erosion.* Soil Conditioners 7: 77-82.
6. Armbrust, D.V., and L. Lyles. 1985. *Equivalent wind erosion protection from selected growing crops.* Agronomy Journal 77: 703-707.
7. Armbrust, D.V., W.S. Chepil, and F.H. Siddoway. 1964. *Effects of ridges on erosion of soil by wind.* Soil Science Society of America Proceedings 28: 557-560.
8. Bagnold, R.A. 1943. *The physics of blown sand and desert dunes.* William Morrow & Co . New York, New York.
9. Bates, C.G. 1911. *Windbreaks: Their influence and value.* Bulletin 86. Forest Service, U.S. Department of Agriculture, Washington, D.C. 100 pp.
10. Black, A.L., and F.H. Siddoway. 1971. *Tall wheatgrass barriers for soil erosion control and water conservation.* Journal of Soil and Water Conservation 26: 107-110.

11. Black, A.L., and J.F. Power. 1965. *Effect of chemical and mechanical fallow method on moisture storage, wheat yield, and soil erodibility.* Soil Science Society of America Proceedings 29: 465-468.
12. Budyko, M.I. 1958. *The heat balance of the earth's surface.* Nina A. Stepanova, Translator. U.S. Department of Commerce, Washington, D.C. 259 pp.
13. Carreker, J.R. 1966. *Wind erosion in the Southeast.* Journal of Soil and Water Conservation 11: 86-88.
14. Chepil, W.S. 1944. *Utilization of crop residues for wind erosion control.* Scientific Agriculture 24: 307-319.
15. Chepil, W.S. 1945. *Dynamics of wind erosion: III. The transport capacity of the wind.* Soil Science 60: 475-480.
16. Chepil, W.S. 1946. *Dynamics of wind erosion: V. Cumulative intensity of soil drifting across eroding fields.* Soil Science 61: 257-263.
17. Chepil, W.S. 1950. *Properties of soil which influence wind erosion: II. Dry aggregate structure as an index of erodibility.* Soil Science 69: 403-414.
18. Chepil, W.S. 1954. *Seasonal fluctuations in soil structure and erodibility of soil by wind.* Soil Science Society of America Proceedings 18: 13-16.
19. Chepil, W.S. 1955. *Effects of asphalt on some phases of soil structure and erodibility by wind.* Soil Science Society of America Proceedings 19: 125-128.
20. Chepil, W.S. 1956. *Influence of moisture on erodibility of soil by wind.* Soil Science Society of America Proceedings 20: 288-292.
21. Chepil, W.S. 1957. *Width of field strip to control wind erosion.* Technical Bulletin No. 92. Kansas Agricultural Experiment Station, Manhattan.
22. Chepil, W.S. 1958. *Soil conditions that influence wind erosion.* Technical Bulletin No. 1185. U.S. Department of Agriculture, Washington, D.C. 40 pp.
23. Chepil, W.S. 1959. *Wind erodibility of farm fields.* Journal of Soil and Water Conservation 14: 214-219.
24. Chepil, W.S. 1960. *Conversion of relative field erodibility to annual soil loss by wind.* Soil Science Society of America Proceedings 24: 143-145.
25. Chepil, W.S., F.H. Siddoway, and D.V. Armbrust. 1962. *Climatic factor for estimating wind erodibility of farm fields.* Journal of Soil and Water Conservation 17: 162-165.
26. Chepil, W.S., and N.P. Woodruff. 1954. *Estimations of wind erodibility of field surfaces.* Journal of Soil and Water Conservation 9: 257-265.
27. Chepil, W.S., and N.P. Woodruff. 1959. *Estimations of wind erodibility of farm fields.* Production Research Report No. 25. Agricultural Research Service, U.S. Department of Agriculture, Washington, D.C. 21 pp.
28. Chepil, W.S., and N.P. Woodruff. 1963. *The physics of wind erosion and its control.* Advances in Agronomy 15: 211-302.
29. Chepil, W.S., N.P. Woodruff, and A.W. Zingg. 1955. *Field study of wind erosion in western Texas.* SCS-TP-125. U.S. Department of Agriculture, Washington, D.C. 60 pp.

30. Chepil, W.S., N.P. Woodruff, F.H. Siddoway, D.W. Fryrear, and D.V. Armbrust. 1963. *Vegetative and nonvegetative materials to control wind and water erosion.* Soil Science Society of America Proceedings 27: 86-89.
31. Chepil, W.S., N.P. Woodruff, F.H. Siddoway, and L. Lyles. 1960. *Anchoring vegetative mulches.* Agricultural Engineering 41: 754-755, 759.
32. Chepil, W.S., and R.A. Milne. 1941. *Wind erosion of soils in relation to size and nature of the exposed area.* Scientific Agriculture 21: 479-487.
33. Claflin, L.E., D.L. Stuteville, and D.V. Armbrust. 1973. *Windblown soil in the epidemiology of bacterial leaf spot of alfalfa and common blight of beans.* Phytopathology 63: 1417-1419.
34. Cole, G.W., L. Lyles, and L.J. Hagen. 1983. *A simulation model of daily wind erosion soil loss.* Transactions, American Society of Agricultural Engineers 26: 1758-1765.
35. Daniel, H.A., and W.H. Langham. 1936. *The effect of wind erosion and cultivation on the total nitrogen and organic matter content of soil in the Southern High Plains.* Journal of American Society of Agronomy 28: 587-596.
36. Dickerson, J.D., N.P. Woodruff, and E.E. Banbury. 1976. *Techniques for improving survival and growth of trees in semiarid areas.* Journal of Soil and Water Conservation 1: 63-66.
37. Dregne, H.E. 1976. *Soils of the arid regions.* Elsevier Scientific Publishers Co., New York, New York. 237 pp.
38. Dudley, F.L. 1959. *Progress of research on stubble mulching in the Great Plains.* Journal of Soil and Water Conservation 14: 7-11.
39. Englehorn, C.L., A.W. Zingg, and N.P. Woodruff. 1952. *The effects of plant residue cover and clod structure on soil losses by wind.* Soil Science Society of America Proceedings 16: 29-33.
40. Erickson, D.A., P.C. Deutsch, D.L. Anderson, and T.A. Sweeney. 1984. *Wind driven interactive wind erosion estimator.* Agronomy Abstracts 247.
41. Fenster, C.R., G.A. Wicks, and D.E. Smika. 1973. *The role of ecofallow for reducing energy requirements for crop production.* Agronomy Abstracts 122.
42. Ferber, A.E. 1969. *Windbreaks for conservation.* Agriculture Information Bulletin 339. Soil Conservation Service, U.S. Department of Agriculture, Washington, D.C.
43. Fisher, P.S., and E.L. Skidmore. 1970. *WEROS: A Fortran IV program to solve the wind erosion equation.* ARS 41-174. Agricultural Research Service, U.S. Department of Agriculture, Washington, D.C. 13 pp.
44. Food and Agriculture Organization, United Nations. 1960. *Soil erosion by wind and measures for its control on agricultural lands.* Development Paper No. 71. Rome, Italy.
45. Food and Agriculture Organization, United Nations. 1979. *A provisional methodology for soil degradation assessment.* Rome, Italy.
46. Frank, A.B., D.G. Harris, and W.O. Willis. 1977. *Growth and yields of spring wheat as influenced by shelter and soil water.* Agronomy Journal 69: 903-906.
47. Fryrear, D.W. 1963. *Annual crops as wind barriers.* Transactions, American Society of Agricultural Engineers 6: 340-342, 352.

48. Fryrear, D.W. 1969. *Reducing wind erosion in the Southern Great Plains.* MP-929. Texas A&M University, College Station.
49. Gillette, D.A., I.H. Blifford, Jr., and C.R. Fenster. 1972. *Measurements of aerosol size distribution and vertical fluxes of aerosols on land subject to wind erosion.* Journal of Meteorology 11: 977-987.
50. Gore, R. 1979. *The desert: An age-old challenge grows.* National Geographic 156: 594-639.
51. Hagen, L.J. 1976. *Windbreak design for optimum wind erosion control.* In: *Shelterbelts on the Great Plains, Proceedings of the Symposium.* Publication No. 78. Great Plains Agricultural Council, Lincoln, Nebraska. pp. 31-36.
52. Hagen, L.J., E.L. Skidmore, and J.D. Dickerson. 1972. *Designing narrow strip barrier systems to control wind erosion.* Journal of Soil and Water Conservation 27: 269-270.
53. Hagen, L.J., L. Lyles, and E.L. Skidmore. 1980. *Application of wind energy to Great Plains irrigation pumping. Appendix B: Method of determination and summary of Weibull parameters at selected Great Plains weather stations.* ATT-NC-4. Science and Education Administration, U.S. Department of Agriculture, Washington, D.C. 20 pp.
54. Hagen, L.J., and N.P. Woodruff. 1973. *Air pollution from dust storms in the Great Plains.* Atmospheric Environment 7: 323-332.
55. Halsey, C.F., W.F. Detmer, L.A. Cable, and E.C. Ampe. 1983. *SOILEROS—a friendly erosion estimation program for the personal computer.* Agronomy Abstracts 20.
56. Hare, F.K. 1983. *Climate and desertification: A revised analysis.* Report No. 44. World Climate Programme, Geneva, Switzerland.
57. Harris, R.F., G. Chesters, and O.N. Allen. 1966. *Dynamics of soil aggregation.* Advances in Agronomy 18: 107-169.
58. Hayes, W.A. 1965. *Wind erosion equation useful in designing northeastern crop protection.* Journal of Soil and Water Conservation 20: 153-155.
59. Hayes, W.A. 1966. *Guide for wind erosion control in the northeastern states.* Soil Conservation Service, U.S. Department of Agriculture, Washington, D.C.
60. Hayes, W.A. 1972. *Designing wind erosion control systems in the Midwest Region.* RTSC-Technical Note, Agronomy L1-9. Soil Conservation Service, U.S. Department of Agriculture, Washington, D.C.
61. Hoag, B.K., and G.N. Geiszler. 1971. *Sunflower rows to protect fallow from wind erosion.* North Dakota Farm Research 28: 7-12.
62. Hurt, R.D. 1981. *The dust bowl: An agricultural and social history.* Nelson Hall, Chicago, Illinois. 214 pp.
63. Jaenicke, R. 1979. *Monitoring and critical review of the estimated source strength of mineral dust from the Sahara.* In: Christer Morales (ed.) *Saharan Dust. Mobilization, Transport, Deposition.* John Wiley & Sons, New York, New York. pp. 233-242.
64. Johnson, G.L. 1978. *Economic design of wind electric systems.* Institute of Electrical and Electronic Engineers Transactions, Power Apparatus Systems PAS-97(2): 554-562.
65. Johnson, V. 1947. *Heaven's tableland: The dust bowl story.* Farrar-Straus, New York, New York. pp. 155-157.

66. Justus, C.G., W.R. Hargraves, and A. Mikhail. 1976. *Reference windspeed distributions and height profiles for wind turbine design and performance evaluation applications*. Technical Report, Contract No. E(40-1)-5108. Georgia Institute of Technology, Atlanta.
67. Lyles, L. 1974. *Speculation on the effect of wind erosion on productivity*. Special Report to Task Force on Wind Erosion Damage Estimates. U.S. Department of Agriculture, Washington, D.C.
68. Lyles, L. 1975. *Possible effects of wind erosion on soil productivity*. Journal of Soil and Water Conservation 30: 279-283.
69. Lyles, L. 1983. *Erosive wind energy distributions and climatic factors for the West*. Journal of Soil and Water Conservation 38: 106-109.
70. Lyles, L., and B.E. Allison. 1980. *Range grasses and their small grain equivalents for wind erosion control*. Journal of Range Management 33: 143-146.
71. Lyles, L., and B.E. Allison. 1981. *Equivalent wind-erosion protection from selected crop residues*. Transactions, American Society of Agricultural Engineers 24(2): 405-408.
72. Lyles, L., D.V. Armbrust, J.D. Dickerson, and N.P. Woodruff. 1969. *Spray-on adhesives for temporary wind erosion control*. Journal of Soil and Water Conservation 24: 190-193.
73. Lyles, L., and J. Tatarko. 1982. *Emergency tillage to control wind erosion: Influences on winter wheat yields*. Journal of Soil and Water Conservation 37: 344-347.
74. Lyles, L., J. Tatarko, and J.D. Dickerson. 1984. *Windbreak effects on soil water and wheat yield*. Transactions, American Society of Agricultural Engineers 27: 69-72.
75. Lyles, L., N.F. Schmeidler, and N.P. Woodruff. 1973. *Stubble requirements in field strips to trap windblown soil*. Research Publication 164. Kansas Agricultural Experiment Station, Manhattan. 22 pp.
76. Malin, J.C. 1946. *Dust storms—part one, two, and three. 1850-1860, 1861-1880, 1881-1900, respectively*. The Kansas Historical Quarterly 14: 129-144, 265-296, 391-413.
77. Mannering, J.V., and C.R. Fenster. 1983. *What is conservation tillage?* Journal of Soil and Water Conservation 38: 141-143.
78. McCalla, T.M., and T.J. Army. 1961. *Stubble mulch farming*. Advances in Agronomy 13: 125-196.
79. McMartin, W., A.B. Frank, and R.H. Heintz. 1974. *Economics of shelterbelt influence on wheat yields in North Dakota*. Journal of Soil and Water Conservation 29: 87-91.
80. Moldenhauer, W.C., and E.R. Duncan. 1969. *Principles and methods of wind erosion control in Iowa*. Special Report No. 62. Iowa State University, Ames.
81. Olmstead, L.B. 1946. *The effect of long-time cropping systems and tillage practices upon soil aggregation at Hays, Kansas*. Soil Science Society of America Proceedings 11: 89-92.
82. Page, J.B., and C.J. Willard. 1946. *Cropping systems and soil properties*. Soil Science Society of America Proceedings 11: 81-88.

83. PEDCO-Environmental Specialists, Inc. 1973. *Investigations of fugitive dust-sources, emissions, and control.* Report prepared under Contract No. 68-02-0044, Task Order No. 9. U.S. Environmental Protection Agency, Office of Air Quality Planning and Standards, Cincinnati, Ohio.
84. Read, R.A. 1964. *Tree windbreaks for the Central Great Plains.* Agriculture Handbook 250. U.S. Department of Agriculture, Washington, D.C.
85. Secretariat of the United Nations Conference on Desertification. 1977. *Desertification: Its causes and consequences.* Pergamon Press, New York, New York.
86. Siddoway, F.H., W.S. Chepil, and D.V. Armbrust. 1965. *Effect of kind, amount, and placement of residue on wind erosion control.* Transactions, American Society of Agricultural Engineers 8: 327-331.
87. Skidmore, E.L. 1965. *Assessing wind erosion forces: Directions and relative magnitudes.* Soil Science Society of America Proceedings 29: 587-590.
88. Skidmore, E.L. 1982. *Soil and water management and conservation: Wind erosion.* In: V.J. Kilmer (ed.) *Handbook of Soils and Climate in Agriculture.* CRC Press, Boca Raton, Florida. pp. 371-399.
89. Skidmore, E.L. 1983. *Wind erosion calculator: Revision of residue table.* Journal of Soil and Water Conservation 38: 110-112.
90. Skidmore, E.L. 1986. *Wind-erosion climatic erosivity.* Climate Change 9: 195-208.
91. Skidmore, E.L., and F.H. Siddoway. 1978. *Crop residue requirements to control wind erosion.* In: W.R. Oschwald (ed.) *Crop Residue Management Systems.* Special Publication No. 31. American Society of Agronomy, Madison, Wisconsin. pp. 17-33.
92. Skidmore, E.L., J.B. Layton, D.V. Armbrust, and M.L. Hooker. 1986. *Soil physical properties as influenced by residue management.* Soil Science Society of America Journal 50(2): 415-419.
93. Skidmore, E.L., and L.J. Hagen. 1977. *Reducing wind erosion with barriers.* Transactions, American Society of Agricultural Engineers 20: 911-915.
94. Skidmore, E.L., L.J. Hagen, D.G. Naylor, and I.D. Teare. 1974. *Winter wheat response to barrier-induced microclimate.* Agronomy Journal 66: 501-505.
95. Skidmore, E.L., L.J. Hagen, and I.D. Teare. 1975. *Wind barriers most beneficial at intermediate stress.* Crop Science 15: 443-445.
96. Skidmore, E.L., M. Kumar, and W.E. Larson. 1979. *Crop residue management for wind erosion control in the Great Plains.* Journal of Soil and Water Conservation 34: 90-96.
97. Skidmore, E.L., N.L. Nossaman, and N.P. Woodruff. 1966. *Wind erosion as influenced by row spacing, row direction, and grain sorghum population.* Soil Science Society of America Proceedings 30: 505-509.
98. Skidmore, E.L., and N.P. Woodruff. 1968. *Wind erosion forces in the United States and their use in predicting soil loss.* Agriculture Handbook No. 346. U.S. Department of Agriculture, Washington, D.C.
99. Skidmore, E.L., P.S. Fisher, and N.P. Woodruff. 1970. *Wind erosion equation: Computer solution and application.* Soil Science Society of America Proceedings 34: 931-935.

100. Skidmore, E.L., W.A. Carstenson, and E.E. Banbury. 1975. *Soil changes resulting from cropping.* Soil Science Society of America Proceedings 39: 964-967.
101. Staple, W.J., and J.H. Lehane. 1955. *The influence of field shelterbelts on wind velocity, evaporation, soil moisture, and crop yields.* Canadian Journal of Agricultural Science 35: 440-453.
102. Svobida, L. 1940. *An empire of dust.* Caxton Printers, Ltd., Caldwell, Idaho. 203 pp.
103. Thornthwaite, C.W. 1931. *Climates of North America according to a new classification.* Geographical Review 25: 633-655.
104. U.S. Department of Agriculture, Soil Conservation Service. 1984. *National Resources Inventory.* Washington, D.C.
105. U.S. Department of Agriculture, Soil Conservation Service. 1988. *National Agronomy Manual,* 190-V. Washington, D.C.
106. Williams, J.R., C.A. Jones, and P.T. Dyke. 1984. *A modeling approach to determining the relationship between erosion and soil productivity.* Transactions, American Society of Agricultural Engineers 27: 129-144.
107. Wilson, L. 1975. *Application of the wind erosion equation in air pollution surveys.* Journal of Soil and Water Conservation 30: 215-219.
108. Woodruff, N.P., and D.V. Armbrust. 1968. *A monthly climatic factor for the wind erosion equation.* Journal of Soil and Water Conservation 23: 103-104.
109. Woodruff, N.P., and F.H. Siddoway. 1965. *A wind erosion equation.* Soil Science Society of America Proceedings 29: 602-608.
110. Woodruff, N.P., J.D. Dickerson, E.E. Banbury, A.B. Erhart, and M.C. Lundquist. 1976. *Selected trees and shrubs evaluated for single-row windbreaks in the Central Great Plains.* NC-37. Agricultural Research Service, U.S. Department of Agriculture, Washington, D.C.
111. Woodruff, N.P., L. Lyles, F.H. Siddoway, and D.W. Fryrear. 1972. *How to control wind erosion.* Agriculture Information Bulletin No. 354. Agricultural Research Service, U.S. Department of Agriculture, Washington, D.C. 22 pp.
112. Woodruff, N.P., W.S. Chepil, and R.D. Lynch. 1957. *Emergency chiseling to control wind erosion.* Technical Bulletin 90. Kansas Agricultural Experiment Station, Manhattan.
113. Zingg, A.W. 1951. *A portable wind tunnel and dust collector developed to evaluate the erodibility of field surfaces.* Agronomy Journal 43: 189-191.
114. Zingg, A.W. 1951. *Evaluation of the erodibility of field surfaces with a portable wind tunnel.* Soil Science Society of America Proceedings 15: 11-17.
115. Zingg, A.W. 1953. *Wind-tunnel studies of the movement of sedimentary materials.* In: *Proceedings, Fifth Hydraulic Conference, Iowa Institute of Hydraulic, Research Bulletin 34.* John Wiley & Sons, New York, New York. pp. 111-135.
116. Zingg, A.W., and N.P. Woodruff. 1951. *Calibration of a portable wind tunnel for the simple determination of roughness and drag on field surface.* Agronomy Journal 43: 191-193.

Chapter 12

METHODS for INVESTIGATING BASIC PROCESSES and CONDITIONS AFFECTING WIND EROSION

E. L. Skidmore, L. J. Hagen, D. V. Armbrust, A. A. Durar, D. W. Fryrear, K. N. Potter, L. E. Wagner, T. M. Zobeck

INTRODUCTION

Although past wind erosion research has been reasonably accepted and implemented, we need to improve our technology for measuring erosion and monitoring soil transport; replace empirical annual soil-loss models with process-based event models; better understand the soil aggregation process as influenced by inherent soil properties, soil management, cropping sequence, climate, and the changes in aggregate status seasonally and during an erosion event; determine probabilities and establish confidence limits for predicting duration and intensity of meteorologic conditions conducive to erosion; and develop user friendly prediction systems that operate on personal computers for conservation planning and resource inventory.

Because it is not feasible to incorporate the new research and technology into existing models or variations thereof, the U.S. Department of Agriculture (USDA) has charged a group of scientists to develop a new "Wind Erosion Prediction System," WEPS (*53, 55*). In this chapter, various members of that USDA core team review some of the methods used for investigating basic processes and conditions affecting wind erosion.

WIND DATA BASES AND SIMULATION

The need to know about the wind has prompted several studies, particularly by those interested in wind as a source of energy (*35, 57, 89*) and those concerned with erosion of soil by wind (*75, 76, 98, 100, 104, 105, 125*).

Skidmore (*98*) computed wind erosion force vectors from frequency of occurrence of direction by wind speed groups. The wind erosion force vectors were used to compute monthly magnitudes of wind erosion forces, prevailing wind erosion direction, and preponderance of wind erosion forces in the prevailing wind erosion direction. These factors indicate, respectively, potential need for wind erosion protection, proper orientation of erosion control measures, and relative merits of proper orientation of the control methods. They were furnished by month for locations throughout the United States (*106*). The resulting handbook has since been used for conservation planning and wind erosion prediction. The prevailing wind erosion direction and preponderance data are included in the recent SCS *National Agronomy Manual* (*106*). In that manual, magnitude of wind erosion forces was presented in an erosive wind energy distribution format as developed by Bondy et al. (*9*) and Lyles (*75*).

Although such wind analyses were essential for conservation planning and wind erosion prediction with the wind erosion model of Woodruff and Siddoway (*124*), they are not adequate for the evolving wind erosion technology (*53, 54, 55*). Skidmore and Tatarko (*103*) developed a wind data base suitable for use in the stochastic approaches in the current wind erosion modeling effort. Their procedure was to use monthly wind speed and wind direction summaries to calculate wind speed distribution parameters for each of 16 directions. In addition, ratios of maximum to minimum hourly wind speed, and hour of maximum wind speed, wind direction, and hourly wind speed were stored in the data base.

A stochastic wind speed/direction simulator was developed which uses data from the data base (*104*). Stochastic wind direction is determined by comparing a random number with the cumulative wind direction distribution. Once wind direction is simulated, Equation 1 is solved to determine daily mean wind speed, UMEAN

$$\text{UMEAN} = c\left\{-\ln[1 - F(u) - F_0)/(1 - F_a)]\right\}^{1/k} \qquad [1]$$

where c and k are monthly Weibull scale and shape wind speed distribution parameters (*118*); F(u) is cumulative Weibull distribution function and has a value between 0 and 1.0 assigned to it from random number generation, and F_0 is the frequency of calm periods. Wind speed at any hour of the day [U(I)] can be simulated from

$$U(I) = UMEAN + 0.5(UMAX - UMIN)\cos[2\pi(24 - HRMAX) + I)/24] \quad [2]$$

where UMAX and UMIN are daily maximum and minimum wind speeds, and HRMAX is the hour of the day when wind speed is maximum.

An output of a few simulations is illustrated in Table 12.1. Wind speed was printed every two hours for each simulation. If wind speed at any time exceeded 8 m/s, then it was flagged by a "yes" in the last column of Table 12.1. This means that wind speed is high enough to cause erosion from an unprotected surface of highly erodible particles, and an erosion submodel should be activated.

PARTICLE DETACHMENT AND TRANSPORT

Detachment and transport of particles during wind erosion are controlled by a number of static and dynamic forces interacting with a wide range of particle sizes in a turbulent flow. This discussion will present a brief overview of the various processes that occur during wind erosion and then review some of the laboratory methods used to investigate them.

Processes

The various processes that compose wind erosion are difficult to study at a point in space and time, so measurements usually represent either quasi-steady state or time integrations of detachment and transport processes over some finite space. One can visualize wind erosion for a finite space near the surface on a typical agricultural soil (control volume) as the conservation of mass of two species (saltation and creep) with two sources (emission of loose particles and abrasion of particles from clods and crust) and two sinks (trapping of saltation/creep and suspension of fine particles) (Figure 12.1). As wind erosion proceeds, rearrangement of the soil surface also occurs. Simplified surfaces are often used in laboratory studies so that individual processes can be investigated. This can be accomplished by manipulating surface roughness and selecting both composition and size of erodible and nonerodible fractions.

The forces acting on a loose grain on a particle bed at the threshold of motion are illustrated in Figure 12.2. For particles below 10 um, forces of adhesion dominate in keeping a particle on the bed, whereas for particles above 100 um, weight dominates (*85*). Single spheres much larger than sand grains have been instrumented with transducers and placed on a bed of similar spheres to study drag and lift forces (*19, 30*).

Table 12.1. Wind direction and wind speed simulation for March & July, Lubbock, Texas (*104*).

WIND	HOUR OF DAY											
DIRECTION	1	3	5	7	9	11	13	15	17	19	21	23	EROSION
MARCH	Windspeed (m/s)												
13	3.3	3.2	3.3	3.7	4.1	4.6	5.0	5.1	5.0	4.6	4.1	3.7	NO
11	4.6	4.4	4.6	5.1	5.7	6.4	6.9	7.0	6.9	6.4	5.7	5.1	NO
13	2.7	2.6	2.7	3.0	3.4	3.8	4.1	4.2	4.1	3.8	3.4	3.0	NO
4	6.2	5.9	6.2	6.8	7.7	8.6	9.3	9.5	9.3	8.6	7.7	6.8	YES
9	6.9	6.7	6.9	7.7	8.7	9.7	10.4	10.7	10.4	9.7	8.7	7.7	YES
11	8.0	7.7	8.0	8.9	10.0	11.2	12.0	12.4	12.0	11.2	10.0	8.9	YES
10	5.4	5.2	5.4	6.0	6.8	7.6	8.1	8.3	8.1	7.6	6.8	6.0	YES
12	1.7	1.7	1.7	1.9	2.2	2.4	2.6	2.7	2.6	2.4	2.2	1.9	NO
5	3.4	3.3	3.4	3.7	4.2	4.7	5.1	5.2	5.1	4.7	4.2	3.8	NO
7	7.3	7.0	7.3	8.1	9.2	10.2	11.0	11.3	11.0	10.2	9.2	8.1	YES
JULY													
5	2.4	2.1	2.0	2.0	2.1	2.4	2.7	2.9	3.1	3.1	2.9	2.7	NO
9	7.2	6.4	6.0	6.0	6.4	7.2	8.1	8.9	9.4	9.4	8.9	8.1	YES
9	4.6	4.1	3.8	3.8	4.1	4.6	5.2	5.7	6.0	6.0	5.7	5.2	NO
9	8.0	7.2	6.7	6.7	7.2	8.0	9.1	10.0	10.5	10.5	10.0	9.1	YES
6	5.7	5.1	4.7	4.7	5.1	5.7	6.4	7.0	7.4	7.4	7.0	6.4	NO
10	8.2	7.3	6.8	6.8	7.3	8.2	9.2	10.2	10.7	10.7	10.2	9.2	YES
7	3.6	3.2	2.9	2.9	3.2	3.6	4.0	4.4	4.6	4.6	4.4	4.0	NO
9	5.0	4.5	4.1	4.1	4.5	5.0	5.7	6.2	6.5	6.5	6.2	5.7	NO
12	3.0	2.6	2.5	2.5	2.6	3.0	3.3	3.7	3.9	3.9	3.7	3.3	NO
9	5.6	4.9	4.6	4.6	4.9	5.6	6.3	6.9	7.2	7.2	6.9	6.3	NO

Directions are clockwise starting with 1 - north, 2 - north, north east, etc.

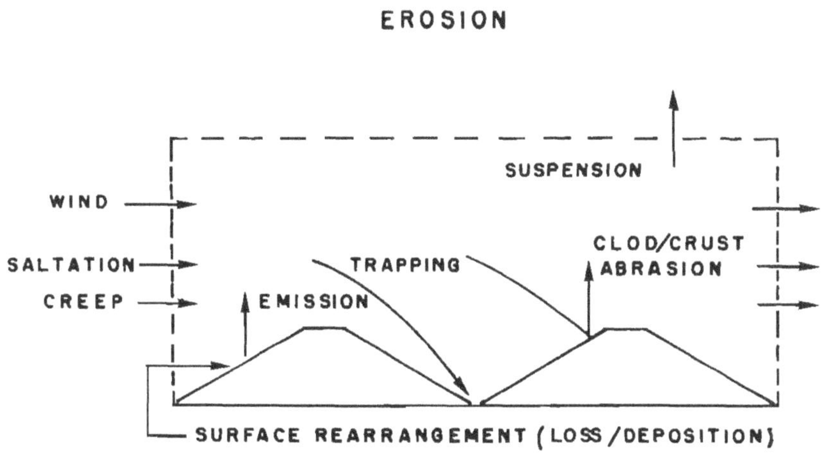

Figure 12.1 Diagram of wind erosion processes on a bare soil.

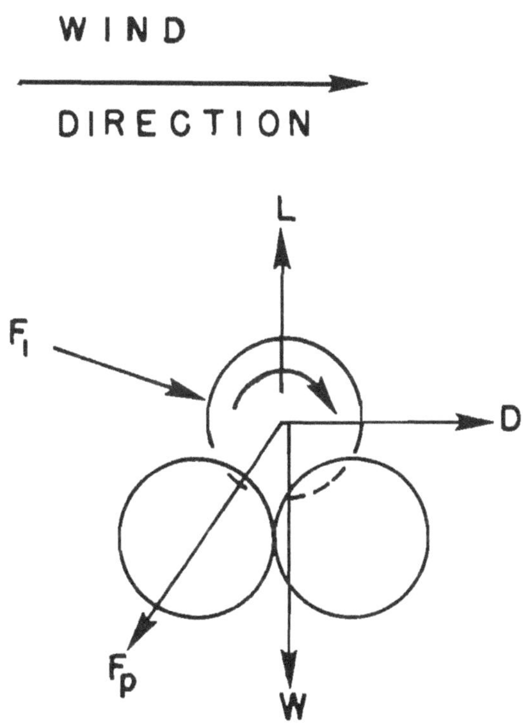

Figure 12.2 Diagram of forces acting on a single particle at the threshold of motion including aerodynamic lift (l), drag (D), moment (M), weight (W), interparticle cohesive force (F_p), and force of impact from an incoming particle (F_I).

Wind tunnels

Environmental wind tunnels are often used to study detachment and transport. These are generally characterized by a long test section designed to develop a turbulent boundary layer whose velocity profile and turbulence characteristics match those of the modeled atmospheric boundary layer. These also often include an adjustable top that can be sloped to remove the pressure gradient down the working section. Even with long test sections, however, it is sometimes necessary to augment boundary layer depth with upstream turbulence generators (*108*). Attempts to adapt short, aerodynamic tunnels to atmospheric simulation have met with mixed success (*84*).

Maintaining similitude while modeling the effects of blowing particles is a problem because of the large number of variables. The variables are generally arranged in dimensionless groups and, in a true model, all the groups have the same value in the model as in the full scale (*66, 68*). Frequently, it is impossible to maintain a true model, such as with snow drifting near windbreaks, and a distorted model must be used. In this case, the degree of distortion must be varied in the experiments and extra care is needed to extrapolate the results to full scale. Recent results show that the Froude number is one of the most important dimensionless parameters when simulating particle transport (*64*).

Threshold

Monitoring initial motion of the particles also has been accomplished by visual observation. However, improved precision is gained by using instrumentation, such as the laser optical system described by Nickling (*83*). He also described an analytical interpretation of the threshold velocity from the sensor output. A transducer with a vertical array of piezoelectric crystals attached to an electronic counter is another instrument that has been used to sense both threshold velocities and momentum of particles (*48*). Finally, low saltation rates have been measured with saltation sand catchers and a saltation threshold discharge defined as 0.001 kg $m^{-1}s^{-1}$ (*78*).

On complex surfaces composed of erodible particles and nonerodible elements, the shearing stress is distributed between the erodible and nonerodible elements (*74*). Even for complex surfaces, however, a surface threshold velocity can be measured. Surface threshold velocities for simulated stubble (*76*), aggregate cover (*73*), and tillage ridges (*56*) have been investigated. Another technique in threshold studies is to use a narrow range of erodible particles and let the wind uncover buried nonerodible elements.

When removal ceases, one can infer that the friction velocity in the sheltered zones is equal to the erodible particle threshold velocity.

Saltation

The saltation process can be investigated by a number of methods. Trajectories of individual saltating particles have been theoretically predicted by numerical integration of the equations of motion. When Magnus lifting force due to particle spin was included, theory agreed well with observations (*120, 121*). Experimental measurements of particle trajectories were obtained from analysis of high-speed motion pictures at 2000 to 10,000 frames/s. In the latter study, particle speeds passing a constant height also were measured using a velocimeter, which consisted of a light source and two phototransistors. The instrument was based on a design modified from Schmidt (*95*). Numerical models of saltating particle trajectories also have been combined with measurements of vertical profiles of the saltation flux to predict such parameters as particle launch velocities and surface dislodgement rates (*107*).

The progress of saltating particles moving downwind can be observed for groups of particles by using dyed grains (*122*) or for an individual particle by using a radioactive tracer (*8*). Methods to prepare, observe, and separate variable size and density sands for saltation studies have been presented by Gerety and Slingerland (*44*). Transport capacity of saltating particles by wind has been measured both outdoors and in wind tunnels using sand catchers. Various empirical formulas have been developed relating the transport capacity to friction velocity (*49*).

Erosion

A convenient way to study abrasion is to place trays of soil in a wind tunnel and blow abrader from an upwind source across them. The abrasion flux from unit surface area (G_{ab}) for a surface composed of n classes of target materials has been modeled as

$$G_{ab} = \left(\sum_{i=1}^{n} F_i \, C_{ai} \right) q \qquad [3]$$

where C_{ai} is an abrasion coefficient with units (L^{-1}), F_i is the fraction of saltation impacting the ith material, and q is saltation discharge ($ML^{-1}T^{-1}$) (*54*). Abrasion coefficient depended mainly on the abrasion resistance of target clods or crust and was only slightly dependent on wind speed or type of abrader. By placing various fractions of non-abradable clods on a crusted surface, values for F_i were calculated after measuring the other three parameters in Equation 3.

AGGREGATE STABILITY

Dry soil-aggregate stability and aggregate-size distribution are primary factors affecting soil susceptibility to wind erosion. Aggregate density, to a lesser extent, affects soil erodibility but is much less variable than stability and size distribution. Soil-aggregate stability and size distribution vary widely in time and space. From Table 12.7 (*124*), it is seen that nonaggregated sandy soils with only one percent of the sand having diameters greater than 0.84 mm are 10 and 100 times more erodible than aggregated soils with 53 and 77 percent of their aggregates greater than 0.84 mm, respectively. Similarly, dry aggregate stability may differ a hundredfold between soils. This section and the next one discuss measuring both of these transient soil properties.

Various methods based on different principles have been used to evaluate dry aggregate stability. Principles include relative aggregate size reduction from applied forces, rupture stress, and energy consumed in size reduction. In the relative size reduction technique, the aggregates were subjected to external forces in several ways: (a) placing in metal cylinders that were inverted end-over-end 20 times (*14*); (b) repeated rotary sieving (*16*); and (c) vigorous sieving with flat sieves (*112*). For rupture stress measurement, aggregates were diametrically loaded between parallel plates (*91, 92, 102*).

Energy consumed in size reduction has been measured using several different methods. The desire to know the work required to subdivide aggregates into smaller units, led Marshall and Quirk (*79*) to use a drop-shatter method. Air-dried samples were shattered by dropping them onto a concrete floor from various heights. The kinetic energy was dissipated by impact with the hard surface. Others (*36, 46, 50*) have used the drop-shatter technique to establish the relationship between the energy imparted to the soil and the degree of fragmentation.

Skidmore and Powers (*102*) measured the energy consumed in crushing an aggregate by integrating the area under the force against a distance curve. Boyd et al. (*10*) developed a soil-aggregate crushing-energy meter, for measuring the energy consumed in crushing an aggregate.

Skidmore et al. (*103, 104*) evaluated four different measures of aggregate stability. The names and unit of measurement for each of these methods were: crushing-energy/surface-area (J/m^2), crushing energy (J/kg), rupture stress (kPa), and initial break force (N). Crushing-energy/surface-area represents the work done in crushing an aggregate divided by the new surface area exposed, which gave energy per unit of surface area. The surface area was calculated by using the arithmetic mean of each of the sieve size fractions and assuming that the aggregates were spherical. The crushing energy was calculated by dividing the work done in crushing an aggregate by the mass of the aggregate being crushed. The rupture stress was calculated by dividing the initial break force by the cross-sectional area of the aggregate. This required an independent measurement of aggregate density. The initial break force was simply the force on the aggregate at initial fracture.

Coefficient of variation for crushing energy (J/kg) < crushing energy/surface area (J/m2) < rupture stress (kPa) < initial break force (N). These four methods correlated reasonably well. When regressed on each other, the coefficients of determination ranged from 0.90 to 0.97. Sample numbers required to estimate the true mean within 25 percent of the mean at 0.05 level of significance were 10, 12, 20, and 22 for crushing energy, crushing-energy/surface area, rupture stress, and initial break force, respectively.

In general, the ranking of simplicity-of-measurement is in the same order as sample numbers required to estimate the mean value. The initial break force is easiest to measure but requires the greatest number of measurements. For the crushing-energy/surface-area method, work required to break aggregate bonds and the newly created external surface area are measured. This method also had the greatest range, more than two orders of magnitude, between the soft and hard soils. One of the drawbacks of this method is the tedium of measuring the new surface area exposed as a result of the crushing.

The energy consumed in size reduction of the crushing-energy method was also measured, but instead of dividing by new surface exposed, the crushing energy was divided by the mass of the aggregate, thus making it more important to crush to the same end point each time. In spite of that, this method required the fewest aggregates for an estimate of the mean. The measurement is extremely simple but does require special equipment for measuring the energy (*10*). The crushing energy method is now used routinely in several laboratories.

In a separate experiment, Skidmore et al. (*101, 102*) evaluated abrasive soil loss as influenced by aggregate stability. Soil loss correlated well with the aggregates' resistance to crushing. Log abrasive soil loss was linearly related to aggregate stability (crushing energy method), yielding a coefficient of determination of 0.99.

DRY AGGREGATE SIZE DISTRIBUTION

Tilled surface soils are composed of clods, soil aggregates, and particles of various sizes. The relative amounts of these constituents, on an air- or oven-dried mass basis, by size class, make up the dry aggregate size distribution (DASD).

The DASD is determined by sieving the dry soil. Early sieving methods used manually or mechanically agitated flat sieves (*29*). Chepil and Bisal (*22*) proposed a standard sieving technique employing a nested set of rotary sieves. The rotary sieve technique has now been generally accepted as the standard technique to determine DASD for wind erosion assessment.

At least once each decade during the past half century, wind erosion researchers have contributed new information on rotary sieves and sieving for determining soil aggregate size distribution and stability. These papers include Chepil and Bisal (*22*), a rotary sieve method; Chepil (*15*), an improved rotary sieve; Chepil (*21*), a compact rotary sieve; Lyles et al. (*77*), a modified rotary sieve; and Fryrear (*39*), a rapid rotary sieve.

Chepil and Bisal (*22*) recommended the rotary sieve over hand sieving mainly because of the wide variance in results obtained among different operators. They attributed this to differences in personal judgment as to the amount of shaking required. Variance between two operators was reduced by using the mechanically operated rotary sieve.

Results obtained by the various models of rotary sieves did not agree well. Lyles et al. (*77*) tested the original, improved, compact, and modified rotary sieves for sieving accuracy with a nonabrasive stone, sand, and gravel mixture and found average errors of 9.4, 32.7, 13.0, and 2.3 percent, respectively. The improved accuracy of the modified rotary sieve was achieved by giving major consideration to mesh length, the primary factor controlling the time sieve material remains on the mesh area.

Even with stable material, size separation by sieving is inherently inaccurate, because of the difficulty of defining unambiguously the point at which to stop the sieving. Part of the difficulty arises because the "near mesh" particles require many encounters with the sieve surface in order to pass through (*51*). With material like soil aggregates, which become abraded or attrited during sieving, not only is defining the end point more difficult but fragile aggregates as from sandy soils may disappear during sieving.

Several alternative methods have been proposed to calculate geometric mean diameter, GMD, and geometric standard deviation, GSD. In his original paper, Gardner (*43*) suggested a graphical approach. Because aggregates are generally log-normally distributed, plots of DASD on log-probability paper, with the log of the sieve diameter as the ordinate, yield straight lines. The GMD is the diameter at 50 percent oversize, and the GSD is calculated as the ratio

$$\text{GSD} = \text{size at 50 percent/size at 15.9 percent} \qquad [4]$$

Gardner (*43*) cautioned that the antilog of GSD has no statistical meaning when determined in this manner.

GMD and GSD also may be calculated from

$$GMD = \exp\left[\sum_{i=1}^{n} m_i \ln d_i\right] = \pi\,(d_i)^{m^i} \qquad [5]$$

$$GSD = \exp\left[\Sigma_{m_i}(\ln d_i)^2 - (\ln GMD)^2\right]^{1/2} \qquad [6]$$

where m_i is the mass fraction in each aggregate size class i, d_i is the geometric mean diameter of class i, and π is product operator.

Hagen et al. (*58*) showed that for aggregates that are size distributed log-normally, the mass fraction of aggregates whose diameters are greater or less than some diameter may be represented by use of the error function of the normal distribution curve. This technique requires only two sieve cuts, from which the geometric mean diameter and geometric standard deviation are calculated. Because these two parameters describe the size distribution of log-normally distributed aggregates, the mass fraction of aggregates greater than some user-selected diameter can be calculated easily.

SURFACE ROUGHNESS: SHELTER ANGLE DISTRIBUTION

The importance of soil surface roughness with respect to wind erosion is well documented (*23, 38, 41, 124*). Several surface roughness indices have been developed that summarize cloddiness contributions to surface roughness (*2, 69, 93*), but it has been difficult to apply these indices to wind erosion processes.

Recently, a surface microrelief index was developed based upon a shelter angle concept, the minimum angle from horizontal a particle must descend in order to strike a given location (Figure 12.3) (*87*). The descent angle for a given point was determined by calculating the maximum angle from horizontal between a vertex point and adjacent measured points on the soil surface in a given direction. The angle for a vertex point and an adjacent point was calculated as the arctangent of the ratio between elevation differences and the horizontal distance separating the two points. This procedure was repeated, retaining the same vertex point, for all other points along a transect in a given direction within a 0.3 horizontal distance, termed the influence zone. The maximum angle was retained and defined as the shelter angle, for that individual vertex. This procedure was repeated, incrementing the vertex point and influence zone one point, across the

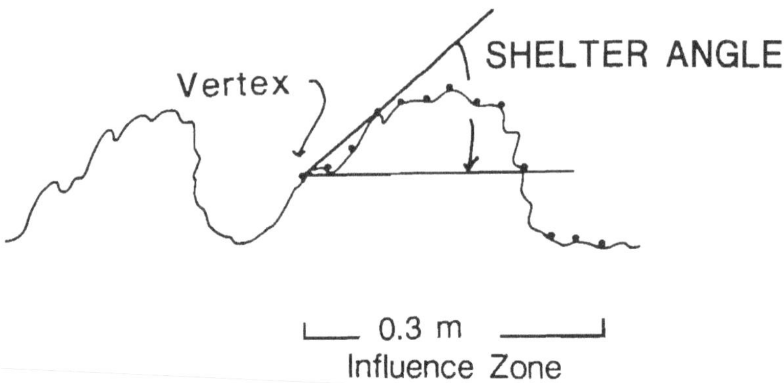

Figure 12.3 Surface roughness schematic illustrating the shelter angle and influence zone concepts (*87*).

transect. Therefore, each measured point has a shelter angle associated with it, except those points near the edge of the plot where the influence zone extended beyond the plot. For points with a negative shelter angle, which indicates that the vertex had a greater elevation than any other point in the influence zone, the shelter angle was set to zero. Therefore, low shelter angles are relatively exposed, so a particle could descend at a very small angle from horizontal and strike the vertex. Large shelter angles would be relatively protected.

The shelter angle for a given vertex represents only that point. To represent the entire surface, many shelter angles are considered by plotting the cumulative shelter angle distribution (CSAD), the fraction of shelter angles less than a given angle versus shelter angle. From the CSAD, the fraction of the measured points with shelter angles equal to or less than an arbitrary angle is easily determined and provides an estimate of the fraction of the surface susceptible to abrasion. In contrast, the fraction of the surface with large shelter angles is likely to be more effective in soil trapping.

Orientation is an important factor for wind erosion control as surface protection, which results from both cloddiness and ridge roughness, varies depending on wind direction relative to the ridge. The effects of ridge and clod roughness can be determined by calculating shelter angles in different directions, i.e., parallel and perpendicular to tillage. An example of shelter angle distributions parallel and perpendicular to sweep tillage with 74-mm ridges is presented in Figure 12.4. It was assumed that the effect of ridging would be minimal in the parallel direction and the CSAD would represent clod roughness. The effect of ridge roughness was determined by calculat-

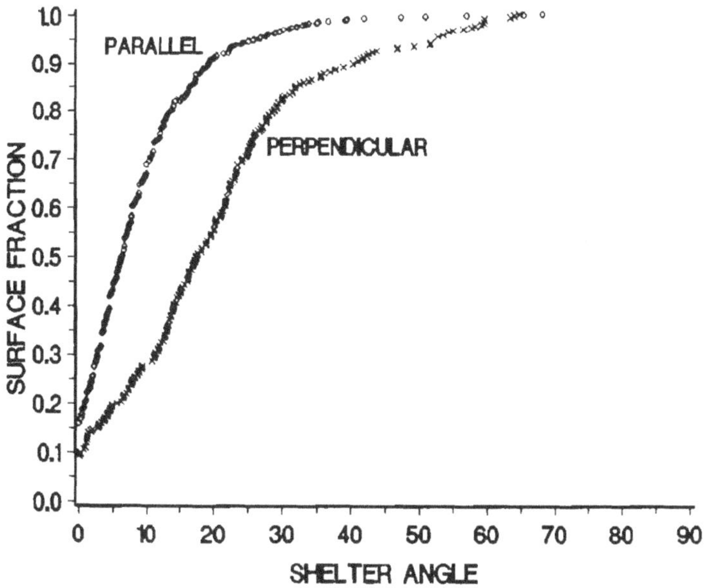

Figure 12.4 Cumulative shelter angle distributions calculated parallel and perpendicular to tillage for a surface with 74-mm ridges (*87*).

ing shelter angles for transects perpendicular to tillage which represents both clod and ridge roughness effects. Therefore, the ridge effect is the difference in the two curves. The fraction of the surface with shelter angles less than 15 degrees for directions parallel and perpendicular to tillage are 80 and 40 percent, respectively.

SURFACE ROUGHNESS: DIGITIZATION OF PIN METER PHOTOS

Parameters that reflect the degree of surface roughness are usually computed from surface elevation measurements. The actual surface elevation data can be obtained through a variety of methods involving image analysis techniques (*90*), microprocessors (*88, 115*), laser systems (*62, 94*), or photography (*119*).

One of the simplest devices to measure surface elevation for calculating shelter angle distribution is a transect, surface profile, pin meter (TSPPM). A typical TSPPM consists of a row of equally spaced pins that contact the soil surface when a measurement is to be taken. The relative pin elevations are then recorded either manually, electronically, or photographically (Figure 12.5).

Manual recording of field pin elevation data is time consuming and prone to human errors. Electronically recording field elevation data is usually faster and less likely to introduce human-induced errors during the actual recording operation. This is an attractive solution when considering the fact that the elevation data can be stored in digital form in the field. However, automated electronic measurement requires that a data acquisition system, sensors, and power source be available at the measurement site. Therefore, the cost of acquisition, maintenance, and repair must be considered when implementing this type of system, especially when multiple field instruments are necessary. Also, electronic measurement devices may require special instruction and training in the operation and maintenance of the units that may not be possible or feasible.

The third method, photographic techniques, can be as fast as electronic means for recording field pin elevation data, but are less costly and generally do not require special handling and maintenance considerations by the field technician. The disadvantage of photographic techniques is that the data still must be digitized into discrete relative elevation values. This digitization step can be performed either manually or through some automated or semi-automated process.

Recently, a PC-based computer program has been developed (*117*) to automate the digitization of transect, surface profile, pin meter elevation data encoded on photographs (available from USDA-ARS, Wind Erosion Research Unit, Manhattan, KS 66506-4006). The Pin Meter Program (PMP) eliminates the tedious, time-consuming process used in manual digitization methods and reduces the amount of operator judgment required in less automated digitization processes. The PMP is designed to simplify the digitization step and provide more consistent results.

The PMP uses a computer-based hand or page digitization scanner to scan the photograph of a transect, surface profile, pin meter field data. This digitized image is then processed using image analysis techniques to locate the pin edges and thereby compute the pin elevation data. The PMP can then produce the data in the desired format and units as selected by the operator. The data can be displayed on the monitor, printed, and/or stored on disk. This elevation data can then be used to obtain the surface roughness parameters of choice. However, the PMP does require a minimum set of computer hardware and peripherals. Those requirements are

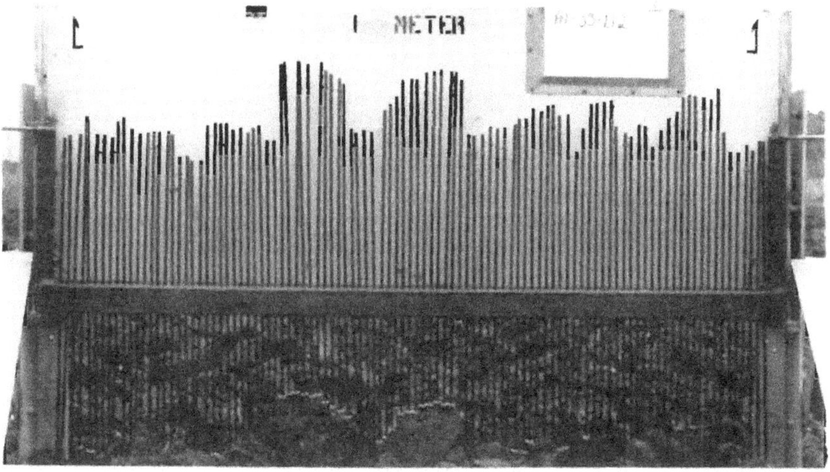

Figure 12.5 Photo of pins in contact with soil surface.

1. An IBM®[1] PC/XT/AT or compatible microcomputer with 640 kilobytes of main memory.
2. MS-DOS or PC-DOS version 2.1 or higher operating system.
3. Additional computer memory supporting the Lotus/Intel/Microsoft (LIM 4.0) Expanded Memory Standard (EMS). Approximately 1.5 megabytes of EMS memory is necessary for a 7.6 by 12.7 cm (3 x 5 in) photo digitized at 300 dpi. The actual amount of EMS memory required is dependent upon the size of the cropped digitized image and the scanner digitizing resolution used.
4. An EGA or VGA graphics adapter with corresponding color monitor is required. Only the 16 color 640 x 350 pixel (EGA) or 640 x 480 pixel (VGA) graphics modes are supported.
5. A Microsoft®[2] compatible mouse with the corresponding software driver.

---

[1] IBM® is a registered trademark of International Business Machines, Inc.
[2] Microsoft® is a registered trademark of Microsoft Corporation.

6. Digitizing scanner. The scanner software must be capable of producing single plane (black and white) digitized images that can be stored in the PC Paintbrush®[3] (.PCX) graphics file format. A relatively inexpensive hand digitizing scanner[4], as compared to more expensive page scanners used during the development of PMP, has provided satisfactory results.
7. A hard disk is not required but highly recommended for convenient storage and retrieval of image and surface profile data files.

Measurement accuracy estimates were made using 20, 101-pin TSPPM photos (3 x 5 in). Each photo was scanned five times at 300 dpi (118 dpts/cm). The standard deviations for the individual pins in each photo were pooled to obtain an estimate of repeatability (standard deviation) of 0.7 mm. This repeatability value may vary with operator, scanner resolution, and contrast quality and size of the TSPPM photos.

SURFACE CRUSTING

Immediately after tillage, the soil surface is usually in a loose, cloddy, unconsolidated condition that is resistant to wind erosion. Heavy rainfall will often produce a consolidated zone at the soil surface, called a surface crust or seal, that is more compact and mechanically stable than some parts of the soil below it (*26*). Soil crusts are generally resistant to erosion by wind alone but they can be rapidly eroded when blowing soil particles are present in the windstream.

Surface crusts have long been the object of scientific investigation, particularly in regard to their effects on plant emergence and water infiltration. Cary and Evans (*13*) have presented an excellent review of crust formation and management. A review of the effects of crust formation on crop establishment has been presented by Awadhwal and Thierstein (*5*). A recent international symposium on surface crusting included sessions on the genesis and morphology of crusts; effects of crusts on seedling emergence, water and gas transfer, and water erosion; methodology to characterize crusts; and crust management (*12*). Few studies have considered the effects of surface crusts on wind erosion.

---

[3]PC Paintbrush® is a registered trademark of Zsoft Corporation.

[4] The actual hand scanner used was DFI Handy Scanner model HS-3000 by Diamond Flower Electric Instrument Co., (USA) Inc., 2455 Port Street, West Sacramento, CA 95961, USA. The use of trade names in this publication does not imply endorsement of the products named.

When crusts are present, the mechanically more stable crust is less susceptible to abrasion by blowing soil than the soil beneath the crust. Crusted soils have been found to erode at a rate of approximately one-sixth to one-half the rate of uncrusted soil (*19*). A determination of the ability of crusts to withstand the abrasive action of blowing soil, the abrasive resistance, is an important factor in the characterization of soil crusts for wind erosion assessment. Crust abrasive resistance has been measured in the laboratory (*14, 33, 126, 127*) and in the field (*24, 28*) using wind tunnel tests.

We describe a common laboratory procedure employed to measure crust abrasive resistance. The procedure requires a rainfall simulator and wind tunnel. Soils are placed in trays approximately 5 cm deep, 30 cm wide, and 100 to 150 cm long. The trays are then subjected to simulated rainfall to create a crusted condition and air-dried or oven-dried at approximately 50°C to simulate field conditions. The bottoms of the trays are perforated to allow water movement through the soil. The trays are then placed in a wind tunnel and abraded using natural aeolian sand or sieved sandy soil. Abrader that is composed of sieved soil material should include portions of the fine and/or medium sand fractions, 100 to 500 um in diameter. Abrader is introduced onto the floor of the wind tunnel using an abrader feeding device that drops a known amount of abrader on the floor of the tunnel through a calibrated orifice. The abrader rapidly reaches maximum velocity using this procedure.

Alternatively, a known amount of abrader can be simply placed on the floor on the up-wind end of the wind tunnel. The latter procedure is suggested only in relatively long wind tunnels, where the initially motionless abrader has time to reach maximum velocity before striking the crusted soil surface. The trays of crusted soil are initially weighed, abraded, and reweighed. The soil loss (mass per unit area) is computed as the difference in tray weight.

In general, the cumulative tray soil loss is linearly related to the cumulative amount of abrader that has passed over the crust. Typical abrasion curves for three crusted soils are illustrated in Figure 12.6. The coefficient of abrasion is calculated from these data by relating the cumulative amount of tray soil loss, as the dependent variable, to the cumulative amount of abrader passing over the crusted soil, as the independent variable, using linear regression analysis. The slope of the line is the coefficient of abrasion. Departures from straight line trends indicate that the amount of abrasion is changing through time, which is often caused by partial destruction of the crusted surface (*14*) or gradient in abrasive resistance of consolidated zone.

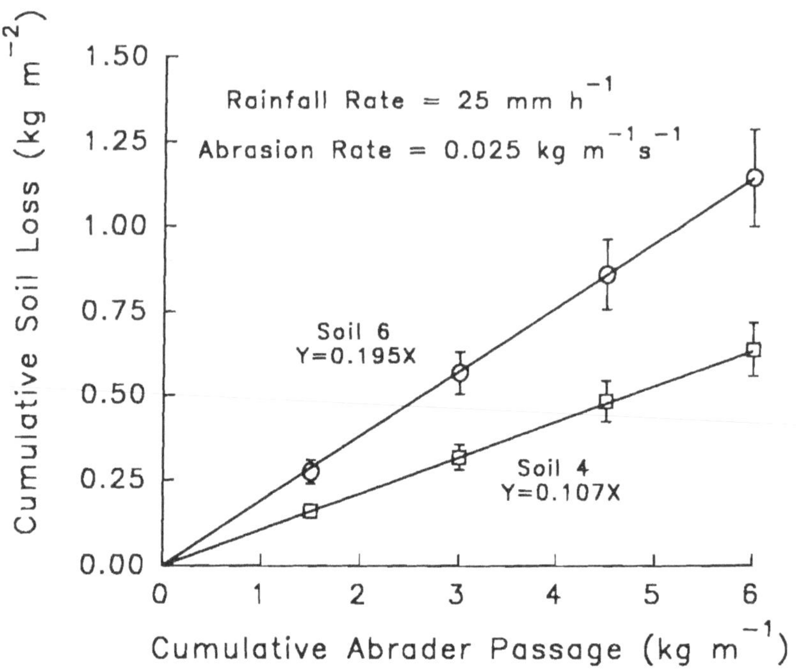

Figure 12.6 Relation of cumulative soil loss and cumulative abrader passage (*127*).

LOOSE SURFACE PARTICLES

Loose particles on a weakly crusted surface are easily eroded by wind. They impact significantly on initiation and perpetuation of wind erosion. Particles between 0.1 and 0.5 mm are easily moved by wind in saltation. The impact force of saltating particles on the soil surface dislodges other soil particles and abrades large soil aggregates into an erodible size (*52*). Chepil (*19*) discussed the interactive relationship between crust strength and abrader quantity on wind erosion. Soils that develop strong crusts, a surface "armor", with little loose abrader material present on the surface are resistant to wind erosion. Other soils that form weak crusts erode readily even though relatively small amounts of abrader are present initially.

Loose material may be generated by a number of mechanisms other than direct abrasion, including freeze-thaw, freeze-dry, and rainfall. We describe here a vacuum method to measure the amount of loose material resting on a crusted surface after a rainfall event. The system had two sedimentation chambers or separators with a combined efficiency of 90 percent (Figure 12.7). The first separator was a modified army ammuni-

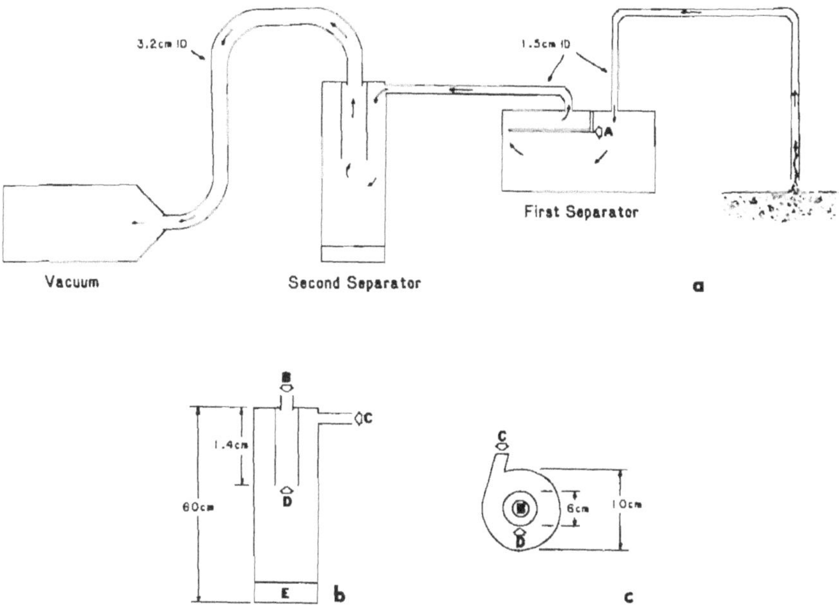

Figure 12.7 Schematic of the vacuum collection system used to collect loose materials. (a) Cross section of the entire system; (b) detail of vertical cross section of the second separator (A—air outlet, B—air inlet, C—tubular guard, and D—bottom cap); (c) detail of horizontal cross section of the second separator (*126*).

tion can with an L-shaped baffle that separated the intake and outflow tubes. The first separator retained most of the sample, allowing relatively clean air to enter the top of the second separator. The second separator was a conventional reverse-flow cyclone with a cylindrical tubular guard that extended below the level of the air outlet to restrict sediment movement. The sample was collected by removing the bottom cap (Figure 12.7). The vacuum was generated by a light-weight vacuum cleaner (110V AC; 1.0 peak horse power motor) powered by a portable generator. Because the surface was not brushed or disturbed during the vacuuming process, the material collected was loose on the soil surface.

Potter (*86*) used this system to develop functions relating quantity of loose particles on the soil crusts after a rainfall event to amount of rain and percent sand in the soil. He found that the variability among soils was large. Coefficients of variation (CV) ranged from 9 to 124 percent and were generally as large or larger for the finer textured soils as for the coarser textured soils. Soil texture had the greatest influence on the amount of loose material on the crusted soil surface. The quantity of loose material was small for the finely textured soils, but increased rapidly as sand content increased. About 30 times as much loose material occurred on the coarsest textured soil (0.091 kg m^{-2}) compared to the finest textured soil (0.003 kg m^{-2}) in Potter's (*86*) study. The vacuum collection system worked satisfactorily to sample loose granular materials on the crusts of the soils tested.

SOIL WETNESS

Soil wetness, particularly at the soil-atmosphere interface, greatly influences soil erodibility by wind. Therefore, an accurate evaluation of soil wetness is a prerequisite for any reliable wind erosion prediction scheme. Chepil (*17*) showed that the resistance of the uppermost soil particles against wind erosion, on the average, was proportional to the equivalent soil water content squared, when equivalent water content was defined as the ratio of the amount of water held in the soil to the amount held by the same soil at -1500 Joules/Kg soil water potential. This was the basis for developing a climatic factor in the wind erosion equation used to predict the amount of soil that will erode from a given agricultural field and to determine the conditions necessary to reduce potential erosion to a tolerable amount (*27*).

The water content at the interface of the soil and the atmosphere is constantly changing. It may change rapidly because of rainfall, irrigation, and/or hail, or slowly because of evaporation, drainage, and in the form of dew or soft-landing snowflakes. The water thus received at the soil surface may run off or infiltrate into the soil and be distributed in the soil volume according to the laws that govern saturated and unsaturated moisture flow, or it may evaporate from the surface.

When the evaporation rate from the soil exceeds the soil's ability to transmit water to that surface, the amount of water in the surface soil can change rapidly. Idso et al. (*63*), using albedo measurements, found that surface volumetric water of a bare Avondale loam soil at Phoenix, Arizona, decreased from 0.20 to 0.07 cm^3/cm^3 between 1300 and 1500 h two days after the soil was irrigated in July. This process was slower in December when the surface soil did not dry until 10 days after irrigation; the volumetric water content decreased from 0.16 to 0.06 cm^3/cm^3 between 1000 and 1500 h. This is significant considering that the amount of water retained at -1500 Joules/Kg water potential for the Avondale loam is 0.15 m^3/m^3 which indicates the soil surface can attain wind-erodible dryness in a few hours.

Van Bavel and Hillel (*113*) presented a dynamic simulation model for calculating the water content and temperature profiles of a bare soil with global radiation, wind speed, air temperature, and dew point temperatures as the only time-dependent input data. Their model can simulate the simultaneous flow of heat and water in a vertical soil profile.

Hillel (*59*), in Computer Simulation of Soil-Water Dynamics, described the formulation of several models simulating soil physical processes. One of these, the model for nonisothermal evaporation of soil water, was modified by Skidmore and Dahl (*100*) to examine the behavior of surface-soil drying as influenced by atmospheric evaporativity, soil hydraulic properties, and initial water content. They examined soil-water dynamics of the surface cm of soil as influenced by evaporativity, soil hydraulic properties, and initial water content. Diurnal wind speed, global irradiance, and air and dew point temperatures for selected days were used to calculate nonisothermal evaporation from hypothetical clay, loam, and sand soil. Surface soil dried quickly to wind-erodible dryness (-1500 joules/kg), especially for meteorological conditions conducive to wind erosion. When the soil was wet (-10 joules/kg) and the weather was typified by medium-high irradiance and strong winds, sand, loam, and clay soils dried to wind-erodible dryness by the first, second, and third days, respectively. When the soil was somewhat drier (-100 joules/kg) initially, it dried to wind-erodible dryness in a few hours. The quick drying of bare surface soil must be considered in the study of the processes that occur at or near the surface.

Lascano and van Bavel (*70, 71*) verified a similar model on bare soil. Using soil hydraulic properties and standard meteorological data as input, their calculated profiles compared favorably to actual measurements. The predicted water content values were within one standard deviation of the average measured values. Bristow et al. (*11*) added a residue component to model dynamic aspects of mass and energy transfer in a soil-residue-atmosphere system.

Although the simulation models that have been developed recently seem to be capable of predicting water content profiles, most require detailed information about soil hydraulic properties and much computer time—maybe more than feasible for the wind erosion modeling. Furthermore, these models fail to predict soil wetness at the soil-atmosphere interface which can be significantly different from that below the surface.

Both Chepil et al. (*27*) and Skidmore (*98*) have assumed that on a long-term basis, equivalent surface water content was approximated by the ratio of precipitation to potential evaporation. If effective precipitation exceeds potential evaporation, the soil remains wet. On the other hand, if potential evaporation far exceeds precipitation, the soil surface must be dry most of the time.

Another estimator of soil wetness is the ratio of actual evaporation to potential evaporation. Water will evaporate from the soil surface at the potential rate only when the surface is wet. When the soil cannot supply water to the surface fast enough to keep up with potential evaporation rate, the surface dries, and the evaporation rate declines below the potential rate. When actual evaporation rate is very slow compared to the potential rate, the surface must be dry. Therefore, the ratio of actual evaporation to potential evaporation is an indicator of the dryness of the soil surface.

To meet the unique requirements of wind erosion prediction technology, a computer simulation of soil water dynamics is being developed and validated for the HYDROLOGY submodel of the Wind Erosion Research Model (*34*).

The HYDROLOGY submodel maintains a continuous soil water balance by accounting for daily amounts of infiltration, snow melt, soil evaporation, plant transpiration, runoff, and deep percolation. The potential evapotranspiration, which is calculated by a combination equation, is partitioned between bare soil evaporation and plant transpiration on the basis of plant leaf area index. Additionally, soil water redistribution by water flow up and down between soil layers is evaluated on an hourly basis using Darcy's flow equation. The first objective of the HYDROLOGY submodel is to estimate water contents of the different soil layers. Potential rates of soil evaporation and plant transpiration are adjusted to actual rates on the basis of soil water availability. Deep percolation from the soil profile is estimated to be equal to the conductivity of the lowermost soil layer, assuming a unit hydraulic gradient at the lower boundary.

Furthermore, the unique feature of this submodel is its ability to estimate soil wetness at the soil-atmosphere interface on an hourly basis as needed by the EROSION submodel. This is accomplished by extrapolating water content from the three uppermost simulation layers, using a second order polynomial equation, and by interpolating the relationship between the evaporation ratio and equivalent surface soil water content.

The submodel was originally tested using the soil water and climatic data from a 1971 bare soil evaporation experiment conducted on an Avondale loam (fine-loamy, mixed [calcareous], hyperthermic Typic Torrifluvent) at the U.S. Water Conservation Laboratory, Phoenix, Arizona (*65*). Further testing was conducted during July-August 1991 on a Pullman clay loam (fine, mixed, thermic Torrertic Paleustoll) at the USDA-ARS Conservation and Production Research Laboratory at Bushland, Texas.

The simulated evaporation results agreed well with the measured daily evaporation rates from lysimeters during the duration of the two experiments. Furthermore, the submodel reasonably estimated the soil water content profiles, particularly soil wetness at the soil-atmospehre interface as compared with measured water contents from the uppermost 2 and 5 millimeters of the soil in the Texas and Arizona experiments, respectively.

Based on our limited testing, the HYDROLOGY submodel is adequate in predicting soil wetness as needed for wind erosion modeling.

CANOPY STRUCTURE

Plant growth simulation models range from the relatively simple, such as SUCROS (*114*), to the extremely complex process-orientated SIMCOTT II (*81*) and from the universal models that simulate many different plants, such as EPIC (*123*), to models that simulate the growth of only one particular species (*3, 60, 67, 117*) or only a specific plant part (*61*). The one common characteristic of all these models is the main output, which is dry weight of the biomass produced. Dry weight is often required in soil erosion modeling.

Canopy structure characteristics of height, structure and flexibility of individual plants, size and arrangement of plant parts, and the number of plants occupying a given area determine its aerodynamic roughness, thus affecting the friction velocity at the top of the canopy (*96*). The distribution of leaf and stem area with height is needed to determine the depletion of the friction velocity through the plant canopy and, thus, the velocity near the soil surface (*6*). Distinguishing between stem and leaf area is necessary because leaves tend to streamline with the wind flow and have a drag coefficient (C_d) of about 0.1, whereas stems remain rigid and have a C_d of about 1.0. Thus, on a unit area basis, stems are about 10 times more effective than leaves in depleting the friction velocity.

Following is an approach used by Armbrust and Bilbro (*4*) for predicting canopy structure.

Thirty plants were sampled every seven days from emergence to maturity. Plants were cut at the soil surface, placed in sealed plastic bags with wet paper towels, and returned to the laboratory where the following measurements were made:

1. Plant height—base of plant to top leaf curve or top of head.
2. Stem length—base of plant to top leaf collar.
3. Stem diameter—2 cm above base and at top leaf collar, including sheath.
4. Peduncle length—above top leaf collar to bottom of head.
5. Peduncle diameter—above top leaf collar and at bottom of head.
6. Length and width of head, if present—width at midpoint of length.

7. Dry weight of head—only that part of head emerged from leaf sheath.
8. Dry weight of sections of stems and leaves equal to one-fifth of plant height and leaf area of each section.

Canopy cover was measured weekly at solar noon (1134 CDT + or - 1 hr) in the field on three areas, each 2 m by 0.76 m, selected immediately after planting, and five positions in each area were marked as locations for canopy cover determination by the meter stick method (*1*).

Armbrust and Bilbro (4) fit equations to the natural growth function equation

$$Y = a\left(1 - e^{bx}\right) \qquad [7]$$

where Y = parameter to be estimated, X = total above-ground dry weight, a = asymptotic value of Y, and b = rate at which Y approaches "a" for several crops. They found that stem area of grain sorghum remained in the bottom 80 percent of the plant height. Average leaf and stem areas versus normalized height is shown in Figure 12.8.

FIELD SCALE WIND EROSION MEASUREMENT

Wind erosion has proven difficult to measure in the field. Attempts have been made to measure change in soil depth resulting from wind erosion (*26, 45*). A change in soil depth of 1.0 mm for a one-year period represents excessive wind erosion on many soils and contributes a significant load to the atmospheric aerosol. It is difficult to measure even 1.0 mm change in depth because the surface is rough and the elevation of the soil is not constant. It may vary several mm because of change in water content or change in bulk density resulting from tillage or compaction.

Redistribution of the fallout radionuclide cesium-137 has been used to estimate net sediment redistribution from wind erosion (*111*). This technique is also limited to long-term studies. Long-term trends in erosion, although useful, do not provide detailed information about wind erosion processes. To investigate erosion processes, measurements encompassing a single storm are needed, or even better, measurements of erosion continuously within the storm event.

Various techniques have been used to measure short-term wind erosion. These include columns of cake-pans and psammagraphs (*32*); tipping-bucket gauge (*80*); vertical tube sand traps (*72*); adhesive tape (Mainquet, Reims, France, personal communication); and numerous vertical slot samples (*7, 18, 82*). Fryrear et al. (*42*) outlined equipment and procedure for measuring field erosion.

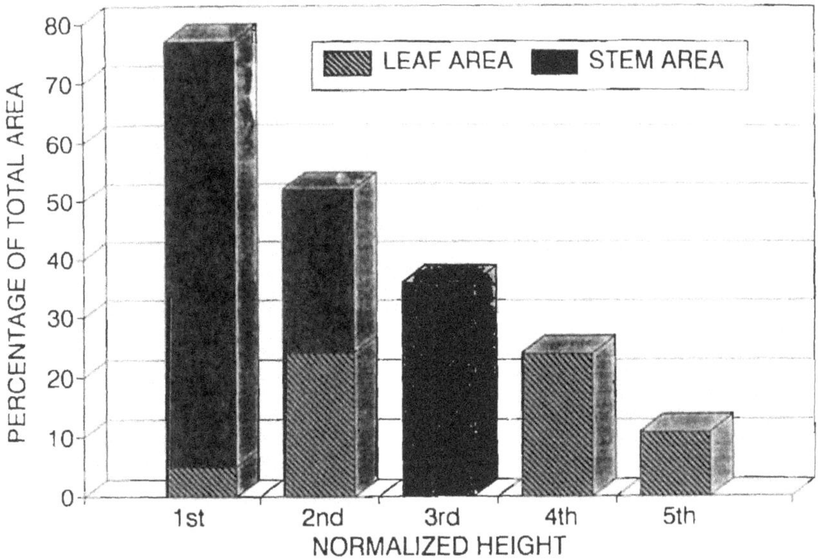

Figure 12.8 Grain sorghum leaf and stem area distribution with normalized plant height. Average for all sample dates from emergence to maximum canopy cover.

Creep and saltation

Because wind eroded soil is transported in various modes, field instruments must be designed to provide samples of eroding soil in the separate transport modes. The creep component of the soil flux can be sampled using an exposed horizontal slot or circular opening that leads to a buried container. If a slot is used, it should be able to rotate so it remains normal to the wind direction. The important consideration is to avoid surface protuberances that prevent the sampler from collecting a representative sample of the moving soil. A small fraction of saltation flux is also trapped by creep samplers, but this can be estimated by sieving the trapped sample.

Vertical slot samplers are generally used for trapping saltating particles. Their overall efficiency is the product of the entrance efficiency and the trapping efficiency for particles that have passed the entrance. An early slot sampler was developed by Bagnold (*7*), but it did not rotate into the wind. Tests on a slot sampler similar to Bagnold's showed it was only about 60 percent efficient (*48*). Chepil (*18*) modified Bagnold's sampler to rotate into the wind. Later, trapping efficiency was improved by venting catchers (*37*) and also by making the sampler in a wedge shape to increase the area for air flow between the inlet slot and the vents (*82*).

Stout and Fryrear (*110*) developed an improved slot sampler for saltation and creep particles. Features of the improved sampler included a vented, wedge-shaped inlet section that was 98 percent efficient for trapping saltation in tunnel tests. In addition, the base at the soil surface consisted of a nonrotating outer portion and a rotating inner portion, so that the rotating portion could not contact the soil surface as in earlier samplers. Because the sampler is 0.2 m tall, it must be combined with other samplers to completely sample the saltation zone. Stout's sampler is currently being used in the WEPS field studies.

Suspension

Accurate sampling of suspended particles requires an isokinetic sampler, i.e., one in which inlet velocity and surrounding air velocity are equal. Sampling errors caused by nonisokinetic sampling depend on both particle size and level of deviation from isokinetic flow (*31*). To achieve near-isokinetic sampling, horizontal inlet tubes coupled to a filter and powered suction device have been used (*25, 47*). Steen (*109*) developed a passive sampler shaped as a diffuser to induce isokinetic inlet flow, but then used power to provide electrostatic trapping of the particles. In general, these samplers were not suitable for untended operation near the surface in an eroding field where high particle concentrations are the norm. Fryrear (*40*) developed a low-cost, wedge-shaped, passive sampler that rotates into the wind, dubbed the BSNE. The sampler provides near-isokinetic inlet conditions and separates particles from the flow by using a combination of particle settling, impact, and a screened outlet. The BSNE sampler efficiency averaged over 89 percent on samples of washed sand and sieved soil. For particles less than 0.050 mm diameter, the efficiency decreased. However, trapped samples can be sieved and mathematically adjusted to account for effects of known trapping efficiencies. A combination of Stout and BSNE samplers allows measurement of soil flux profiles from the surface to a height of several meters.

Threshold

Most wind erosion models require an accurate estimate of threshold conditions and erosion duration during field tests. Thus, a method to determine both the onset and cessation of saltation is needed. To accomplish this task, a piezoelectric quartz crystal impact sensor has been developed for field use (*48*). The sensor, called SENSIT, counts particle impacts at a fixed height near the surface. Coupled with wind speed and direction sensors, SENSIT allows onset, duration, and direction of wind erosion to be monitored.

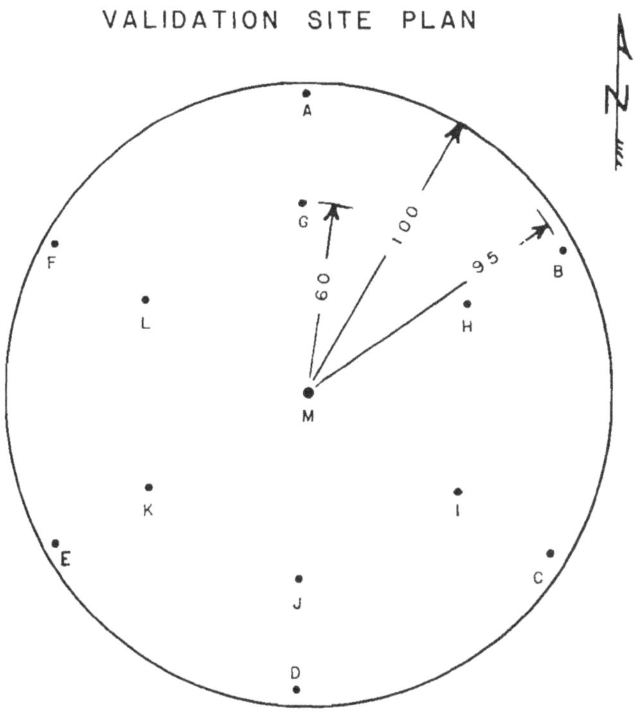

Figure 12.9 Plan for locating 14 clusters (A through N) of samplers in a 200-meter diameter circular field.

Field installation

For a typical uniform field with a well defined non-eroding boundary, the following equipment and instruments were used by Fryrear et al. (*42*).

Meteorological tower instrumented with the following

- four anemometers, 0.20, 0.50, 1.0, and 2.0 m
- two air temperatures, 0.2 and 2.0 m
- one solar radiation
- one soil temperature, 0.02 m
- one wind direction, 2.5 m
- one tipping bucket rain gauge
- one relative humidity sensor, 2.0 m

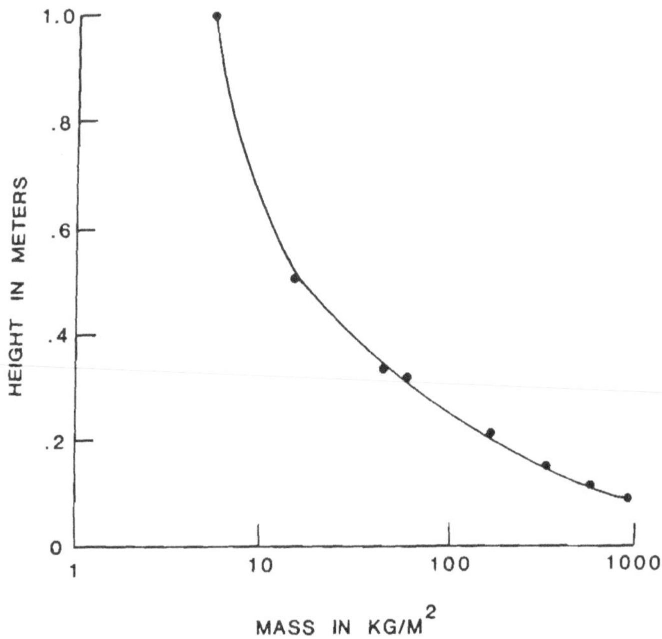

Figure 12.10 Vertical distribution of airborne mass to a height of 1 meter on March 11, 1988, 134 meters from an upwind non-eroding surface at Big Spring, Texas.

Erosion samplers

- 14 clusters BSNE samplers at 0.05, 0.10, 0.20, 0.5, and 1.0 m above the soil surface (located as illustrated in Figure 12.9)
- one surface creep (0.0 to 0.003 m height, 0.005 m wide opening) with saltation (0.003 to 0.02)
- one weighing BSNE sampler at 0.02 m height
- one SENSIT (threshold) at 0.02 m height

This array and selection of samplers enabled one technician to maintain a site. After every dust storm, all samplers were emptied, the contents were transferred to metal containers, and temporal soil field conditions that had changed since the last measurements were recorded. Typical vertical and horizontal distributions of eroded material collected in the above described arrangement of samples are shown in Figures 12.10 and 12.11.

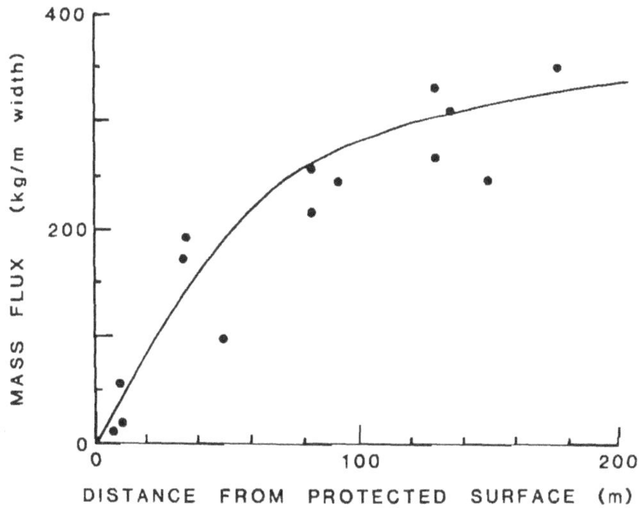

Figure 12.11 Horizontal distribution of eroded mass across a 200-meter-wide field at Big Spring, Texas, March 11, 1988.

REFERENCES

1. Adams, J.E., and G.F. Arkin. 1977. *A light interception method for measuring row crop ground cover.* Soil Science Society of America Journal 41: 789-792.
2. Allmaras, R.R., R.E. Burwell, W.E. Larson, and R.F. Holt. 1966. *Total porosity and random roughness of the interrow zone as influenced by tillage.* Conservation Research Report 7. U.S. Department of Agriculture. 22 pp.
3. Arkin, G.F., R.L. Vanderlip, and J.T. Ritchie. 1976. *A dynamic grain sorghum growth model.* Transactions, American Society of Agricultural Engineers 19: 622-626, 630.
4. Armbrust, D.V., and J.D. Bilbro. 1988. *Prediction of canopy structure for wind erosion modeling.* Paper No. 88-2559, International Winter Meeting, American Society of Agricultural Engineers, December 13-16, Chicago, Illinois.
5. Awadhwal, N.K., and G.E. Thierstein. 1985. *Soil crust and its impact on crop establishment: A review.* Soil Tillage Research 5: 289-302.
6. Bache, D.H. 1986. *Momentum transfer to plant canopies: influence of structure and variable drag.* Atmosphere and Environment 20: 1369-1378.
7. Bagnold, R.A. 1943. *The physics of blown sand and desert dunes.* London, Methuen, 265.
8. Barndorff-Nielsen, O.E., J.L. Sensen, H.L. Nielsen, K.R. Rasmussen, and M. Sorensen. 1985. *Wind tunnel tracer studies of grain progress.* Proceedings, International Workshop on Physics of Blown Sand. Department of Theoretical Statistics, University of Aarhus, Aarhus, Denmark. 2: 243-252

9. Bondy, E., L. Lyles, and W.A. Hayes. 1980. *Computing soil erosion by periods using wind energy distribution.* Journal of Soil and Water Conservation 35: 173-176.
10. Boyd, D.W., E.L. Skidmore, and J.G. Thompson. 1983. *A soil-aggregate crushing-energy meter.* Soil Science Society of America Journal 47: 313-316.
11. Bristow, K.L., G.S. Campbell, R.J. Papendick, and L.F. Elliott. 1986. *Simulation of heat and moisture transfer through a surface residue-soil system.* Agriculture and Forest Meteorology 36: 193-214.
12. Callebaut, F., D. Gabriels, and M. DeBoodt. 1986. *Assessment of soil surface sealing and crusting.* Proceedings, Symposium in Ghent, Belgium, 1985. State University of Ghent.
13. Cary, J.W., and D.D. Evans (eds.). 1974. *Soil crusts.* University of Arizona Agricultural Experiment Station Technical Bulletin 214. Tucson.
14. Chepil, W.S. 1951. *Properties of soil which influence wind erosion: V. Mechanical stability of structure.* Soil Science 72: 465-478.
15. Chepil, W.S. 1952. *Improved rotary sieve for measuring state and stability of dry soil structure.* Soil Science Society of America Proceedings 16: 113-117.
16. Chepil, W.S. 1953. *Field structure of cultivated soils with special reference to erodibility by wind.* Soil Science Society of America Proceedings 17: 185-190.
17. Chepil, W.S. 1956. *Influence of moisture on erodibility of soil by wind.* Soil Science Society of America Proceedings 20: 288-292.
18. Chepil, W.S. 1957. *Width of field strips to control wind erosion.* Kansas Agricultural Experiment Station Technical Bulletin 92. p. 16.
19. Chepil, W.S. 1958. *Soil conditions that influence wind erosion.* U.S. Department of Agriculture, Agricultural Research Service Technical Bulletin 1185. U.S. Government Printing Office, Washington, D.C.
20. Chepil, W.S. 1960. *Conversion of relative field erodibility to annual soil loss by wind.* Soil Science Society of America Proceedings 24: 143-145.
21. Chepil, W.S. 1962. *A compact rotary sieve and the importance of dry sieving in physical soil analysis.* Soil Science Society of America Proceedings 26: 4-6.
22. Chepil, W.S., and F. Bisal. 1943. *A rotary sieve method for determining the size distribution of soil clods.* Soil Science 56: 95-100.
23. Chepil, W.S., and R.A. Milne. 1941. *Wind erosion of soils in relation to the size and nature of the exposed area.* Sci. Agric. 21: 479-487.
24. Chepil, W.S., and N.P. Woodruff. 1954. *Estimations of wind erodibility of field surfaces.* Journal of Soil and Water Conservation 9: 257-265.
25. Chepil, W.S., and N.P. Woodruff. 1957. *Sedimentary characteristics of dust storms: II. Visibility and dust concentration.* American Journal of Science 255: 104-114.
26. Chepil, W.S., and N.P. Woodruff. 1963. *The physics of wind erosion and its control.* Advances in Agronomy 15: 211-302.

27. Chepil, W.S., F.H. Siddoway, and D.V. Armbrust. 1962. *Climate factor for estimating wind erodibility of farm fields.* Journal of Soil and Water Conservation 17: 162-165.
28. Chepil, W.S., N.P. Woodruff, and A.W. Zingg. 1955. *Field study of wind erosion in western Texas.* U.S. Department of Agriculture Technical Bulletin SCS-TP-125.
29. Cole, R.C. 1939. *Soil macrostructure as affected by cultural treatments.* Hilgardia 12: 429-472.
30. Coleman, N.L., and W.M. Ellis. 1976. *Model study of the drag coefficient of a streambed particle.* Proceedings 3rd Federal Interagency Sedimentation Conference, March 22-25, Denver, Colorado.
31. Davies, C.N. 1968. *The entry of aerosols into sampling tubes and heads.* British Journal of Applied Physics Ser. 2. (1): 921-923.
32. DePloey, J. 1980. *Some field measurements and experimental data on wind-blown sands.* In: M. DeBoodt and D. Gabriels (eds.) *Assessment of erosion.* John Wiley and Sons. pp. 541-553.
33. Diouf, B., E.L. Skidmore, J.B. Layton, and L.J. Hagen. 1990. *Stabilizing fine sand by adding clay: Laboratory wind tunnel study.* Soil Tech. 3: 21-31.
34. Durar, A.A., and E.L. Skidmore. 1989. *The hydrologic components of the wind erosion research model.* Agronomy Abstracts. Madison, Wisconsin. p. 278.
35. Elliot, D.L., C.G. Holladay, W.R. Barchet, H.P. Foote, and W.F. Sandusky. 1987. *Wind energy resource atlas of the United States.* DOE/CH 10093-4. Available from National Technical Information Service, Springfield, Virginia.
36. Farrell, D.A., E.L. Greacen, and W.E. Larson. 1967. *The effect of water content on axial strain in a loam soil under tension and compression.* Soil Science Society of America Proceedings 31: 445-450.
37. Fryberger, S.G., T.S. Ahlbrandt, and S. Andrews. 1979. *Origin, sedimentary features, and significance of low-angle aeolian "sand sheet" deposits, Great Sand Dunes, National Monument and vicinity, Colorado.* Journal of Sedimentary Petrol. 49(3): 733-746.
38. Fryrear, D.W. 1984. *Soil ridges-clods and wind erosion.* Transactions, American Society of Agricultural Engineers 18: 445-448.
39. Fryrear, D.W. 1985. *Determining soil aggregate stability with a rapid rotary sieve.* Journal of Soil and Water Conservation 40: 231-233.
40. Fryrear, D.W. 1986. *A field dust sampler.* Journal of Soil and Water Conservation 41(2): 117-120.
41. Fryrear, D.W., and E.L. Skidmore. 1985. *Methods of controlling wind erosion.* In: R.F. Follett and B.A. Stewart (eds.) *Soil Erosion and Crop Productivity.* American Society of Agronomy, Crop Science Society of America, Soil Science Society of America, Madison, Wisconsin. p. 443-457.
42. Fryrear, D.W., J.E. Stout, L.J. Hagen, and E.D. Vories. 1990. *Wind erosion: field measurement and analysis.* Transactions, American Society of Agricultural Engineers 3: 155-160.

43. Gardner, W.R. 1956. *Representation of soil aggregate-size distribution by a logarithmic-normal distribution.* Soil Science Society of America Proceedings 20: 151-153.
44. Gerety, K.M., and R. Slingerland. 1983. *Nature of the saltating population in wind tunnel experiments with heterogeneous size-density sands.* In: M.E. Brookfield and T.S. Ahlbrant (eds.) *Developments in Sedimentology* 38: 115-151, Elsevier, New York.
45. Gibbens, R.P., J.M. Tromble, J.T. Hennessy, and M. Cardenas. 1983. *Soil movement in mesquite dunelands and former grasslands of southern New Mexico from 1939 to 1980.* Journal of Range Management 36: 145-148.
46. Gill, W.R., and W.F. McCreery. 1960. *Relation of size of cut to tillage tool efficiency.* Agricultural Engineering 41: 372-374, 381.
47. Gillette, D.A., I.H. Blifford, and C.R. Fenster. 1972. *Measurements of aerosol size distributions and vertical fluxes of aerosols on land subject to wind erosion.* Journal of Applied Meteorology 11: 977-987.
48. Gillette, D.A., and P.H. Stockton. 1986. *Mass momentum and kinetic energy fluxes of saltating particles.* In: William G. Nickling, (ed.) *Aeolian Geomorphology.* Allen and Unwin, Boston, pp. 35-56.
49. Greeley, R., and J.D. Iversen. 1985. *Wind as a Geological Process.* Cambridge University Press, Cambridge.
50. Grossman, R.B., J.B. Fehrenbacker, and A.H. Beavers. 1959. *Fragipan soils of Illinois: I. General characterization and field relationships of Hosmer silt loam.* Soil Science Society of America Proceedings 23: 65-70.
51. Gupta, V.S., D.W. Fuerstenau, and T.S. Mika. 1975. *An investigation of sieving in the presence of attrition.* Power Technology 11: 257-271.
52. Hagen, L.J. 1984. *Soil aggregate abrasion by impacting sand and soil particles.* Transactions, American Society of Agricultural Engineers 27: 805-816.
53. Hagen, L.J. 1988. *New wind erosion model developments in the U.S. Department of Agriculture.* Proceedings of the 1988 Wind Erosion Conference. Texas Tech. University, Lubbock. pp. 104-116.
54. Hagen, L.J. 1991. *Wind erosion mechanics: Abrasion of aggregated soil.* American Society of Agricultural Engineers 34(4): 891-837.
55. Hagen, L.J. 1991. *A wind erosion prediction system to meet user needs.* J. Soil and Water Conserv. 46: 105-111.
56. Hagen, L.J., and D.V. Armbrust. 1989. *Aerodynamic roughness and saltation trapping efficiency of tillage ridges.* Transactions, American Society of Agricultural Engineers 35: 1179-1184.
57. Hagen, L.J., L. Lyles, and E.L. Skidmore. 1980. *Application of wind energy to Great Plains irrigation pumping.* Advances in Agricultural Technology, USDA-AAT-NCO-4, 20 pp.
58. Hagen, L.J., E.L. Skidmore, and D.W. Fryrear. 1987. *Using two sieves to characterize dry soil aggregate size distribution.* Transactions, American Society of Agricultural Engineers 30: 162-165.
59. Hillel, D. 1977. *Computer simulation of soil-water dynamics.* International Development Research Center, Ottawa, Canada, 214 pp.

60. Holt, D.A., R.J. Bula, G.E. Miles, M.M. Schreiber, and R.M. Peart. 1975. *Environmental physiology, modeling, and simulation of alfalfa growth. I. Conceptual development of SIMED.* Purdue University Agricultural Experiment Station Research Bulletin No. 907. Lafayette, Indiana.
61. Hoogenboom, G., and M.G. Huck. 1986. *ROOTSIMU V4.0: A dynamic simulation of root growth, water uptake, and biomass partitioning in a soil-plant-atmosphere continuum: update and documentation.* Alabama Agricultural Experiment Station Agronomy and Soils Department Series No. 109, 84 pp.
62. Huang, C., I. White, E.G. Thwaite, and A. Bendeli. 1988. *A noncontact laser system for measuring soil surface topography.* Soil Science Society of America Journal 52: 350-355.
63. Idso, S.B., R.D. Jackson, R.J. Reginato, B.A. Kimball, and F.J. Nakayama. 1975. *The dependence of bare soil albedo on soil water content.* Journal of Applied Meteorology 14: 109-113.
64. Iverson, J.D. 1986. *Small scale wind tunnel modeling of particle transport—Froude number effect.* Aeolian Geomorphology 19-33.
65. Jackson, R.D. 1973. *Diurnal changes in soil water content during drying.* In: *Field Soil Water Regime.* Soil Science Society of America Special Publication No. 5, pp. 37-55.
66. Jensen, M. 1958. *The model-law for phenomena in natural wind.* Ingenioren 2: 121-128.
67. Johnson, I.R., T.E. Ameziane, and J.H.M. Thornley. 1983. *A model of grass growth.* Ann. Bot. 51: 599-609.
68. Kind, R.J. 1976. A *critical examination of the requirements for model simulation of wind-induced erosion/deposition phenomena such as snow drifting.* Atmos. Environ. 10: 219-227.
69. Kuipers, H. 1957. *A relief meter for soil cultivation studies.* Netherlands Journal of Agricultural Science 5: 255-262
70. Lascano, R.J., and C.H.M. van Bavel. 1983. *Experimental verification of a model to predict soil moisture and temperature profiles.* Soil Science Society of America Journal 47: 441-448.
71. Lascano, R.J., and C.H.M. van Bavel. 1986. *Simulation and measurement of evaporation from a bare soil.* Soil Science Society of America Journal 50: 1127-1132.
72. Leatherman, S.P. 1978. *A new aeolian sand trap design.* Sedimentology 25: 303-306.
73. Logie, M. 1982. *Influence of roughness elements and soil moisture and resistance of sand to wind erosion.* In: D.H. Yaalon (ed.) *Aridic Soils and Geomorphic Processes.* Catena Supp. 1., 162-173. Braunschweig.
74. Lyles, L. 1976. *Wind patterns and soil erosion in the Great Plains.* In: *Proceedings of the Symposium "Shelterbelts on the Great Plains,"* Great Plains Agriculture Council Publication No. 78. pp. 22-30.
75. Lyles, L. 1983. *Erosive wind energy distributions and climatic factors for the West.* Journal of Soil and Water Conservation 38: 106-109.

76. Lyles, L., and B.E. Allison. 1976. *Wind erosion: The protective role of simulated standing stubble.* Transactions, American Society of Agricultural Engineers 19: 279-283.
77. Lyles, L., J.D. Dickerson, and L.A. Disrud. 1970. *Modified rotary sieve for improved accuracy.* Soil Sci. 109: 207-210.
78. Lyles, L., R.L. Schrandt, and N.F. Schmerdler. 1974. *How aerodynamic roughness elements control sand movement.* Transactions, American Society of Agricultural Engineers 17: 134-139
79. Marshall, T.J., and J.P. Quirk. 1950. *Stability of structural aggregates of dry soil.* Australian Journal of Agricultural Research 1: 266-275.
80. McCauley, J.F., C.S. Breed, P.J. Helm, G.H. Billingsley, and D.J. MacKinnon. 1987. *Monitoring desert winds.* In: John F. McCauley and Jack N. Rinker (eds.) *A workshop on desert processes, September 24-28, 1984 - Report on the conference.* U.S. Geological Survey Circular 989. pp. 6-9.
81. McKinion, J.M., D.N. Baker, J.D. Hesketh, and J.W. Jones. 1975. *Part 4-SIMCOT II: a simulation of cotton growth and yield.* In: *Computer simulation of a cotton production system.* ARS-S-52. p. 27-82.
82. Merva, K.R., and G. Peterson. 1983. *Wind erosion sampling in the North Central Region.* Paper No. 83-2133, Summer Meeting, American Society of Agricultural Engineers, Bozeman, Montana.
83. Nickling, W.G. 1988. *The initiation of particle movement by wind.* Sedimentology 35: 499-511.
84. Peterka, J.A., and J.E. Cermak. 1974. *Simulation of atmospheric flows in short wind tunnel test sections.* Report No. CER73-74JAP-JEC32, Fluid Dynamics and Diffusion Laboratory, Department of Civil Engineering, Colorado St. University, Ft. Collins.
85. Philips, M. 1980. *A force balance model for particle entrainment into a fluid stream.* J. Phys. D: Appl. Phys. 13: 221-233.
86. Potter, K.N. 1990. *Estimation of wind-erodible materials on newly crusted soils.* Soil Sci. (in press).
87. Potter, K.N., T.M. Zobeck, and L.J. Hagen. 1990. *A microrelief index to estimate soil erodibility by wind.* Transactions, American Society of Agricultural Engineers 33(1): 151-155.
88. Radke, J.K., M.A. Otterby, R.A. Young, and C.A. Onstad. 1981. *A microprocessor automated rillmeter.* Transactions, American Society of Agricultural Engineers 24: 401-404, 408.
89. Reed, J.W. 1975. *Wind power climatology of the United States.* SAND 74-0348. Sandia Laboratory, Albuquerque, New Mexico, 163 pp.
90. Rice, C., B.N. Wilson, and M. Appleman. 1988. *Soil topography measurements using image processing techniques.* Computers and Electronics in Agriculture 3: 97-107.
91. Rogowski, A.S., and D. Kirkham. 1976. *Strength of soil aggregates: influence of size, density, and clay and organic matter content.* Med. Fac. Landbouww. Rijksuniv. Gent 41: 85-100.
92. Rogowski, A.S., W.C. Moldenhauer, and D. Kirkham. 1968. *Rupture parameters of soil aggregates.* Soil Science Society of America Proceedings 32: 720-724.

93. Römkens, R.J.M., and J.Y. Wang. 1986. *Effect of tillage on surface roughness.* Transactions, American Society of Agricultural Engineers 29: 429-433.
94. Römkens, R.J.M., J.Y. Wang, and R.W. Darden. 1988. *A laser microrelief meter.* Transactions, American Society of Agricultural Engineers 31: 408-413.
95. Schmidt, R.A. 1977. *A system that measures blowing snow.* U.S. Department of Agriculture Forest Service Research Paper RM-194.
96. Shaw, R.H., and A.R. Pereira. 1982. *Aerodynamic roughness of a plant canopy: A numerical experiment.* Agric. Meteo. 26: 51-65.
97. Skidmore, E.L. 1965. *Assessing wind erosion forces: directions and relative magnitudes.* Soil Science Society of America Proceedings 29: 587-590.
98. Skidmore, E.L. 1986. *Wind erosion climatic erosivity.* Climate Change 9: 195-208.
99. Skidmore, E.L. 1987. *Wind-erosion direction factors as influenced by field shape and wind preponderance.* Soil Science Society of America Journal 51: 198-202.
100. Skidmore, E.L., and L. Dahl. 1978. *Surface soil drying: simulation experiment.* Abstracts for Commission Papers, 11th Congress. International Society of Science, Edmonton, Canada, 1: 246-247.
101. Skidmore, E.L., J.B. Layton, and L.J. Hagen. 1990. *Dry soil-aggregate stability: measuring and relating to wind erosion.* In preparation.
102. Skidmore, E.L., and D.H. Powers. 1982. *Dry soil-aggregate stability: energy-based index.* Soil Science Society of America Journal 46: 1274-1279.
103. Skidmore, E.L., and J. Tatarko. 1991. *Wind in the Great Plains: speed and direction distributions by month.* pp. 195-213. In: J.D. Hanson, M.J. Shaffer, and C.V. Cole (eds.) *Proceedings of the Sustainable Agriculture Symposium.* Fort Collins, Colorado. USDA-ARS-89.
104. Skidmore, E.L., and J. Tatarko. 1990. *Stochastic wind simulation for erosion modeling.* Transactions, American Society of Agricultural Engineers 33(6): 1893-1899.
105. Skidmore, E.L., and N.P. Woodruff. 1968. *Wind erosion forces in the United States and their use in predicting soil loss.* Agricultural Research Service, U.S. Department of Agriculture. Agriculture Handbook No. 346.
106. Soil Conservation Service, U.S. Department of Agriculture. 1988. *National Agronomy Manual.* Second Edition. 190-V Issue I (Part 500-509) SCS, USDA, Washington, D.C.
107. Sorensen, M. 1985. *Estimation of some aeolian saltation transport parameters from transport rate profiles.* Proceedings of the International Workshop on Physics of Blown Sand 1: 141-190. Department of Theoretical Statistics, University of Aarhus, Aarhus, Denmark.
108. Standen, N.M. 1972. *A spire array for generating thick turbulent shear layers for natural wind simulation in wind tunnels.* Report LTR-LA-94, National Aeronautical Establishment, Ottawa, Canada.
109. Steen, B. 1977. *A new simple isokinetic sampler for the determination of particle flux.* Atmos. Environ. 11: 623-627.
110. Stout, J.E., and D.W. Fryrear. 1989. *Performance of a wind blown particle sample.* Transactions, American Society of Agricultural Engineers 32: 2041-2045.

111. Sutherland, R.A., T. Kowalchuk, and E. DeJong. 1991. *Cesium-137 estimates of sediment redistribution by wind.* Soil Sci. 51: 387-396.
112. Toogood, J.A. 1978. *Relation of aggregate stability to property of Alberta soils.* In: W.W. Emerson, R.D. Bond, and A.R. Dexter (eds.) *Modification of Soil Structure.* John Wiley and Sons, New York. pp. 211-215.
113. van Bavel, C.H.M., and D.I. Hillel. 1975. *A simulation study of soil heat and moisture dynamics as affected by a dry mulch.* Proceedings, Summer Computer Simulation Conference 1975. pp. 815-821.
114. van Keulen, H., F.W.T. Penning de Vries, and E.M. Drees. 1982. *A summary model for crop growth.* In: F.W.T. Penning de Vries and H.H. van Laar (eds.) *Simulation of plant growth and crop production.* Simulatio Monogr., Wageningen, Netherlands. p. 87-97.
115. Van Ouwerkerk, C., M. Pot, and K. Boersma. 1982. *Electronic microreliefmeter for seedbed characterization.* Soil Tillage Research 2: 81-90.
116. Wagner, L.E., and Yiming Yu. 1991. Transactions, American Society of Agricultural Engineers 34: 412-416.
117. Wann, M., C.D. Raper Jr., and H.L. Lucas Jr. 1978. *A dynamic model for plant growth: a simulation of dry matter accumulation for tobacco.* Photosynthetica 12: 121-136.
118. Weibull, W. 1951. *A statistical distribution function of wide applicability.* J. Appl. Mech. 18: 293-297.
119. Welch, R., T.R. Jordan, and A.W. Thomas. 1984. *A photogrammatic technique for measuring soil erosion.* Journal of Soil and Water Conservation 39: 191-194.
120. White, B.R. 1982. *Two-phase measurements of saltating turbulent boundary layer flow.* Multiphaser Flow 8: 459-473.
121. White, B.R., and J.C. Schulz. 1977. *Magmes effect on saltation.* J. Fluid Mech. 81: 497-512.
122. Willets, B.B., and M.A. Rice. 1985. *Wind tunnel tracer experiments using dyed sand.* Proceedings, International Workshop on Physics of Blown Sand, 2: 225-242. Department Theoretical Statistics, University of Aarhus, Denmark.
123. Williams, J.R., C.A. Jones, and P.T. Dyke. 1984. *A modeling approach to determining the relationship between erosion and soil productivity.* Transactions, American Society of Agricultural Engineers 27: 129-144.
124. Woodruff, N.P., and F.H. Siddoway. 1965. *A wind erosion equation.* Soil Science Society of America Proceedings 29: 602-608.
125. Zingg, A.W. 1949. *A study of the movement of surface wind.* Agr. Engin. 30: 11-13, 19.
126. Zobeck, T.M. 1989. *Fast-vac, a vacuum system to rapidly sample loose granular material.* Transactions, American Society of Agricultural Engineers 32: 1316-1318.
127. Zobeck, T.M. 1991. *Abrasion of crusted soils: Influence of abrader flux and soil properties.* Soil Science Society of America Journal 55: 1091-1097.

INDEX

D

Q–R

T

Y–Z

For Product Safety Concerns and Information please contact our EU
representative GPSR@taylorandfrancis.com
Taylor & Francis Verlag GmbH, Kaufingerstraße 24, 80331 München, Germany

www.ingramcontent.com/pod-product-compliance
Ingram Content Group UK Ltd.
Pitfield, Milton Keynes, MK11 3LW, UK
UKHW022206190926
475702UK00003B/231

* 9 7 8 1 1 3 8 4 6 8 6 0 3 *